Writing History in the Age of Biomedicine

Writing History in the Age of Biomedicine

Roger Cooter with Claudia Stein

Yale
UNIVERSITY PRESS
New Haven & London

Published with assistance from the foundation established in memory of Philip Hamilton McMillan of the Class of 1894, Yale College.

Copyright © 2013 by Yale University.
All rights reserved.
This book may not be reproduced, in whole or in part, including illustrations, in any form (beyond that copying permitted by Sections 107 and 108 of the US Copyright Law and except by reviewers for the public press), without written permission from the publishers.

Yale University Press books may be purchased in quantity for educational, business, or promotional use. For information, please
e-mail sales.press@yale.edu (US office) or sales@yaleup.co.uk
(UK office).

Designed by Sonia Shannon.
Set in Garamond type by IDS Infotech, LTD.
Printed in the United States of America.

Library of Congress Cataloging-in-Publication Data
Cooter, Roger.
   Writing history in the age of biomedicine / Roger Cooter with Claudia Stein.
     page cm
   Includes bibliographical references and index.
   ISBN 978-0-300-18663-5 (clothbound : alk. paper)
   1. Medicine—History—20th century.   I. Stein, Claudia. II. Title.
   R149.C66 2013
   808.06'661—dc23
   2012047822

A catalogue record for this book is available from the British Library.

This paper meets the requirements of ANSI/NISO Z39.48–1992 (Permanence of Paper).

10 9 8 7 6 5 4 3 2 1

*The Wellcome Trust
Prince of Patrons*

CONTENTS

	Preface	*ix*
	Acknowledgments	*xiii*
1.	The End? *History-Writing in the Age of Biomedicine (and Before)*	1
2.	Anticontagionism and History's Medical Record	41
3.	"Framing" the End of the Social History of Medicine	64
4.	The Turn of the Body	91
5.	Coming into Focus *Posters, Power, and Visual Culture in the History of Medicine*	112
6.	Visual Objects and Universal Meanings *AIDS Posters, "Globalization," and History*	138
7.	The Biography of Disease	160
8.	Inside the Whale *Bioethics in History and Discourse*	170
9.	Cracking Biopower	183
10.	The New Poverty of Theory *Material Turns in a Latourian World*	205
	Notes	*229*
	Bibliography	*285*
	Index	*337*

PREFACE

This book is a personal statement of my core belief in the function of history-writing as a form of critical politics for addressing the present. In fact, it's a restatement of this conviction since most of its chapters have been previously published. Today, however, critical history-writing is under threat. As elaborated here in the hitherto unpublished first and final chapters, the space for understanding its function, practice, and relevance has been undermined. Our world is deeply invested, economically as much as culturally, in the belief that the present is all there is and that history-writing has no purpose beyond the entertainment industry and the utilitarian politics of neoliberalism. Our times, as one historian recently put it, are "aggressively unhistorical."[1] In part this can be attributed to the fact that biological neo-Darwinian explanations and understandings have become hegemonic and quotidian. Especially frightening for critical history-writing and, indeed, for the survival of the very idea of history, is the tendency today in the social sciences and humanities to accept biological, and especially neurobiological, explanations for what it is to be human. Biology, besides thinking that it stands outside the historical formation of all knowledge, does not conceive of itself as just one way of understanding humanness. It sees itself as *the* way—more "real" than history-writing or any other subject in the humanities because ostensibly more useful and economically efficient. But all human knowledge is made in history, including that of the natural sciences. It is vital, therefore, to uphold critical history as an equally important form of knowledge and inquiry into what it means to be human. This is my belief.

But if all knowledge is historical, what about critique itself? Does it stand apart? This book suggests that it, too, cannot escape the rule of history. Although a critique of our so-called biomedical times currently preoccupies me, the bringing together of these essays is not directed to that end. My purpose, rather, is to demonstrate the changing nature of historical critique, and what this might mean for the future of history-writing. Through the illustration of my own involvements in the business, these essays highlight the various ways in which its content and form have changed over time in relation to the moments of its undertaking—a confirmation of the view that history is always written in the present. Collectively, they posit the hope that through greater awareness of our own historicity as historians, and the consequent need for us to adopt more reflexive modes of engagement, our enterprise might be revitalized and even, against all the odds, be re-secured as an essential discipline for the human understanding of humans. To this end, *Writing History in the Age of Biomedicine* is an invitation for historians to turn their analytical tools upon themselves as much as upon the objects of their study, while not forgetting that they themselves, in their values and morals as much through their heuristic tools, are historically fashioned. To be aware of one's historicity, and therefore deny the "objectivity" of historical knowledge is, I believe, the precondition to making history-writing engaged and political.

All of the essays, bar one, were written over the past decade as commentaries on postmodern and post-postmodern tropes in academic history-writing (four of them coauthored with Claudia Stein). Focused on the historiography of medicine, the body, bioethics, biopower, and neurobiology, they comment on the various turns that historians have made, from the social and the cultural, to the somatic, the visual, the spatial, the ethical, and, now, to the "material" and "posthuman." The exception is an essay written in the 1970s on the social construction of medical knowledge (Chapter 2), which is included in order to make apparent some of the conceptual and methodological shifts that came later. All, including the exception, were conceived as critical inquiries into the politics of these moves, particularly as manifested (or not) in the history of medicine. They were also deliberately pitched to arouse some degree of constructive discomfort or to circumvent the complacency that inhibits the quest for what is always, ultimately, self-understanding. Together they constitute a kind of contour map of recent trends and trajectories in the discipline of history, or the various points of its passage. At the same time they contribute didactically to the need to reclaim academic history's political role as a critical epistemology conducted through reflection on the dialectics of epochal change.

To enhance these aims and render the whole greater than the sum of its parts, I have added prefaces to the essays. These elaborate on the conditions of possibility for their being written in the first place. Along with the

Introduction (Chapter 1), which also charts my own place in the changing landscape for history-writing, the prefaces explore the intertwined socioconomic, political, conceptual, and epistemic constraints upon the practice past and present. Among other things, they mark how intellectual positions are never just intellectual positions, but are always a part of wider interpersonal relations exercised in and through work environments. I was fortunate in my career to experience several such places, each with its own particular dynamics (on a quasi-permanent basis, in Cambridge, Calgary, Halifax, Nova Scotia, Oxford, Manchester, Norwich, and London). How those different contexts bear upon what I produced seems to me important to bring out, if only because my passage through these places also constitutes a part of a fast-fading history of our times, and of the men and women within different branches of academic history who strove to shake and shape it.

At one and the same time, then, this book is a chronicle of historical critique and a retrospective reflection on the shifting nature of its conduct. To a degree, it is also a call-to-arms for a change of view in the regard and practice of academic history-writing. As emphasized in the first and final chapters, it is an encouragement to self-reflexive critique as a means to counter what I see as the two major forces now constraining it—the first, the new persuasion to neurobiological understandings of humanness, and the second, the postmodern identification of historical practice as entirely "subjective" in being steeped in, and wholly reliant upon, the moral values and norms of its times. In its conception as an exercise in self-reflection, this book suggests a possible solution to this double bind: the active locating of the historian in the problem, or the abandonment of the belief that the historian can continue to stand on the outside of what is to be analyzed, as if capable of autonomous "objective" judgment.

As some readers will be aware, this has not always been my position. In fact it is relatively new to me, something I was helped to only over the course of the past decade or so. In effect, *Writing History in the Age of Biomedicine* is the story of how I came to it. Through its Introduction and the prefaces to its chapters, no less than through its chronologically ordered illustrations of historical critique, it documents my changing methodological and historiographical interests in order to reveal how the separation between the political and the personal, or the intellectual and the subjective, is not possible except through the denial of the historian's self. As such, these essays sketch what might be described as a conceptual-cum-epistemic journey. In this case it is one instanced through the *Werdegang* of a social and cultural historian of ideas in science and medicine who has moved (not without some effort, it will be seen) from a broadly neo-Marxian to a broadly neo-Foucauldian identity.

Because the Introduction and the prefaces to the chapters have been written from the furthest reach of this journey and reflect upon it, the narrative cannot be regarded as having had preconceived destination. Nor should it now be conceived as disclosing an evolution to some greater historical meaning. There is no meaning to historical writing precisely because there is no end to, or escape from, the ever-being-remade present in which it is composed. Thus it is always provisional and imperfect. At most the historian can attempt to delineate some of the shifting limits to it by attending to the struggles, strategies, and tactics that, as Foucault observed, render "meaningless history" nevertheless intelligible and susceptible to analysis down to the smallest detail.[2] *Writing History in the Age of Biomedicine* aspires to just this, in much the same way as its individual chapters were originally conceived.

Because nine out of ten of them were written over a relatively short period (some more or less simultaneously), there is a degree of overlap in the way in which they approach their objects of inquiry, as well as in the viewpoints expressed and the kinds of sources drawn on to illuminate crucial shifts in historiographical discourse. I make no apology for this; it is largely unavoidable and possibly even advantageous, much as looking at a painting from different vantage points or, to be more exact here, looking at different paintings from approximately the same point, with attention, above all, to shifts in that "approximation." Wider changes in politics and ideology, and in language, discourse, and thinking, are only subtly mediated through intellectual endeavor. If they were not, if they were immediate or seismic, there would be little need for the kind of exploration and historicization undertaken here. It is precisely because of their occluded nature that they invite it. Hence, overlap in the essays' conceptual architecture should be taken as illustrative and instructive in itself—indeed, as I suggest in Chapter 1, can comprise a fit subject for theorization if approached from the perspective of underlying epistemic warrants. Beyond light editing, therefore, I have mostly only altered the previously published essays where there was duplication of some textual material, inserting cross-references where necessary. To attempt more would be to editorialize my historical self out of the picture, the opposite of what this book intends.

ACKNOWLEDGMENTS

These essays would be considerably diminished were it not for the inspiration, encouragement, and critical feedback of dozens of friends and colleagues over the years. Very much, their investments of time and intelligence constitute a part of the book's narrative. I cannot thank them enough. Some, such as Bill Luckin, Chris Lawrence, and Illana Lowey, have been at my elbow all along, while others more recently have played a decisive role. Among the latter, I am especially grateful to Rhodri Hayward, Stephen Casper, Suzannah Biernoff, Rob Kirk, Mike Sappol, Max Stadler, Roger Smith, Sander Gilman, and, above all (for reasons assigned in Chapter 1), Claudia Stein. I'm also obliged to Steve Fuller for forcing me to challenge what exactly it is that I'm trying to challenge, and why. Not least, I am grateful to the two anonymous referees for Yale University Press for spurring me on and helping me to sharpen my reflections.

It is a pleasure to thank the many others who lent their expertise in the original preparation of these essays. Among them, in chapter order, are Dorothy Porter, Katherine Angel, Rebecca Earle, Graham Mooney, and Norberto Serpente; Stella Butler, John Pickstone, Sam Shortt, and Bob Young; John Arnold, Elsbeth Heaman, Rickie Kuklick, Eve Seguin, Sally Sheard, and Steve Sturdy; Adam Bencard and Javier Moscoso; Renate Stein, Martin Gorsky, Julie K. Brown, Guy Attewell, Jürgen Döring, Sir Nick Partridge, and Mike Laycock; Skúli Sigurdsson, John Harley Warner, and Beth Linkner; and Logie Barrow, Mark Hailwood, Marion Hulverscheidt, Paolo Palladino, and Joan W. Scott.

I must also express my gratitude to the journals and collections in which eight of the essays first appeared, not forgetting the often generous help of their editors and technicians. Chapter 2 appeared in *The Problem of Medical Knowledge: Examining the Social Construction of Medicine,* edited by Peter Wright and Andrew Treacher (Edinburgh: Edinburgh University Press, 1982); Chapter 3 in *Locating Medical History: The Stories and Their Meanings,* edited by Frank Huisman and John Harley Warner (Baltimore: Johns Hopkins University Press, 2004); Chapter 4 in *Arbor Ciencia, Pensamiento y Cultur*

186 (2010): 393–405; Chapter 5 in *Medizinhistorisches Journal* 42 (2007): 180–209; Chapter 6 in *Medical History* 55 (2011): 85–108; Chapter 7 in the *Lancet* 375 (2010): 111–12; Chapter 8 in *Social History of Medicine* 23 (2010): 662–72; and Chapter 9 in *History of the Human Sciences* 23 (2010): 109–28. I am grateful to all the publishers for permission to reprint the papers (with only minor changes to text and titles in most cases).

For consummate professionalism, guidance, and unfailing good humor, I am especially grateful to Jean Thomson Black and her editorial team at Yale University Press, including my meticulous copyeditors, Joyce Ippolito and Laura Jones Dooley.

Last, but hardly least, I acknowledge with the utmost pleasure the charitable body to which this volume is dedicated. Most of these essays would never have seen the light of day had it not been for its generous patronage for almost the whole of my career.

1

The End?
History-Writing in the Age of Biomedicine (and Before)

Some of the most important watersheds in human history have been associated with new applications of technology in everyday life: the shift from stone to metal tools, the transition from hunting and gathering to settled agriculture, the substitution of steam power for human and animal energy. Today we are in the early stages of an epochal shift that will prove as momentous as those other great transformations. This time around, however, the new techniques and technologies are not being applied to reinventing our tools, our methods of food production, our means of manufacturing. Rather, it is we ourselves who are being refashioned.
—Michael Bess, "Icarus 2.0"

There is now a substantial and rapidly expanding literature depicting the transformation of our times through biomedicine, biotechnology, and neurobiology.[1] Written mainly by scholars in sociology, anthropology, and science studies, it reveals how the new life sciences have come fundamentally to challenge our understanding of being human. "Biocitizenship" and "neurological identity," it claims, have replaced social citizenship in the course of molecular biology and neurobiology's reduction of human life entirely *to* its biology.[2] A "neuromolecular gaze" is said to have come to reconfigure our moods, desires, and personalities.[3] And through visualizations made possible by molecular diagnostic technologies such as DNA sampling, functional magnetic resonance imaging (fMRI), microarrays, microfluidics, "lab-on-a-chip," and protein staining, the human body has been reconstituted "at the critical interval between perception and knowledge," or between how we see it and how we think it.[4] This reconstitution, linked to biometrics and other information technology, is held to have produced a new anatomical body for what many scholars now refer to as "posthuman medicine."[5] No longer are we simply the potential victims of applications of biology *to* society (sociobiology).[6] In a world in which humanness is flattened to the biological, the salience of the social disappears altogether, while humanness becomes, as one social scientist has put it, a folk category in a neo-Darwinian world.[7] Such are some of the alleged features of our "posthuman age" and some of the ways in which our "neuroculture futures" are coming to be imagined.[8]

Illuminating much of it, Nikolas Rose, author of *The Politics of Life Itself* (2007) and "The Human Sciences in a Biological Age" (2012), has pointed to how peoples' conception of themselves has been transformed from hardly even half a century ago. No longer do we conceive of ourselves largely in terms of a "deep interior psychological space" but rather, increasingly, in terms of a self-evident biology.[9] Other social scientists have expanded on this, elaborating, for example, how today's hugely profitable sale of products for "erectile dysfunction" rests on an understanding of "impotence" not as a psychological problem, but a physiological one—a comprehension powerfully solidified through "the problem's" instant pharmacological solution.[10] This example is also but one of how the ideal of the perfect body armored against all the corruptions to which flesh is heir has come substantially to redefine the practice of Western medicine away from the treatment of illness toward the supposed greater enhancement of physical and mental well-being.[11] Further, it points to how the human body down to its molecular level has come to be exploited and sold as a valuable commodity. Newly "materialized" in this sense, bodies and brains have become shimmering objects not just for research, explication, and sensational popular exhibition, but for vast capital accumulation.[12] Ethics themselves have been "somaticized," or made to support the commoditization and commercialization of body parts in their global exchange.[13] Little wonder, then, that the medical profession's "selfless humanity"—its stamping of medicine as "the greatest benefit to mankind"—has been all but consigned to the dustbin of history along with the literature that lent it support.[14]

Yet—and here's the rub—through the very act of observing this professed new "biological age," its ethnographers also market it. In their determinist claims for the new bio-knowledges and practices, they unwittingly perform *for* the professional, commercial, and political interests invested in the claims. This includes (less wittingly perhaps) their own academic investments. At the very least, their observations may be said to be more than what they seem; as a character in Richard Powers's novel *Galatea 2.2* wryly observes, "You think observation doesn't have an ideological component?"[15] Nor, of course, are these scholars today's only "observers"; the new age of biology is also substantiated by the many popularizers of neurobiology, including those who use it to buttress claims for social justice and other liberal values, and those who enhance its cultural valance by stressing how human nature is now to be understood in terms of this or that neurochemical marker.[16]

I do not mean to suggest that the image of a new biological and neurobiological order is a fiction, or that the economic and political power attributed to the new life sciences is but a fabrication of the self-interested and/or deluded. It is far from it, in my opinion, even if a neurobiological "revolution" remains yet to be proved, and even if the general public and most scholars in the humanities may yet have little awareness of the claim that perception, cognition, identity construction, "the social," and "life itself" have all undergone a refashioning as a result of the new life sciences.[17] That a new generation of scholarship is "coalescing around the assertion that the very grounds of life itself are changing" is only right; the new biomedical theories and practices clearly *do* demand careful exploration and analysis.[18] But is not such attention at the same time an elaborate *representation* of our times, in the sense of something that both helps to make up a regime of truth and performs for it, culturally and existentially? "Truth," as Nietzsche said, "is a power that renders itself true by prevailing," and the representation of our times as neurobiological surely does just that.[19] Instructively, the particular game of truth that Nietzsche observed in the nationalistic context of the newly unified German state was that of an excessive valorization of the past. Through "a consuming historical fever," he observed, the dead were burying the lives of the living.[20] Today, in stark contrast, the living might be said to be burying the dead (the past) through a consuming *a*historical preoccupation with the bio-present as all there is and biology as all we are. One of the many consequences of this is an obliviousness to, or denial of, the practice of history as a means to understand the present and guide the future. Instead, the past is trivialized; packaged into a sellable media commodity, and in this way, by no coincidence, effectively fitted to today's entrepreneurial prizing of innovation, or the "sloughing off of yesterday," and starting over.[21] Academic history, beyond its use in political propaganda, is dismissed as having no practical (that is, market) value, let alone intrinsic value. What's "old" is unwanted, like last week's iPod or mobile phone. Instead, we are encouraged to the new, to a future of never-ending growth and "economic progress"—an ideology now all the more forcefully preached after being caught (yet again) with its pants down. At best, the study of the past can only contribute to an understanding of decay, the stuff perceived as holding back the present and the future. In the English-speaking world, and perhaps in the Westernized world as a whole, the prevailing view is that present-centric economistic thinking about the future is all that matters.

Thus, to cut to the chase, it is not the accuracy, extent, or otherwise of the representation of our times as neurobiological and posthuman that should immediately draw our attention, but rather how this truth or how this whole field of observation with all its ahistoricity and disdain of academic history has come into being. How, in temporally specific terms, did it get constituted and naturalized? What legal, moral, ethical, philosophic, economic, political, and other institutional strategies or technologies of power, as Foucault called them, were involved in enabling this to happen, so as to permit a new governance of the self and others? And how (to the extent they have) did older values, technologies, practices, political and economic ideas and ideologies, notions of rights, and so on get retooled for this truth to become so shiny, believable, and powerful? What has been closed off, or is *being* closed off and tarnished, by this truth coming to prevail?

Such questions are historical, although not conventionally so, and not just because they query the present. They reflect a distinct kind of historical practice, sometimes called postmodern, constructivist, or representationalist. Honed over the past thirty years or so by critical theorists exploring the constitution of modernity, representations theory encompasses a multitude of different means and methods, but at its core suggests that there is no reality beyond what is represented through language, which is shaped historically.[22] Its scholars, including now many historians, have elaborated the socio-political and historical epistemology of concepts and categories as basic to modern science and modern thought as "objectivity," "empiricism," and "experiment."[23]

They have exposed and contested the idea of would-be universal, historically transcendent concepts and essentialist categories, such as "consciousness," "biology," "the body," "the social," and even "history" itself (as a universalizing and homogenizing metanarrative).[24] They have challenged the naïve acceptance of facts as nothing more than facts. (Hence, as here, the frequent intrusion of scare-quotes around words and concepts to indicate that they can no longer be taken for granted.) And they have called into question and de-privileged, if not wholly inverted, the exalted valuation of science and its method as *the* standard for all other social and cultural enterprises.[25] In short, they have demonstrated that everything is historically contingent, including the reasoning and rationality involved in representations of "the present," "the past," and "the future."[26]

Representations theory and practice owes much to Foucault, whose "genealogical method" directed attention to the conditions of possibility for the emergence and unfolding of any kind of knowledge and power.[27] Impelled by a desire to comprehend the evaluative frameworks of the present (as well as the past within what he referred to as the "history of the present"), Foucault's genealogical method encourages the de-centering of any object of investigation.[28] It urges resisting the temptation to follow well-established modes of investigation, such as the compulsion in academic history to the linear tracing of causes and effects, "forerunners," "origins," "culminations," "inevitable outcomes," and so on. As far as possible, the approach permits standing on the outside of any object of inquiry and investigative practice to explore the nature and exercise of the theories, concepts, and categories that sustain it. What, it asks, are the multiple sources of power that make things seem the way they seem? What are the practices and theories, and the combined conceptual and epistemological, juridical, and institutional strategies that create a particular field of truth in which the object of any investigation is to be perceived? Basic to Foucault's genealogical method, and that of representations theory in general, is the understanding held by Marx in his early writings that nothing is ever outside of history, including our conception of ourselves and our most abstract categories, not absenting the very historical knowledge on which the modernist or Enlightenment project to demystify the past was based.[29] But in representations theory, Marx's view of history is to be seen as itself a representation—one heavily reliant on Hegel and with a particular understanding of itself as a conscious process that consciously supersedes itself. Thus, from the perspective of representations theory, Marx's view of history constitutes a technology of power, or a means by which a particular representation of the world is sustained.

A representations approach to the would-be neurobiological present encourages directing attention to features that do not come into focus through its ethnography, anthropology, and sociology. It incites us to read outside or against such texts—above all, to see them as *a*historical in denying any relation to the past of their objects of investigation. Among other things, it raises for question the insistence in these texts that the refashioning of the understanding of "life" necessitates the radical overhaul of long-held conceptual distinctions and divisions of labor between academic disciplines (especially between the natural and human sciences)—the politics of which I will come back to in the third and final part of this chapter. It also encourages questioning the epistemological implications of returning to biological essentialist explanations and conceptualizations for what it is to be human. Further, it invites exploring how an emphasis on the would-be vital importance of the (represented) biologized and biotechnologized present devalues the worth of trying to historicize it, even though, paradoxically, the alleged grounds for both the devaluation and the historicization are inherently historical: after all, as Rose and others insist, the present is far more "complex" than anything that has preceded it.[30] Perhaps most importantly, the representations approach permits exposing what might be depicted as the nasty fly in the ointment for the representationalist understanding of our biologized times, the fact that that biologization closes off that very means to its investigation.

Most evidently, this closure is manifested in what is now proudly proclaimed as "non-representational theory," the aspiration of which is to elaborate "multisensual worlds" that are "more-than-human and more-than-textual."[31] Although the genealogy of non-representational theory and its modes of thinking and expression are entirely postmodern, it also embraces "affect theory," which, though similarly postmodern in its origins, draws directly on the neurosciences to make claims about human cognition and identity.[32] Both non-representational and affect theory routinely encourage the intertwining of biology and culture and explicitly call for the "biological understanding of semiconscious cognition."[33] Both convey the belief that there really is a reality beyond representations, which can now be located in the brain. Explicitly, they claim that the days of representations and postmodern theory are gone. "Constructivism" is passé, reports Rose, "no longer are social theories thought progressive by virtue of their distance from the biological."[34] Thus, no less than the analysts and ethnographers of our biological times, the theorists of non-representationalism and affect provide technologies of power for the new game of truth. Moreover, in doing so they endorse neoliberal economics and its faith in modern managerialism. This is clear in their devotion to concepts such as fluidity, exchange, interchangeability, logics of propensity, and "susceptible situations which can be ridden rather than rigidly controlled"—notions all culled from contemporary business management.[35] In this respect they manifest, uncritically, one of the tenets of critical theory as elaborated in the New Historicism of the 1980s and 1990s where representations theory was born—namely, that no critical method or language adequate to describe culture under capitalism can do so without participating in the economy described.[36]

Admittedly, representations has not been every *historian's* cup of tea. Far from it. Above all, this is because it inhibits historians from assuming "objective" detachment from the contexts in which their objects of investigation appear. Indeed, it renders them incapable of acting as analysts of pasts "out there" or "over there," separated from the contexts in which their own questions are framed. The present in which the historian operates, it insists, provides no neutral shelter in which to stand objectively observing. Rather, like all spaces, that of the present is to be seen as historically constituted and hence impossible for the historian to extricate him- or herself from. As some historiographers and critical theorists have suggested, therefore, the only solution for the historian is continually to interrogate him- or herself as the analyzing subject, constantly self-monitoring and destabilizing the historical self.[37] Understandably, many practitioners find this is a frightful prospect. But it need not be; far from marking an end to academic history, it might be regarded as a means to restore a degree of honesty and credibility to a practice now often criticized by insiders and outsiders alike for its pretentions to scientificity or objective truth recovery. At a time when academic history-writing has never been more under threat (at root, ironically, precisely because of the adoption of *a*historical modes of thought), any opportunity for enhancing its credibility should be seized upon. It does mean, though, the abandonment of the idea that the historian stands as if on Mars. And it involves understanding the need for this abandonment on grounds beyond those merely acknowledging that history-writing is always conducted in the present.

It also has to be confessed that the representations or postmodern mode of critical history, no less than non-representational theory, is itself a technology of power. It emerged at a particular moment in time in order to apprehend the present in a particular way (above all by deliberately trying to read against other people's texts, of which all the human world could be said to consist and be made "real"). Its moment was the end of the Cold War when totalizing theories and investigative practices such as those of Marx and Weber no longer seemed adequate to explain the world. But to register the representations mode of critical history as itself a historically contingent technology of power does not mean that we must now consign it to the proverbial dustbin. In my opinion it remains a healthy means to keeping history-writing fit—if nothing else, a way to maintain unfettered curiosity into the temporal specificities of the past and the present. Indeed, the method is possibly the only means to do so in the face of two powerful new forces in historical practice that threaten to impale it: first, a return to neo-empiricist or neo-descriptive history-writing—to the "real"—often posed in opposition to postmodern representations theory,[38] and second, more worrying for its explicit commitment to non-representational and affect theory, the argument that history-writing needs to be de-centered from within as a result of our new location in a biologized posthuman world. Both positions need some elaboration.

The first finds its apogee today precisely in that history-writing that buys directly into the new bio-representations of reality, especially evolutionary brain science. Its practitioners do so in the belief that the new knowledge provides an innovative tool for digging deeper into the understanding of ourselves and our past. What they and fellow neuro-travelers in the humanities and social sciences overlook, however, is that the tool is reliant on the positivist methodology of the natural sciences.[39] Its adoption (which instructs us not only on *what* we are to perceive as "relevant" in our world, but also on *how* we are to perceive it), entails accepting the purportedly ideological-free reductive logic of the natural sciences as based on experimentation and controllable data—the very things that we know from thirty or more years of research in the history of science were politically fashioned in history. Thus the very act of taking up of the neo-realist position means the surrender of a view of the world and its conceptual categories (including biology and the brain sciences) as historically constituted. Wittingly, or unwittingly, its advocates fly in the face of postmodern thinking, or for that matter the young Marx, for they undermine not only the basis for the historical pursuit of the alleged bio-human or posthuman condition as a *representation* of a particular power constellation, but also the rationale for its investigation as such. At the same time, the neo-realist position opens the floodgates to a neo-essentialist historical practice, reinforcing the new scientized representation of reality and submitting to its totalizing grand narrative of evolutionary progress.[40] Lost along the way through the collapse of the historical study of human life to its biology is the idea of history as a form of knowledge and inquiry different from the biological—a form of knowledge with the potential to contribute as an alternative to the now increasingly culturally singular neo-Darwinian grand narrative of science as inherently evolutionary and progressive. Through this collapse, academic history gets overwritten by a biology that assumes itself outside the historical formation of all knowledge, or doesn't see itself as just one way of conceiving what it is to be human.[41]

The second position inheres much of the latter's pre-postmodern thinking, but from a post-postmodern perspective. Inculcated by scholars keen to move beyond the thinking of their postmodern elders, it takes as given that "molecular biological research has opened up an interest in the biological within us."[42] Co-opted *to* this representation, its advocates search for ways to think about the "posthuman condition" that is assumed to derive from this new dispensation, and to think about how it might be historicized. Touted is "the human condition in the post-genomic [world] ... of biology."[43] And, just as post-genomic biology ("proteomics") promotes itself as offering "customized study design," "differential expression profiling," "exchanges," and so on, so (the argument goes) intellectuals must now attend to the "concrete matter of the body, ... spatiality, materiality, affect, experientiality, emotions, the unconscious life of the body, and *the more-than-representational*." As the author of these remarks concludes, "the question of what sort of history we should write and how we should do it then comes down to how we can produce knowledge that is *pertinent for the condition of bio-humanism*. Traditional structures within history as a discipline specifically, and the humanities more generally, simply no longer seem to provide useful or pertinent insights. The task now is ... to figure out how to produce such knowledge" (italics mine).[44] But pertinent to whom, we might ask (mindful of the evident deployment here of the language of neoliberal economics). The issue is lost sight of in part because here too science is unwittingly reinstated as having unmediated power, if not in fact returned to positivistically as something wholly separate from culture. Science is once again transformed into ideology, a source of autonomous power, without being recognized as such. Thus would-be posthumanist historical study itself becomes complicit with the ever-extending investment in the *a*historical bio-construction and understanding of human identity and selfhood—the new politics of life. Within this frame, the task of the historian becomes merely that of following these politics around, further ethnographizing the world according to the new life sciences' field of observation (as if, incidentally, the designation "life sciences" wasn't itself another historical construct, and a rather telling one at that vis-à-vis the "human sciences" and the older, if corrupted in cultural translation, distinction between *Naturwissenschaften* and *Geisteswissenschaften*).[45] Academic history thus becomes installed as another site for the new game of truth, not as a means to its critique.

Posthumanism, by placing humans on a par with animals, plants, and other nonhuman material things, deprives human agency of any special status.[46] It has worrying political implications therefore beyond those merely for writing academic history (discussed in Chapter 10). Compromised, if not assaulted, is the conception of ourselves as unique ethical beings. Although some might argue that this assault was presaged in the "anti-humanist" origins of poststructuralist or postmodernist thinking, in practice, poststructuralists and postmodernists retained an inherently humanist faith. Importantly, they also maintained a healthy skepticism of biological concepts and categories for fear of falling into essentialisms deemed hostile to the possibilities for cultural transformation.[47] The ideology of science they contested.

I agree with the humanist historian of science Roger Smith that, contrary to what posthumanists maintain, the bounds of being human are not known, even if it is only too apparent that the term "human nature" is no longer adequate to the task of its historical analysis.[48] Further, I agree with one of the pioneers of postmodern cultural theory, W. J. T. Mitchell, that humanists are precisely those who cannot accept the obsolescence of the concept of wisdom that we humans alone possess (to the degree we possess it). The concept of wisdom, Mitchell insists, "underlies all critical and theoretical investigation of the species being of humanity." Hence our business in the humanities must continue to be that of "engag[ing] in ethical reflection, analyses of the grounds for the making of wise decisions, responsible interpretations, logical deductions, accurate estimations of aesthetic quality, the critique of religion, culture, politics, the arts, the media, languages, texts, and so on."[49] Sure, this is a faith—one fostered through the Enlightenment—but so what? Is biology, as positivist ideology, any less a faith (and with proven hideous consequences at that)? And are we to forget that biology was also an invention of the Enlightenment, like all the other modern academic disciplines? In effect, the business of critical history now more than ever becomes the realization and defense of the humanities as the custodian of a metanarrative in need of clear positioning against the powerful reductive one of the natural sciences as narrowly based on *a*historically conceived evolutionary theory and notions of progress *in science*. At issue is how to halt the forces now in place for the further degradation of this alternative metanarrative as meted through the universalizing biological understanding of the world—representationally, experientially, and materially. How can historical practice help, and maybe help itself in the process? Although it seems unlikely that we will soon (or ever) witness a new humanist philosophy of history to challenge the discourse of the natural sciences, the business of endeavoring to set it out *as an alternative* can at least provide some clearing for the possibility of the project and its rationale. Defending representationalist postmodern critique, in other words, can in itself be regarded as a form of resistance to rapidly expanding *a*historical biological representation.

Required in order to initiate this, however, is a better appreciation of the constructedness of all historical thought and the norms and virtues on which it runs. In effect, what is needed is attention to the historical epistemology of history-writing and the historical ontology of its authors, that is (as the historian of science Lorraine Daston defines "historical epistemology"), to "the history of the categories that structure our thought, pattern our arguments and proofs, and certify our standards for explanation."[50] It is precisely this, though, that tends to be resisted by Anglo-American historians. Instead, they prefer to draw a sharp and inherently evaluative distinction between their historical practice and the discussion of its premises. Very much this is a part of the problem when it comes to seeking to defend the practice of history-writing today; the defense can not be mounted for it knows not on what it rests. In the anglophone world, R. G. Collingwood in the 1930s and 1940s and E. H. Carr in the 1960s were lonely exceptions to this kind of concern.[51] In their different ways, both made apparent that history-writing is always historiographical, not only because it is always informed by the politics and ideologies of the moment in which historians choose their topics and interpretations (or, more accurately, their topics and interpretations choose them), but above all, because history-writing is always underpinned by and infused with reigning philosophical and/or metaphysical beliefs—be it late nineteenth-century positivism or twentieth-century liberal notions of self-creation and self-governance.[52] Following Collingwood and Carr, we might argue that what is required today is less an understanding of the contemporary sociopolitical shaping of the writing of the past than a comprehension of the unspoken subjectivities, or below-the-radar values and ideals of the present through which history-writing is articulated.[53] Arguably, it is these unstated assumptions and values that saturate, girdle, and guide the professional enterprise as a whole, including its making-up of narrative structures and supportive "fact" and "fiction" dichotomies.[54] These underpinnings are best styled epistemological, or as having to do with the nature of knowledge and understanding and the subtle ways in which it is generated and sustained in different epochs.

To subscribe to this view and ask how history-writing embodies the epistemic moment of its production is not to concede to "present*ism*," the long-denounced "temporal imperialism" that naïvely reads the past from the assumed-to-be-knowable values, interests, and assumptions imbibed from the present. That kind of presentism, even in its more recent ideological castigation as a kind of teleology in reverse that dis-estranges a past that should always be estranged, is not one that sees history-writing (or any other form of intellectual activity) as always inside its epistemic present and, as such, always embodying and reproducing the values and virtues of its conceptual or metaphysical moment.[55] Nor is it one with much respect for the now fairly conventional view that history, like meaning, is made, not found, and that that which is made has itself already been shaped by the terms on which we seek further to shape it. Rather, the epistemological presentism suggested here entails admitting that the study of the past, and the very objects for its study, invariably come into existence within epistemes (to use Foucault's term) that permit and condition the basis of thought.[56] To unveil this thinking in history—the Nietzschean idea that our thoughts are never our own, or that the questions we pose and the solutions we find are always of the epoch in which we live—it is necessary to plumb the specific values, virtues, assumptions, prejudices, and beliefs to which history-writing commits and through which it mediates accommodations to the present. We need only add that such virtues and values also morph over time and, hence, are themselves as provisional as all history-writing.

The task might be depicted as an invitation for historians to turn their analytical tools upon themselves as much as upon the objects of their study, while not forgetting that the tools, too, have been historically fashioned and are far from neutral. This is not simply a matter of "knowing thyself" for its own sake; it is comprehending the historian's own moral categories, subjectivity, and epistemic envelopment in order that he or she can act politically within any historically shaped present. Without such awareness the historian is stymied, stuck in the prison-house of epistemology as it were, unable to either understand the nature of his or her own intellectual actions and agency or formulate positions to mobilize critically and politically in defense of them. One of the consequences of this failing, as we shall see, is that others then easily come to decide the worth and business of academic history, while still others inside the academy and even inside the humanities can point to the discipline's contemporary irrelevance.[57]

Historians, while often brilliant at fitting ideas into their appropriate historical context, usually resist reflecting on their own projects in relation to the multiple contexts in which they conduct them.[58] Not present to the fact that history-writing (including its choice of methodologies) is always already *in* history, *in* theory, *in* epistemology a priori, they unwittingly treat the present *a*historically. They disclaim responsibility for adopting any such understanding on the grounds of dealing only with an ostensibly objectified past. They may not always treat the past as a place where the facts of history can be left to speak for themselves, but typically they regard it as a place in which the present that the past is written through is factual, real, and value free.[59] Justifications for history-writing, from Herbert Butterfield in the 1930s to Arthur Marwick, Mary Fulbrook, and Wolfgang Mommsen more recently, are relentless in their insistence that historians should use concepts that are as value neutral as possible. Yet, all the while, they assert such normative values as "the deployment of empathy with respect to a variety of voices and positions" instead of "partisan appeal to the audience"; the need to participate in controversies about the past "in as clear and 'disengaged' a manner as possible, irrespective of personal political and moral views"; and so on and so on.[60] Such are the virtues (the "good intentions," as Hayden White once dubbed them) that govern professional history-writing.[61] But they constitute nothing less than an ideology professing to be non-ideological. And it is this ideology that is extolled above all when "theory" is negatively juxtaposed to the presumed-to-be-value-free historical pursuit of "objective reality." Indeed, "theory" is often overtly disdained, while attention to a believed-to-be-objectively-observable "context" that does not politically involve the historian's self is insisted on as the discipline's highest and morally self-evident goal.[62] Instead of attending seriously to the "a priori," the historical and historicized present is closed off or ignored altogether. Regarded as a neutral space in which to conduct observations of "the past," the a priori theory-laden epistemic present is innocently reproduced in historical scripts. A part of the problem, as Joan

W. Scott points out, is that historians who continue to believe that moral evaluations lie outside their responsibility "do so without interrogating the meaning of the morality they take to be a shared, self-evident set of beliefs. Indeed, 'ethics' for most of these writers refers to a closed system of evaluations, one in which fixed categories of 'the good' and 'the just' are applied to events and actions in the past. The kind of self-reflective examination of the historian's own moral categories (called for by poststructuralists) is largely absent in this attempt to reconcile ethics and historical objectivity."[63] It is not the past, therefore, that historians need to conquer through their various rationalist engagements, but themselves located in their history and ideology.[64] In short, what requires estranging is the epistemic present that is continually reproduced *a*historically through the study of the past.

This is a very different "why bother?" than that which commonly impels academic and other forms of history-writing: the "how can we know the present without knowing the past?" It flies in the face of it, for assumed in that is the belief that the past as an objective reality can truly be known and, moreover (again normativistically), that there is something inherently worthwhile in seeking it out.[65] Further, it assumes that the tools to it (empiricism and objectivity, which were borrowed from the natural sciences) are not themselves historical constructs.[66] Thus, this rationale only re-inscribes on the practice of history-writing the critical triumph of modern scientific consciousness, perfectly instancing how historians are always implicated in the epistemic present whether they like it or not.

Taken as a whole, the essays in this book commit to a different rationale: that a history-writing that stretches to epistemic self-awareness can critically contribute to alternatives to the practice of history as we find it. Hence, it can open the capacity to reconfigure it and inform, or even precipitate, change. This rationale, which in the latter respect is at least as old as the Enlightenment, is no less idealistic or moralistic than any other.[67] But for that no apology is needed, since nothing is ever outside of historically contingent moral ideals. "Practical ethics" or "risk assessment," to take but two of today's commonly held counterexamples, only assume themselves free of idealism (as if such authorizing philosophies as that of William James's pragmatism or Jeremy Bentham's utilitarianism were not themselves expressions of idealism). For the writing of history today—indeed, for its very survival in the academy—this awareness is strategically important, since those wielding authority in today's utilitarian knowledge economy commonly legitimize it through reference to an efficacy based on pragmatics, and frequently distain history and other branches of the humanities for their lack of it, if not specifically for their supposed idealism. Which brings us to the relevance of studying epistemic virtues, for it is precisely these, as non-visible regulative ideals, that not only run the popular dismissal of academic history as lacking "relevance," but run the practice of academic history as something in which the fronting of these ideals is perceived as irrelevant.

Epistemic virtues have been defined by historians of science Lorraine Daston and Peter Galison as the "norms that are internalized and enforced by appeals to ethical values, as well as to pragmatic efficacy in securing knowledge." As assiduously practiced in the pursuit of natural knowledge, they are the "techniques of shaping the self, as well as of picturing nature."[68] For instance, a late nineteenth-century scientist seeking to render the ideal image of a natural object also has within her- or himself an image of the ideal image maker. Such embodied ideals, argue Daston and Galison, have "much to do with the ways in which epistemology is translated into science."[69] Today, they submit, the principal epistemic virtue running science is not that of objectivity in its high-Victorian positivist sense, but "trained judgment," which they define in terms of its difference from "truth to nature," the epistemic virtue that ran natural knowledge production prior to the mid-nineteenth century. This is different again from the "mechanical objectivity" that ran it thereafter. Whereas "truth to nature" was concerned with unveiling a nature that was obscured, and "mechanical objectivity" with clear-sightedness through technical command, "trained judgment" is the ethical virtue of the person of the scientist as a technical expert capable of interpolation who actively employs subjectivity in that process. Scientists involved with neuroimaging, for instance, color the images they produce not according to some external objective standard but according to their own (shared) aesthetic sensibilities. This is how they clinch arguments among their peers—a carrying-on that would have horrified scientists in the nineteenth century who argued for self-discipline in the possibility for (mechanical) objectivity. Hence the "natural representation of a natural object is no longer the sole object of scientific desire."[70] What is "real" emerges only "from the exercise of trained judgment" with its *subjective* appeals to "resemblances" rather than ideals of truth. Since the 1950s, apparently, what has been increasingly prized within scientific imagining has not been objective truth, or "the blind sight of mechanical objectivity," but the epistemic virtue of trained judgment extending from the unconscious intuitive self of the scientist.[71] The current objective of science is no longer to explain "a real world" but to "reveal *how* to know [it]." Thus, "subjective" judgment (interpretation, resemblance, discrimination, intuition, and so on) has supplanted the old notion of (mechanical) objectivity—a point to hold on to, as much for any historical approach to twenty-first-century science, medicine, and technology as for the problematic of the epistemic present embodied in historical practice.[72]

It would be difficult systematically to transpose the idea of invisible epistemic virtues in the production of natural knowledge to the production of historical knowledge. And it is noteworthy that Daston and Galison, in common with other historians of science, do not attempt this in their own historical project.[73] They stay outside their own analytical framework for the study of the past, keeping their history-writing apart from the business of historical epistemology. Nowhere do they ever impart the image of the ideal historian that commits them to their own task. Nevertheless, even if academic historians are far more diffuse than scientists in their core commitments, methodologies, and objectives, and indulge in different degrees of self-reflexivity (at the historiographical cutting edge, at least), the transposition of the idea of underlying hopes and habits from scientific to historical enterprises seems worth considering if historians are to become more self-aware of the constructedness of their own thought and, thereby, enable their critical engagement with the new biological regime of truth. It also seems possible in a way loosely parallel to Daston and Galison's historical undertaking; after all, historical knowledge, no less than scientific knowledge, is governed by ideals or by non-visible embodied normative values and epistemic virtues embedded in its practitioners and their institutions.[74] (What else is at stake in debates over historiographical methods, for example?) Furthermore, one of the main points that Daston and Galison raise about the role of epistemic virtues in the history of natural knowledge production seems entirely applicable to history-writing—namely, that existing norms or virtues are never simply superseded by newer ones. Epistemic virtues might sometimes collide, but rarely do they "annihilate one another like rival armies"; instead, "*they accumulate*."[75]

Again, such thinking, along with the growth of interest in historical epistemology more generally, owes much to Foucault.[76] Although Foucault never referred to epistemic virtues, he coveted an interest in the slow and often invisible maneuvers in knowledge and society over time, which, he believed, crisscrossed and connected institution and practices of all sorts.[77] These subtle maneuvers, he argued, were reflected in specific techniques and functions of power. They were to be perceived less as properties than as strategies, which (as Daston and Galison's history of the pursuit of scientific objectivity well illustrates) continuously shift their constellations over time, as well as continue to serve older strategies of power/knowledge.[78] It is not just that regulatory disciplines create norms that become embodied; it is also that what gets embodied produces new kinds of knowledge/power, the knowledge and the power never being separable.

As this idea of the role of epistemic regulatory values in knowledge production draws heavily on Foucault, so it can also be illustrated though the history of postmodernity as the intellectual project that Foucault did so much to inspire, especially through his attention to discourses around biomedicine and the human body. Postmodernity, both as a poststructuralist hermeneutic movement and as a more general outlook on culture, has clearly affected perceptions, valuations, and experiences of the world. Probably to a much greater extent than the new neurobiological truth (to date, at least), it has upset and instilled new ways of viewing and conceptualizing—especially among anglophone audiences (and for historically contingent reasons, as we can now better see).[79] To the extent that historians became attentive to this, they became complicit, if not always knowingly, with Foucault's grand project to problematize and historicize reason, and examine how people "govern themselves and others by the production of [such] truth[s]."[80] Whether they did so unwittingly, reluctantly, or with open arms (ditching ingrained causal and linear understandings for the study of temporally specific conditions of possibility, or abandoning "the chimera of origins" for genealogical approaches), prior ways of knowing were not left pristine; they were transmuted simply by becoming something *in relation to* "the postmodern."[81] As the historian of science Paul Forman has observed, "postmodernity ... has taken ever wider and firmer hold on our globalized psyche. Although we historians may be little schooled in postmodernist theses, we are firmly in the grip of postmodern prejudices. ... As a result, historians today—and not only the younger—reject the possibility of conceptually characterizing any substantial chunk of the past, reject the hierarchic ordering of causes of any historical happening—indeed, the very idea of causes—and, consequently, take complexification as the only legitimate goal of historical research and exposition."[82]

Even to wish postmodernity away, as neo-empiricists do, is to be complicit with it. At the same time, though, to embrace it does not necessarily entail the wholesale abandonment of all formerly held values, virtues, and beliefs and the institutions and practices attached to them—as belied, for instance, by Daston and Galison's self-confessed embrace of the Enlightenment rationale for their own history-writing and, indeed, by their chosen (hence celebrated) subject matter of "objectivity" in knowledge production in science.[83] The fact that the postmodern literary historian Terry Eagleton could also be a Marxist and Catholic critic of postmodernity says much in this regard. For it is only subtly, and over a relatively long span of time—and then only at a level of abstraction—that casts of mind can be seen to change fundamentally and thereby come to render former castings invisible.[84] However many manifestos for history-writing might appear, shakeouts are possible only in hindsight, and then only in part, since the epistemic basis for the triangulation made in hindsight has already shifted in the meantime. "A" does not linearly succeed "B" in the same argument; rather, "B" replaces "A" because everything around "A" comes to be replaced. A different mode of thought comes to distinguish the one from the other, despite retaining much of the language and many of the modes of thought of "A." (Non-representational theory in relation to representations theory is a good example.) In history practice, "A" in the project that nurtures it loses the relevance it has accumulated as "B" acquires new relevance. Rarely, then, do "A" and "B" clash head on, as the advocates of "B" might later purport.

I find this helpful for thinking about historiography—the history of history-writing—and for the historiography of medicine in particular. For one thing, for the history of medicine, it moves beyond mere congratulation for a subfield of medicine turning into a subfield of history. That is to say, it goes beyond praising progress within a framework of narratives of origin and causal change. For another, it permits collapsing the distinction between the history of medicine and the epistemological envelope in which it is written, without having to surrender wholly to language, or sacrifice attention to social, political, and economic contexts (the unfortunate tendency of Daston and Galison's philosophically inclined study of scientific objectivity).[85] More importantly, to think of the historiography of medicine in terms of regulatory constellations of thought that overlap and shift at the same time raises for analysis the intellectual neutralization or *a*historicization of the present in history-writing. It permits the historian consciously to be brought back into history-writing as the embodied bearer of virtues and values that change the nature of history-writing practice over time, just as in science invisible embodied virtues change natural knowledge and scientific epistemology. Histo-knowledge production can thus reunite with its producer. As an approach this offers some escape from the myriad clutches of a modernist historiography in its denial of subjectivity and the problem of objectivity.[86] For the instrumental purpose of turning history into an effective technique for critically engaging with the material and epistemological present, including the biological naturalism of the new life sciences on the one hand, and post-human or postnaturalist philosophy on the other, it is valuable because it takes seriously the knowledge producer—the historical self—and the object of study. It permits revealing from the inside the normative values and epistemic virtues that inform and outlive the period of their production. At a trivial level, it also permits perceiving the "so what?" and "why bother?" of history-writing as themselves embedded and embodied moral claims for historical relevance. Both smart-aleck remarks, after all, lay the charge of wasteful and indulgent—sinful even—at the door of a history-writing regarded as less efficacious or practical than that purported to stem from science and political economy. Thus, the "so what?" and "why bother?" can be seen implicitly to collude with the latter's ideological and epistemic claims.

Auto-Historical Illustration

The essays in this volume track my own overlapping epistemic presents in history-writing. They illustrate how different modalities of thought come to operate and then disappear, surfacing and then evaporating not through radical reformulations, but through gradual and largely unaccounted-for accommodations to shifting political and material conditions, themselves constitutive of different structures and hierarchies of intellectual production. That the essays are my own (albeit co-authored in four instances) may appear to violate a normative value that the German professor of medicine Julius Pagel (1851–1912) regarded as the cardinal one arising from the study of medicine's past among his medical colleagues, that of modesty.[87] I hope not (to express one of my own shared values). As a rule, academic historians do not autobiographize, and collected essays tend to be gathered in festschrifts produced by others.[88] In fact, with a few notable exceptions, historians are unusual in this modesty.[89] For the most part, they retain a pre-postmodern, would-be author-ative, passive voice—distanced, ostensibly objective, and third person. This voice, as George Orwell observed in "Politics and the English Language" (1946), is one of the constituent features of ideological language.[90] Intended to remove a sense of agency, or to suppress the primary position or will of the author, it is that of the would-be disembodied biomedical and scientific author/agent alleging disavowal of bias, interests, and investments in outcomes. Mimicked by historians, it is the voice that permits them to stand outside their own production of knowledge.[91]

If I differ here from most historians in seeking to resist this passive voice and to argue for a more embodied one that reasserts author-ative responsibility, it is not to stake a claim for a postmodern status (passé in any case). My intention is almost to the contrary, inasmuch as the first-person voice proclaimed by postmodernists confessed to the death of the author (albeit in the course of usefully promoting the de-objectification of subjective categories).[92] My object, rather, is to illustrate the union of my own agency and construction within the particular contexts in which these essays were written.

There was little call for this sort of thing when I entered the history of science and medicine in the 1970s. For the most part history-writing was then perceived as an unproblematic tool, a value-neutral means to describe, interpret, explain, and animate "the world we had lost," as the title of one of the popular texts of the time framed it.[93] This of course assumed the acquisition of a historical perspective from "the world we had gained," as a sequel volume titled itself.[94] But the nature of that acquisition was seldom pondered. It couldn't be, for history-writing sought primarily to fulfill an overtly moral and political function: casting light on the past in order to justify and legitimate change in the present (especially egalitarian-minded change). In retrospect, it was a move to accommodate a present that was infused with the epistemic notion of change as virtuous. Behind it, invisible, was the long-standing naturalization of equality, which was imbibed in the history of medicine and social policy from such left-leaning sociologists of health and welfare such as Richard Titmuss and Brian Abel-Smith.[95] Blindly, our research amplified that from which we took succor: a socialist-humanist master narrative. Change to less inequality was good and was put beyond question, while radical or systemic social change to greater equality was regarded as better still.

In the emergent social history of medicine this spirit was endemic; indeed, the social history of medicine was very much the sociological embodiment of it. No one dared accuse it of being a species of retrospective diagnosis (the practice of pressing current knowledge into the service of diagnosing the past, discussed in Chapter 7 in relation to contemporary biological knowledge of disease). That is to say, no one at the time thought the worse of the social history of medicine (and social history in general) for ostensibly diagnosing the political ills of the past by utilizing the sociologisms of the present, and vice versa. Nor was the history of medicine ever historicized as a self-serving product of Enlightenment medical humanism, despite the turn-of-the-century insights on this by Max Weber.[96] The critique of modern medicine was cast in terms of the moral *correction* of medical humanism, not as critical insight on the project itself understood as a historical phenomenon.

This is not to impugn the quality of the research. If there was a tendency to journalistic finger-pointing at particular ghettos of evil (the industries of smoking, asbestos, and health care, for example), there was also reverence for rigorous argument and analysis, and for the disciplinary hallmarks of footnoting and the archive. Behind it all was the animus of a widely and (in itself) uncritically shared sense of social crisis in the Western world, and of the need to engage politically with it. Depending on political orientations, this crisis was either the dreaded or desired meltdown of the social order, the countercultural signature of which was scrawled over everything from the civil rights movement to feminism, anti-psychiatry, the 1968 student protest, the Stonewall Riots, Vietnam, Chairman Mao, Che Guevara, Ho Chi Minh, the Campaign for Nuclear Disarmament, the Pill and sexual revolution, and the wholesale recreationalization of mind-bending drugs. We tripped on our times.

Around the profession of medicine especially crisis seemed manifest, appropriately so perhaps, given the origins of the word in medical theory.[97] The idea that medicine could do no harm was coming seriously unstuck, initially through critiques of psychiatric practice and theory and then, very publicly, through the thalidomide tragedy of the early 1960s. The latter appeared to many observers to burst the bubble of postwar miracle medicine. Critics marched in, buoyed by countercultural talk of revolution. Medicine was identified not just as a central part of the (by definition) conservative "Establishment," but a corrupted system built on a depersonalizing and alienating paternalism. Technology, less than bureaucracy, was held to rule and to turn its practitioners into smug overpaid robots. *Medical Nemesis* (1975) by Catholic priest and polemicist Ivan Illich captured perfectly the mood that its title summed up. Socialists and left-liberals confronted medicine's evil power, while feminists of the same persuasion liberated *Our Bodies, Ourselves* (1973). The new enterprise of bioethics, leavened by scandals on human experimentation such as that on the impoverished black population of Tuskegee, championed patient autonomy, the celebration of which was soon to find its place in the sociology of medicine and eventually—programmatically, if anachronistically by the mid-1980s—in a would-be patient-centered social history of medicine.[98]

This was a different mediation of the present than the subsequent one of the 1990s when the study of the medical marketplace and consumerism accommodated the dominant discourse of identity politics and the rhetoric of consumer choice.[99] From the point of view of values and virtues grating unwittingly and unbeknownst against competing ones, it is telling that those central to making this move (Roy Porter most notably in the history of medicine) did not see themselves as operating out of, or for, anything other than the social ideals of the 1970s.[100] That their history in its choice of subject matter was serving to accommodate the market values of the new age of monetarism would have horrified them. Yet it was neoliberalism and its rhetoric of free choice that was largely running the show. Although Porter's depiction of eighteenth-century Britain as a land of medical consumers and market transactions was driven more by a desire to extend a 1970s critique of consumer capitalism than represent a past that appeared commercially to be much like Maggie Thatcher's present, it nevertheless shored up that present, helping through history-writing to make it real and acceptable.

Of course, here as everywhere, what was "real" was not quite what was claimed; looking back on the 1970s and 1980s, a good deal of "unreal" is also apparent. For example, in the United Kingdom, despite the National Health Service and the curbs it imposed on ethical irresponsibility in medicine of the American sort, the critique of medicine was just as strident as it was in the United States. The pharmaceutical industry was then still tiny and was rather better known for its philanthropy (notably through the funding of the history of medicine by the Wellcome Foundation, as it was then called) than for the ruthless global pursuit of profits. As for the British medical profession, a vulgar Marxist class analysis might have served better than anything more sophisticated. Essentially class structured (as it remains), with an elite of consultants at the top and a proletariat of general practitioners at the bottom, it had yet to endure—mostly at the bottom—the tyrannies of Thatcherite managerialism (internal economies) and, later, evidence-based medicine, and the relentless push by politicians for the wholesale privatization of medical care.[101]

The situation in the history of medicine was further unreal in that the ostensible enemy, the medical profession, often labored happily among us. Just as many of those in the sociology of medicine who set the agenda for the field were themselves medically qualified and employed, so many historians of medicine were, or had been, doctors, midwives, and nurses. They were not writing the crude Whiggish history that the rationale for the professionalizing social history of medicine claimed was the case for medico-centric history. Irvine Loudon, for example, a general practitioner prominent in the College of General Practitioners (established in 1952 and turned "Royal" in 1972) may only have been legitimizing general practitioners (GPs) through his historicization of their institutionalization, but his *Medical Care and the General Practitioner, 1750–1850* (1987), among other works, was a far from facile retrospective valorization. It was more a history of struggle, contestation, and achievement—precisely those E. P. Thompson–like characteristics for the working class that were admired and aspired to by those of us involved the fledgling Society for the Social History of Medicine (established in 1970) and its *Bulletin* (turned professional journal in 1986).[102] In fact, in deeper time, much of the alleged self-interested historicization of medicine by its practitioners never quite came up to scratch. Karl Sudhoff, for instance, the first professor of the history of medicine (Leipzig 1906) was not the simple-minded positivist and predestined Nazi that the rationale for the social history of medicine might have liked.[103] Similarly, from the other side of the political coin, Sudhoff's successor at Leipzig, Henry Sigerist, who later moved to Johns Hopkins, never quite lived up to the model social/socialist historian represented by his left-leaning lionizers. He wrote mostly triumphalist medical narratives, after all.[104] But historical niceties like this didn't much matter in the accommodation of medical history to its times. Social, or socialist, or community medicine was a good thing ipso facto, much as Rockefeller Medicine had to be a bad thing according to our fairly monochromatic social humanist script.[105]

Like many others at the time, I slipped into the enterprise of medical history through the side door of the history of science.[106] This domain, too, was accommodating itself to its countercultural moment, a move that was facilitated in many ways by Thomas Kuhn's *Structure of Scientific Revolutions* (1962) with its encouragement to a view of the relativity of scientific truth.[107]

Through radical Marxist and "bourgeois" liberal social constructivist projects intended to extend that view, science studies, as they came to be known, heated up long before American hacks hyped the science wars in the 1990s.[108] For myself, these "changing perspectives in the history of science" came not as a welcome release from "old perspectives" in the field (of which I knew next to nothing), but as an escape from conventionalized social history.[109] In 1972 I completed a dissertation on the history of Irish life in the northeast of England, but it did little to inspire me in any lasting way.[110] If I had been passionate about Irish politics this surely would not have been the case, but I'd been brushed by the campus radicalism of my undergraduate days in Vancouver in the late 1960s, my father's brand of brewery socialism, and the Marcusian existentialist geography of one of my flatmates. When I met the socialist historian of science and activist Robert M. Young in Cambridge in the spring of 1972, therefore, I was only too receptive to his suggestion to apply my historical training to the topic of the power of popular science, and to the popularization of nineteenth-century phrenology in particular.

Young was then at the forefront of the assault on the old internalist history of science characterized by its philosophical idealism and lack of socio-political contextualization and awareness. Through landmark studies of Darwin and Malthus he demonstrated to a generation of young scholars that science was constitutive of the social relations embedded in its construction. Natural knowledge was not just *in* or acting *upon* culture, it *was* culture.[111] With scarcely a nod to Ludwik Fleck (Kuhn's inspiration), let alone to Foucault, the history of modern positivist science, he showed, was part and parcel of the politically evaluative dichotomies constructed between nature/culture, science/society, culture/science, subjectivity/objectivity, and fact/value that served (from this perspective) to mediate the hierarchical relations of capitalism.[112]

These were among the insights that animated the seminars and workshops in the history of science at the time. For me they were exhilarating, not least because of the intellectual company of those discussing them in the seminars at the Department for the History and Philosophy of Science in Cambridge: among others, Ludmilla Jordanova, Roger Smith, Jim and Ann Secord, Karl Figlio, Mary Hesse, Roy Porter, Jim Moore, Martin Rudwick, and later, Simon Schaffer. But they were also intimidating to one who had never heard of Kuhn and hadn't a clue as to what phlogiston theory and such like might be (and was afraid to ask). Yet somehow, they contributed all the while to the primitive desire in me to know how and why people (above all myself) came to think the way they did. I had no idea then that Foucault's project on how people govern themselves was grappling with essentially the same issue, as indeed had Nietzsche over a century before.[113] Also dimly dawning on me was the notion that, while history-writing served better than the guesswork of psychology to answer such questions, it, no less than science itself, might be a positivist conduit through which the present passed in mediated form. Could the new history of science no less than the old be a cultural product shaped by social assumptions and epistemic virtues? More tellingly so than anthropology, sociology, and all the other subjects in the humanities that came to be scientized in the nineteenth century (literature excepted), history was not just history, perhaps (no more than sociology was just sociology), but ideology and epistemology in the guise of a neutral tool for interpreting the past.[114] The new history of science, by taking for granted that the power of science in contemporary society justified its critical study, in fact contributed to the collapse of the "two cultures" that C. P. Snow famously sought to defend. Our critical faculties simply did not stretch far enough to see it. They couldn't, in part because our social humanist faith had not been radically shaken. Up to this point, it sat confidently beside, or even within, the evolutionary and progressivist metanarrative of science and scientific socialism. That faith was celebrated through histories of science that focused on such socialist humanists in science as Julian Huxley, Joseph Needham, and J. D. Bernal—biologists, public intellectuals, and historians all.[115] Such studies occluded the performative work of historical, sociological, and literary studies of science.[116] They pointed instead to what was then far more compelling to many of us around Young: stimulus to thinking about the politics of science.

But through phrenology? Admittedly, even then, in the late 1970s, it seemed slightly bizarre that the history of one of modern science's most obscure and tarnished subjects should be having this effect on me, while the pursuit of Anglo-Irish relations did not, despite that the latter were then literally exploding all around us. Today, I might attribute this perversity to always preferring the study of less-obvious manifestations of power to more blatant ones—to power/knowledge cooked through science in culture rather than experienced bluntly at the end of a gun (which any good journalist ought to be able to describe). I might invoke Hannah Arendt on how authority precludes the use of external means of coercion and is "incompatible with persuasion, which presupposes equality and works through a process of argumentation."[117] But at the time, the inclination had more to do with the rush that came from engaging with a subject that seemed hot-wired to the world as I found it (and still find it): hoisted by science and scientism.

My doctoral dissertation on the rise and fall of phrenology in nineteenth-century Britain was broadly a means to illustrate that science-Truth was socially and culturally made, not found. It challenged the assumptive metaphor in most history of science up to that time, the inevitable progress and triumph of a truth called science, and hence the "inevitable" quashing of a "pseudo-science" like phrenology. More programmatically, I argued against the idea that social explanations accounted only for "bad" or "pseudo" science as opposed to "normal science"—a line of thinking that was foundational to the then emergent field of social studies of science wherein "rejected science" was to be treated symmetrically with "successful science." Such was a part of an increasingly fashionable histo-sociological enterprise—another form of countercultural politics at intellectual remove. Basic to it was an understanding of scientists as social actors making up knowledge, instead of the older view of natural knowledge as transcendent and self-evident.[118] Later, scholars drawing on discourse theory would add to this, revealing how codes and conventions of narrative used in writing and speech conveyed ideology at the same time as creating and mediating cultural and social practices. In the 1970s, however, the main issue (for me at least) was countering the positivist politics of Karl Popper, who had wanted to mingle "disreputable false science" with "disreputable politics" in order to assert the ideological neutrality of "reputable" science. My attention to phrenology sought to turn that claim on its head, literally.

As much as anything else, my attraction to phrenology was shaped by the unsung and often radical individuals involved in its promotion. This, after all, was the heyday of the socialist history so brilliantly exemplified by E. P. Thompson and enthusiastically taken up in the History Workshop movement under the leadership of the late Raf Samuel. My dissertation, later published as *The Cultural Meaning of Popular Science: Phrenology and the Organization of Consent in Nineteenth-Century Britain,* which I dedicated to Young, was as much a contribution to the then un-studied (and still under-studied) place of science in working-class culture as to a relativist kind of semiotic sociology of scientific knowledge, as one critic referred to it.[119] But by the time I was halfway along, Antonio Gramsci, not Thompson, had become my pole star, as the "organization of consent" in the subtitle was meant to indicate.

Indeed, it was in the face of Thompson's *Poverty of Theory* (1978) that I posed the Italian theorist of hegemony. In doing so, I joined the company of many other historians who were inclining to Gramsci in order further to counter vulgar Marxist economic reductionism and deterministic explanations for social injustice. Among them was Edward Said in his famous critique of *Orientalism* (1978), which also embraced Foucault for his insights on the politics of exclusion. Such writings were hardly theoryless, but they increasingly shied away from or disdained what appeared to be the overblown *a*political jargon of postmodern critical theorists. Hence the emergence of the criticism that Eagleton would come to level at Said: that any objection of his to theory (meaning the embrace of poststructuralism and postmodern modalities) was necessarily also a commitment to some kind of theory.[120] Ultimately, it was to Eagleton's agenda, not Said's, that Gramsci's notion of hegemony was itself to succumb. In postmodern thought, talk of ideological consent was seen as sustained by underlying and inherently universalist and essentialist assumptions about the nature of power and about what it was to be human.[121] In the late 1970s and 1980s I was among those who believed that, for all intents and purposes, "human nature" (unquestionably "real") *was* fixed and uniform, and that (following Marx) all that was required for human fulfillment was to do what the philosophers proverbially had not: change power relations by changing social environments. In this I was on the side of Said and Thompson and other social humanists, including, not least, those associated with the Frankfurt School of Critical Theory (the better grasp of which I obtained after reading Martin Jay's brilliant study).[122]

Other contemporary innovations in thought were also involved in the makeup of my thinking. Particularly potent was the technique of "thick description" extolled by Clifford Geertz. Based on the idea that an event could be read like a text, its attraction was complemented by that of "implicit meanings" extolled by Mary Douglas. The work of both anthropologists was important not least for the use of "cultural" in the title of my study. Douglas was also crucial for a chapter on "The Power of Body" that I contributed to Barry Barnes and Steven Shapin's *Natural Order* (1978)—a book largely inspired by Douglas.[123] Although my contribution now seems to me juvenile, its title and concerns hint at the trope that over the next decade was to preoccupy postmodern-tending scholars, even if my own framing of "power" remained solidly Gramscian (and for that reason *Natural Order*'s oddball).

Many other theorists also stood in the wings, among them an ill-digested Foucault, my attraction to whose *Madness and Civilization* had taken me to Cambridge in the first place.[124] But it was many years before I grasped Foucault's supra-Enlightenment genius. The invaluable complete edition of his occasional writings and interviews was not yet available in any language, and YouTube had yet to make ludicrously apparent the incommensurability between him and America's then leading socialist public intellectual, Noam Chomsky. In the 1970s, I knew incommensurability only from Immanuel Velikovsky's *Worlds in Collision,* while the idea of "cracking" Foucauldian notions of biopower (Chapter 9) was as remote as the moon.[125] Perhaps the most salient feature of this whole period, however, was that our thinking was still largely in cold-war legacy mode. Worlds *could* still be in collision, unlike now when only fragments predominate and, hence, the messy multiplication of incommensurabilities even within the humanities.

From these beginnings, the essays in this volume traverse some of the methodological and historiographical turns that came later (excepting Chapter 2, written in the late 1970s). They exhibit, at the same time as they historicize, the various nondiscrete habitations for history-writing up to and including today's neuro-turn—up to, that is, the final scooping-out of the socialist-humanist narrative and the sedimenting of a neo-positivist one. None deal with the specific historical topics that occupied me over those years: those extending from my interest in phrenology, science popularization, and alternative medicine (mesmerism, bonesetting, and alterity in medicine and medical theory more generally).[126] Nor do any of them draw on the foci that came later: war and medicine, child health, scientific management, doctors in politics, the epistemology of epidemics, disability, death, and the history of the idea of an accident—all of which grew out of my work on the history of the surgical specialism of orthopedics, begun in 1995 after I moved to the Centre for the History of Science, Technology and Medicine in Manchester.[127] Some of the latter interests became a means for me to better understand the role of science and scientific management in the shaping of Western medicine.[128] Perceiving modern medicine as a thoroughly modernist project in the Weberian sense of ever-extending (disenchanting) rationalization was a way to gain purchase on postmodern orientations, or at least to construe modernity and postmodernity as distinctive historical moments. But those particular historical investigations were not conceived in terms of the present project: the historicizing of historical writing around medicine and the body, pre- and post-postmodernity. They were intended mostly as historiographically informed additions to, or revisions of, the extant historical record. Written within a modernist frame and a Marxian intellectual inheritance, they presupposed that one could arrive at a picture of historical reality free from distorting idealizations of one sort or another. I believed that then, and it is mainly because of it that I have chosen not to include samples of that work here, despite that some of it harbored a degree of epistemological reflection. The business of this volume is with the intellectual contexts that were then coming into being, or with the kinds of critical thinking that were coming to be possible (and, to varying degrees, the political conditions of possibility for them). Those contexts, or that thinking, was not consciously ruminated on in the histories I was writing, although there might have been signs of it in the margins. Attention to it became central only after an invitation to a conference on the historiography of medicine held in Maastricht in 1999. And it was only in the early years of the twenty-first century, after I began working on medicine and visual culture with the historian of science and medicine Claudia Stein, that it came to matter more. It moved upstream in the course of what became our mutual enthusiasm for rethinking thinking about history-writing.

Collaborating with Claudia was a very different experience than that I'd enjoyed with historians such as John Pickstone, Steve Sturdy, Mary Fissell, and Steve Pumphrey. It was as fruitful and fun, but it went well beyond plotting the presentation of research findings within reviews of the latest historiography. For her, the question to be pushed was not the historiographical "so what?" that had impelled most of my work, but rather the "so what?" of the practice of history itself, albeit always in the hope of finding an answer that would intellectually satisfy and justify its academic continuance. At the top of her agenda was the question of what the practice of history was supposed to be about. Not content with history-writing as a form of critique, nor with contemporary critical theory for its own sake, her impulse was to the moral worth of it all, or how it could be turned into a device to instruct the rest of mankind. These concerns were animated not by Marx, as they had been for me, but primarily by Foucault, in the conviction that he had never ceased to be political, always refused to get stuck in one intellectual rut, and always underlined his own investment and involvement in the present.

This was the kick in the pants I needed to move beyond mere critical depiction of the effects (or not) of Foucault and postmodernism in the writing of the history of medicine, such as had begun to preoccupy me after the meeting in Maastricht. Why not strive to put Foucault's practice of the history of the present into the present of historical practice, the better to realize one's own contingency? That was the challenge, opening up not just a life in history, but putting a life-in-history perspective into history-writing. In effect, I was becoming a product of a new episteme, in the course of which the writing of "realist history" began to interest me less, while at the same time heightening my political and ideological interest in reading it.

Coming to terms with the rush of new turnings in cultural studies was the first step in our collaboration, although one not easily taken given how much we ourselves were enmeshed in them, more or less unreflexively. Endlessly we articulated, wrestled with, and revised our thinking, which in the process became inextricable. We groped our way to understanding how contemporary theory had moved beyond what it had been in the 1990s (never mind the 1980s), how the currency of postmodernism might be regarded as having become politically tarnished by its complicity with neoliberalism inside and outside the classroom, and how its cynics (similarly complicit) could now charge it as valueless in an intellectual world that was busy reshaping itself around yet more theoretical moves and empirical enterprises drawn, knowingly or unknowingly, from the natural sciences. Hesitantly, we began to discern how late twentieth-century politics and monetarist philosophy were integral to it all, and how biotechnology (built mainly around private enterprise, after all) was, in turn, increasingly integral to the intellectual mediation of that politics and ideology.[129] For me, ever interested in the mystery of the relationship between thought and the material and political conditions surrounding it, this was a pathway to political reanimation.

If this was more challenging than anything I had attempted before, it was largely because it compelled acknowledging the subjective self in the business of critique. Forced was the recognition that all life—not just history-writing—was already inside theory and was capable of expansion only through that awareness. Life in history, no less than "life" in the representation of contemporary bio-reality, could not stand outside it. In a sense, this was to entertain seriously what Spanish filmmaker Pedro Almodóvar insisted: that "anything that is not autobiography is plagiarism."[130] But it was more than that; it was coming to see that if our norms and values were historically fashioned, then we historians had to have some relation to them. But how could the past be brought to the present through one's own values? What could be the basis for our authority to act in socially responsible ways in a world that had already talked the social out of existence? If academic history among paid-up professionals was to take seriously its role in preparing people to think about the future, what might that responsibility entail in terms of the kind of stories to write? At every turn the self was challenged; no longer sufficient was an understanding of the duty of the historian to be an analyst of conflicting versions of reality and responsibility "over there."[131] More urgent was the need to become present to the self in the historical "over here," including the need constantly to question one's conceptual location in ever-shifting times.[132]

Today this agenda seems more than ever necessary in a context in which the continuance of academic history is under threat, along with the pursuit of the humanities in general (at least in much of the anglophone world).[133] The question of the place, purpose, and value of the discipline has never been more stark, with an actual end to history as a financially viable autonomous academic enterprise appearing as a distinct possibility. Neoconservative historians fight back with renewed empiricist vigor, heaping blame on postmodern theorists for undermining the basis of the enterprise by challenging its objectivity. At the same time, many neoliberal historians embrace the neuro-turn as a post-postmodernist alternative, doing so directly or indirectly though attention to affect, the history of emotions, and materiality. But efforts to blame and shame, on the one hand, and the positing of new directions and manifestos, on the other, might equally be regarded as mere fiddling while Rome burns. Both might also be accused of dangerous historical naïveté. More urgent and compelling, surely, is understanding the situation in which academic history now finds itself and attempting to act purposefully in view of it. In order to do so, attention needs to turn to its intertwined material, political, ideological, and epistemic context. Without attention to that it is difficult to see how an effective defense of the discipline can be mounted, or even wanted.

A Call to Arms

The need to defend disciplinarity in history-writing was hitherto a fairly foreign academic matter, or so it vaguely seemed to me. Like others during the 1970s and 1980s, I negatively associated discipline with punishment (the idea that disciplines discipline) and came to regard academic disciplines as relics of modernity. I was not alone in thinking that best of all was a discipline such as the history of science or medicine whose boundaries were only lightly policed. Indeed, it was in part because of the intellectual porosity and eclecticism of these particular areas of study that they had been exempted from the charge of "disciplinary blindness" or "single vision" that some American historians leveled at their colleagues in the wake of challenges from the literary turn.[134] The latter, which was characterized by its own historical turn in the form of the New Historicism heralded by Stephen Greenblatt, was itself a major incentive to the dissolution of disciplinary boundaries, as well as to the growing sense from the late 1970s of the irrelevance of conventional academic history.[135] Given postmodernism's privileging of the body as a site for the discursive analysis of modernity, it helped the history of medicine's exceptionalism that the human body was ostensibly its remit. Similarly, the history of science was exempted for its long entertainment of the problems of objectivity and facticity that were now coming seriously to vex historians in other fields.[136] Although the social history of medicine as a professional activity located in departments of history and sociology was actually more in the rear than in the van of emergent postmodern positions, it could feign otherwise by riding on methodological eclecticism and pluralism (see Chapters 3 and 4). Inherently fearful of postmodern postures for the most part, it continued to stand by the humanist social politics that drove what was increasingly, in effect, populist history in academic dress.

But what the subdiscipline paraded as a legitimizing virtue in the 1980s could not be sustained in the 1990s. For one thing, the cultural bases for its faith in the social-humanist narrative had withered; it was delegitimized socio-economically and politically. Marx, with his perfect descriptions of the dynamics of capitalism, became, as the radical philosopher Slavoj Žižek has quipped, "the poet of commodities" loved on Wall Street.[137] For another, disciplinarity itself came under attack, its identity, virtue, and function called into question and its authority seriously challenged, especially in Anglo-American settings.[138] Thus, the continuance of academic history could no longer be presumed in the way that it had only a few decades before when the American-based historiographer Georg Iggers deflected the challenges stemming from the then fearful literary turn. Iggers rightly concluded that "if history will be written in the future, it will have to take on completely different forms," but he was confident that the discipline would survive by taking on these new forms.[139] To some extent, by the 1980s, it already had through the incorporation of linguistic analyses.[140] We might be living in a "posthistorical age," Iggers submitted, but for all the talk of the fall of objectivity, academic history was not dying. It was simply moving away from the old grand narratives and opening itself to a "diversification of approaches," or to an "expanded pluralism."[141] Today, however, in an alleged posthuman world, Iggers's optimism seems scarcely warranted, and academic history is far from immune from the indivisible politico-economic, ideological, and epistemic forces arranged against it. Whereas for Iggers in the 1990s contending with the literary turn was the order of the day, today, for academic historians, it is the fight for the discipline's very survival.

First and foremost, among the more mundane reasons for this situation is the assault on the worth of the academic discipline by those who govern higher education. This often seems especially so in British academia, where keenly felt is the audit culture that exercises round after round of financial retrenchment.[142] But the situation is scarcely different in North America outside of a few Ivy League colleges.[143] For common to the thinking of the managers of higher education everywhere, as well as those running the research-funding bodies, is the difficulty of comprehending the possible practical value of studying the past. And the more seemingly remote and arcane that past, the harder it is for them to understand.[144] The managers' difficulty is not helped, moreover, by historians' own inability to defend their practice on other than historiographically anachronistic grounds. As one critic has remarked, most scholarly works of history "are written as if to cater only to those who already want to know about a particular subject and they write off the rest of the public. In the way they hang on to outmoded kinds of narrative and analysis, they seem to assume that you should care about what they have to say, but they don't justify that assumption. History is good for you, they imply—but they never say why. And if they don't answer that question, why would anyone else?"[145]

Conventional historians don't help matters by condemning postmodern trends without admitting to the subjectivity of their own practices and their blind faith in objectivity. Antony Beevor, for example, the author of *Stalingrad* (1998), only digs deeper holes for himself by railing at Hilary Mantel's phenomenally successful fictionalization of the life of Thomas Cromwell (*Wolf Hall*, 2009) on the grounds that this "histotainment" and "faction-creep" dangerously corrupts the boundaries between fact and fiction. "You just can't tell what's original any more," he concludes, immune to the irony.[146]

In contrast, self-evident to those running higher education are the economic benefits and practical virtues of business schools and science parks on university sites, since manifest income generation is believed to be the highest measure of utility. These virtues and values—constituting in fact an idealization and ideology of "the practical" over "the abstract," or over "the purely academic" or, indeed, ironically, over "idealism"—have been transported to the humanities in the form of impact indicators to measure the immediate feel-good factor on students-cum-"clients" or "consumers."[147] Bizarrely, everything except the transmission of knowledge and ideas among peers is conceived as capable of such scientific measure. But this is hardly bizarre to the managerially minded, who have every reason to believe that scientific measures—performance indicators, citation indexes, research outputs, and the like—translate into money.[148] As confident products of the neoliberal marketplace values of exchange, growth, and profits, they follow their political orders in a culture that assumes economic utility to be the highest attribute to which a human can aspire.[149]

It would be naïve, therefore, simply to cast blame on those in charge of higher education for the current state of affairs. Like everyone else in the world, including historians, managers operate according to certain values and virtues that are upheld through a whole set of linkages to other practices and institutions. Collective cultural wisdom supports the shift in power they participate in: away from the humanities and toward the applied sciences and management that are within the economic order that grants intellectual dominance to a human nature defined as fundamentally greedy and grasping. More worrying is that academics in the humanities and social sciences follow suit, colluding through an un-self-conscious rescripting and revaluing of their own practices. As the sociologist of science Barry Barnes has noted of recent changes in the estimation of his field of study, " '*useful practices* now have greater value than . . . knowing 'for its own sake.' "[150] A decade or so ago this was not the case, although one of the forces behind it, the wholesale privatization and capitalization of formerly public-sector research in science and biomedicine (largely ignored by sociologists of science), has been rapidly gaining ground since the 1960s. In the sciences today, as various scholars have noted, the beau ideal is no longer the "disinterested" scientist of modernity, but the entrepreneur for whom financial success is the only criteria of merit.[151] Again, one can hear how only money talks, and how it talks loudest in cultures (such as universities turned into retail enterprises) that reproduce its values and tirelessly justify the enormous bureaucracies that go with it. Indeed, the ability to talk money comes close to defining what it is to be human, exactly what neuroeconomics endorses.[152]

The End of History-Writing? 35

The assault on academic history is furthered by those purporting to defend the pursuit of history-writing on populist grounds (persons themselves often married to reductive economic rationales and openly hostile to the kind of history-writing that aspires "to ask awkward questions, [rather than] … validate current assumptions").[153] Never greater has been enthusiasm for family history in local archives, history programs on television, and historical novels and movies of stuttering kings and impotent queens. Such sedations are far from culturally meaningless. They may not be driven nationalistically or idealistically, as in the Third Reich (a society not unlike our own in also being impressively biologized), but they nevertheless provide impressions of worth, tangency, and meaning in a world of surplus valuelessness where nothing, outside of biology, is real (often including photographs of "the real") and where, unlike from the 1960s to the 1980s, there is no shared vision of the future. Paradoxically, in the commoditized world that celebrates endless purchase and exchange, and where the presumption of the self is that of full autonomy and wholly voluntaristic agency, history often gains the appearance in the public sphere of providing unmediated access to moral authority.[154] Such authority is rhetorically cherished for seeming to be more amenable and trustworthy than the knowledge produced by hair-splitting philosophers or, in biomedicine, by morally tainted (but would-be philosophically neutral) bioethicists. Inasmuch as populist history reflects "the erasure of the boundary between the scholarly and the popular," it is, as Forman notes, further indicative of postmodernity's reversal of cultural presuppositions.[155] As he suggests, postmodernity is not the cause of the problem; the reversal is sustained by the material, economic, political, and ideological investments in it. The overall effect, however, is to compound the trivialization of academic history-writing.

What has been less commented on in the analysis of the problems facing academic history-writing today is how scholars in other fields often appropriate history to bolster their own particular views and valuations, and in the process repackage and revalue history's purpose for a wider public. This move is perhaps most pronounced in the sociology of science and biomedicine—domains widely regarded as more relevant to the "real world" than academic history. As noted earlier, Nikolas Rose, for one, inheres history by arguing for the "greater complexity" of our biologized times, and thereby garners support for his and his colleagues' further ethnographic investigations. More straightforwardly, the sociologist of science Steve Fuller uses history to uphold the belief that science is a productive force that generates great things (to the extent of his providing unblushing praise for blatantly biological and neurological reductionist accounts of history-writing).[156] Elsewhere, Fuller uses history to legitimize a new space for biology in sociology. It was, he contends, merely the result of historical contingency that sociology left biology out; hence sociology can now be historically "corrected" by *re*investing in biology.[157] A past that not long ago was upheld by historians as a "contested and colonized terrain" is by these means flattened through an *in*contestable neurobiological "human nature."[158] Such uses of history to naturalize biology, moreover, frequently find their way onto the bandwagons of those further up the power chain in higher education, to those on the funding councils, and occasionally even to a university manager.[159] In a culture that values only knowledge with economic potential, and that hypes science and information technology in terms of "knowledge economy," such strategic uses of history—however cavalier, simplistic, and reductive—command far more respect than "purposeless" academic history. The marketing success of the various recent neurological reductions of history is but a further instance of this trend, for, as indicated above, these works purposively, if un-self-consciously, service the same set of anti-historical (*a*historical) values.[160] Since sociologists of science and the new neurobiological reductionists of history can presumably write all the history that is needed for contemporary culture, why bother with funding professional historians at all—or, for that matter, history learning anywhere? Herein lies the real "end of history," the furtherance of which (through a rewrite of sociobiology no less) is now made by the same neoconservative political scientist who originally authored the simple-minded book by that name.[161]

Oddly, albeit at another level understandable, academic historians assist in this process. Again, this often seems especially so in Britain, where individual and departmental survival is determined by highly competitive grant-income generation. No money, no time for research; no research, no chance of scoring in the government's competitive research assessment exercise (a Thatcherite purchase bespoke from the Harvard Business School) and therefore no chance to bid for grants—indeed, no chance for survival, let alone acquire the public funding that is supposed to result from a high scoring in the assessment exercise ("when times are not tough").[162] Like other academics, historians busy themselves competing against each another in the division of spoils (hence promotion, hence escape from one fraught environment to one slightly less fraught, they hope). Such is their collusion with the degradation of their own intellectual labor. But even if they were not compelled to busy themselves in this way, their belief in their own intellectual autonomy militates against union and protest (unlike many academics in France, Spain, and Germany). Hence, exquisitely they fit themselves to the interests of management, giving up intellectual autonomy to please the employer or the funding patron or agency. Given that they are supposed to be intellectuals, not managers (and when managers not competitive capitalists), this is difficult to pardon. Yet, to be fair, the neoliberal forest in which all this has transpired has been difficult to penetrate. Even scholars with an interest in the world outside the classroom have been hard pressed to find the time or opportunity to bring it inside, or comprehend what exactly it is that has to be brought in and held up to scrutiny. Exhaustion and job survival prevail as much as hapless functionarialism.[163]

Another, not disconnected, path to the degradation of academic history has been the advocacy of interdisciplinarity by those who control the purse strings of higher education. Drawn from the notion of teamwork inculcated in industry and science laboratories from the turn of the twentieth century for purposes of productivity and profits, it has recently become conceptually obese—not least in the areas of health care and biomedicine through such devices as integrative medicine and translational medicine/research.[164] Described by one investigator as, at best, "the result of opportunism in knowledge production," interdisciplinarity or multidisciplinarity is often held to promote novel cross-fertilization in research, something held to be in itself inherently good.[165] (Academic brains, unlike certain cacti and genetically modified plants, apparently, risk none of the dire pestilential consequences of cross-species intermingling.) However, rather than promoting the disciplinarity it appears to presuppose, interdisciplinarity surrenders it to a politically-correct-sounding, and thoroughly "dissensus"-avoiding, pluralism in which, crucially, all parties forfeit responsibility, thus leaving the agronomists of higher education further to extend their culturally nourished economistic and scientistic values and virtues.[166] Which is not to say that interdisciplinarity is necessarily an evil in itself, even if there is little evidence that the vaunted practice has ever produced anything truly novel or intellectually creative outside of public relations firms and the world of advertising. Least of all has it done so through collaborations between those working in the natural sciences and those in the humanities. Indeed, between "the unwashed and the white-coated," as Bruno Latour refers to them, interdisciplinarity appears to be entirely unproductive if judged by quality outputs.[167] Confirmed is only the observation of Max Weber a century ago that mediocrity is all that results from academics of different sorts trying to cooperate.[168] For the most part, interdisciplinarity merely contributes to what Žižek identifies as today's general "prohibition against thinking"[169]—a manager's dream come true.

More crucial is the political effect of the dogma of interdisciplinarity in absenting academics from the exercise of intellectual and ethical responsibility. And in this connection, too, it has to be said that historians (and scholars in the humanities in general) act in naïve complicity through the belief gushingly held by some of them that research today has so many new and exciting ways of being conducted (despite the reality of funding bodies largely setting research parameters). More appalling and profound, historians are complicit in mimicking and swooning to the same (outdated) political epistemology of science that runs the administrators' thinking on interdisciplinarity—a mental act on the part of historians of science and medicine that ought to be treated as criminal given their supposed expertise. To some extent, the problem, at root, is the one identified by Latour, that "when impressed by white coats, humans ... stop 'objecting' to inquiry and abandon any recalcitrance."[170] But even among sociologists of science—perhaps especially among them—there seems a great reluctance to confront the wider and deeper sets of epistemic virtues that inform and validate the managerialized utilitarian culture in which such confrontation is rendered meaningless.

Yet there may still be room for hope. Since interdisciplinarity absents academics from the exercise of intellectual and ethical responsibility, its challenge by asserting disciplinarity might become, in turn, a counter-hegemonic strategy for the reclamation of responsibility. Its assertion can become a political act in the face of the pervasive and aggressive politics for the disappearance of disciplinarity. Gaining the will to it might be encouraged by noting that there is no lack of faith in disciplinarity in the natural sciences. There, through the proliferation of expertise, disciplinarity is celebrated. Indeed, the logic of disciplinarity might be seen as the raison d'être of the natural sciences. Entirely one sided therefore—science sided—is the view that the humanities and social sciences should collaborate with the natural sciences, since no one in the natural sciences feels the slightest need to surrender the assumed superiority of their historically fashioned disciplinary logic.

With regard to the history of medicine and biomedicine, the need to repoliticize through the device of advocating disciplinarity is all the more compelling in the face of one particularly virulent form of would-be institutional species-crossing, medical humanities. As rebranded on American campuses over the past few decades, and now formally promoted by the Wellcome Trust in the United Kingdom, medical humanities aspires to provide a consortium for all manner of parties (and priests) interested in biomedicine: literary scholars, visual artists, playwrights, media buffs, philosophers, sociologists, anthropologists, health economists, jurists, medical ethicists, *and* historians.[171] Since most medical history can hardly be accused of subverting the project of medical humanism, the move to medical humanities (mostly in medical schools) cannot be interpreted as a calculated act on the part of managers to suppress critical histories of medicine and the biosciences. Nevertheless, this is its potential and likely effect, for in this forum history not only is demoted to the status of an unequal partner (unequal given the owners of the forum) but also is understood by those running the show as a means primarily to supply witness to biomedicine's "greatness." Not here encouragement to history-writing that might be "critical … thought-provoking, frustratingly ambiguous and uncertain"—in many of these respects, interestingly, not unlike the nostalgic values behind research in the natural sciences.[172] Even if such history-writing were encouraged, the same could be equally extended to the other parties involved. Marginalized, then, if not entirely lost, is the importance and relevance of history as a critical tool to explicate the present in all its linguistic, epistemological, political, and ideological creation. Instead, the history of medicine (writ large or small) becomes but a market device for biomedicine and medical philanthropy, a public relations exercise much like contemporary bioethics for clinical practice (discussed in Chapter 8). In a social and cultural context of what appears to be ever-shorter collective memory and ever-greater scientization of the humanities, it is entirely likely that this is how the history of medicine will come to be perceived, funded, practiced, and socially valued.[173] To a considerable extent it already has.

Never more than now, then, is there a need for sustained historical unpacking of the episteme in which we write the past to help think the future. Never greater is the need for historically informed wisdom, especially in view of the new bio- and neuro-popular ways of reductively conceiving identity and understanding. Never more than now do historians need to comprehend how they have been corralled, and at what psychological and professional cost. And never more than now do they need to reappraise and revalue the community, identity, and solidarity they have, and which they actually tacitly acknowledge—for example, through the colleagues who have helped them in their research (an act of recognizing and valuing a set of shared assumptions within a community), or through the huge investments they often make in the welfare of their students. But the solidarity that is needed now is far from that of seeking to establish methodological consensus or homogeneity; quite simply, it is that which is necessary for a political voice. If academic history is to survive as something different and distinct from populist apology, and from the sale of the interconnected epistemic virtues of scientists, sociologists, and business managers, its practitioners need more than ever to assert disciplinarity in the face of the counterpolitics outlined above. In the classroom especially, among students at ease with the pervasive sleaze of neoliberal managerialism, the "so what?" of historical disciplinarity is vital to set forth.

But to accomplish this, historians need to do more than merely agree upon "the rules that govern their discourse," which was the strategy that Kuhn adopted in the 1960s for the continued practice of normative science in the face of his exposure of it as socially and culturally contingent.[174] They need to understand how the challenge to disciplinarity and the anxieties and intellectual constraints flowing from it have come to be constituted and embodied in themselves. Above all, they need to understand how the increasingly *a*historically represented world of biomedicine and posthumanism suppresses their voice. Anthropology, sociology, and many other disciplines in the humanities have substantially remade themselves over the past decade or so. Academic history has not, although there have been beckoning voices. If it wishes to do so, it will take more than blind faith in its own bricks and mortar (already being re-consigned), and it will involve more than seeking salvation in the next trendy turn. Indeed, if the next turn is to be that informed by the theories and practices of the social sciences (themselves now informed by the natural sciences and neurobiology in particular), then the game is pretty well up. The crossroads have been reached and there is no going back. Given how the space has shrunk for historians to continue to be their own blind assassins, and is rapidly shrinking still further, the choice is either a new beginning under self-reflexive critical management, or retirement. To this extent, the future of academic history-writing lies in hands of historians themselves. The time for procrastination and pious hope is past.[175]

2

Anticontagionism and History's Medical Record

THIS ESSAY WAS WRITTEN in the late 1970s for a volume setting out what was then the new social constructivist approach to medical knowledge.[i] Drawing on Frankfurt School critical theory, the rediscovery of the early Marx, and the ideas of Antonio Gramsci, it was an attempt to illustrate how knowledge production (held as class production) and power relations were mutually constitutive or interpenetrated. Whereas so-called vulgar Marxism had interpreted knowledge production in terms of an instrumental (assumed direct) relation between it and the repressive power of capitalism, the new approach was to see knowledge as mediated ideology (to the extent that the causally barbed word "influence" became *verboten*). Contexts of capitalist social relations were seen to determine the shape and the form of both the production and the consumption of knowledge. Moreover, no body of knowledge or set of ideas, however abstract, was to be regarded as neutral or value-free; all was held to be ideological or embedded in social relations. But the expression of these relations was mystified or non-apparent to either producers or consumers. This was a new way to think about natural knowledge, the stuff of science and medicine, and in that domain (so central to modern capitalism) an interesting pathway to further unmask the constitution of unequal power relations under capitalism. It felt right at the time.

The early nineteenth-century belief in anticontagionism was my way of elaborating this approach, a way to reveal how a body of knowledge in medicine's past that was of undoubted politico-economic significance could also be read as a mystified mediation of ideology. Reading was key, and close attention to the historical actors' use of language was a part of the technique. This was not the practice of "close reading" later honed by postmodern literary scholars. It had more in common with the work of Marxist literary historian Raymond Williams, whose *Keywords* (1976) notably contained helpful entries on "ideology" and "mediation." For myself, having studied English literature as an undergraduate and imbibed Dryden's injunction to squeeze from every line of poetry all and every conceivable meaning, it was an attractive means to get at what I conceived to be the historical construction of my own being.

It intrigued me that the Marxist historian of medicine Erwin Ackerknecht (1906–88) had published on the idea of anticontagionism as early as 1948, the year I was born.[ii] For me, this made him the pioneer not of the social history of medicine, but of the social history of ideas in medicine—a mantel in some ways comparable to that which Ludwik Fleck, the 1930s pioneer of the epistemology of facticity in science, was just then beginning to acquire among historians and sociologists of science. I knew little about Ackerknecht other than that he had been a leftist intellectual with an interest in anthropology. (Charles Rosenberg's informative essay on him of 2007 was a long way off.)[iii] I liked that he had plowed a lonely furrow at Wisconsin largely independent (or so I mistakenly thought) of the historians working under Henry Sigerist at Johns Hopkins—those other would-be pioneers of the social history of medicine and the social history of ideas in medicine, George Rosen and Oswei Temkin. Doubtless there were culturally conditioned psychological reasons for my wanting to pay homage to Ackerknecht by seeking, as I thought, to extend his insights. Those reasons are perhaps best forgotten, but it is hard to overlook that my critique of his socioeconomic analysis of the early nineteenth-century debate over contagionism was also something of a backhanded compliment. I can see, too, that I was scarcely different from him inasmuch as I was also focusing on the debate to sell a particular politically informed orientation to history-writing. Whereas for him the history of the debate over anticontagionism proved the worth of a Marxian economic determinist approach to history, for me it proved the merits of a quasi-poststructuralist Marxian one. Indeed, unsurprisingly perhaps, I found in the anticontagionists' own writings the kind of thinking I was interpreting them through.

I was not the only product of my times and intellectual context to be attracted to the anticontagionist debate. Although I was unaware of it when I wrote the paper, at more or less the same time Foucault's student François Delaporte was also mounting a strong critique of Ackerknecht's article.[iv] His was from a very different angle than mine, stressing as it did (à la Foucault) how medical modes of analysis, lexicons, practices, and techniques restructured themselves from within fields previously organized along non-"medical" lines. Ultimately, though, these two different approaches—my Marxian-inflected social constructivist one, and Delaporte's non-Marxian Foucauldian discursive one—came to overlap and to some extent meld. There was a certain poetic justice in this in that Ackerknecht's opposition to Francophone "unscientific" medical history led him, in 1964, to turn away from one of the first historians of medicine to take up Foucault, the German Werner Leibbrand (1896–1974). Ironically, Leibbrand was then ostracized by a younger generation of scholars in Germany for having dared to oppose the Marxist Ackerknecht.[v]

For those of us in the United Kingdom involved with the history and sociology of science and medicine, Foucault was recognized as an important thinker, especially for *The Archaeology of Knowledge* (1969), which was translated into English in 1972, and *The Birth of the Clinic: An Archaeology of Medical Perception* (1963), translated (rather badly, it emerged) in 1973. But he tended to be regarded as an auxiliary source, supporting rather than supplanting our interpenetrative, non-crude (but nevertheless causal) notion of knowledge/power and historical change. This was possible, since in those books Foucault was forwarding his "archaeological" method focused on the classification of scientific and medical knowledge. His "genealogical" method (discussed in Chapter 1), which transcended this, emerged later—initially in *Discipline and Punish* (translated in 1977). The classification of scientific and medical knowledge seemed well within our remit, and since Foucault's "archaeology" confined itself to texts and texts only, and had no concern with questions of causality, he appeared not to threaten our inquiries. Wasn't he just another critical thinker further demystifying the world and its ordering?[vi] Locked within our totalizing Marxist episteme, we didn't understand that our fundamentally causal preoccupation with the contextual determination of knowledge/power was irrelevant to Foucault. When this became clearer, through the efforts of the avant-garde in literary and cultural studies, the tendency among historians was to regard Foucault with disdain, or as an irrelevance.[vii] It was, after all, the literary and cultural lot riding Foucault who came also to challenge the disciplinary canons of history-writing, and the discipline itself as an "objective" enterprise. This largely explains why it was sociologists and anthropologists, such as Nikolas Rose and Paul Rabinow, who were to become Foucault's main interlocutors, not historians (many of whom, to this day, rather dismissively classify him as "a philosopher").[viii]

The essay was begun in 1979 when I was a Killam Fellow at Dalhousie University in Halifax, Nova Scotia, and it was completed a year later when I moved to the Wellcome Unit for the History of Medicine in Oxford. These were very different environments, but each was important in its own way for reinforcing the rightness of my worldview that, circularly, I justified through its historicization. At Dalhousie was Trotskyite historian Bryan Palmer, along with labor historians Greg Kealey, Craig Heron, Ian McKay, and several others who were all connected in one way or another to the American socialist historians Herbert Gutman, Christopher Lasch, Eugene Genovese, and David Montgomery. All were devotees of Edward Thompson.[ix] We partied together, pouring socialist scathe on our political times. If we differed intellectually it was only in terms of the moral and political worth of our particular objects of inquiry. I found it hard to convince them (and doubt I did) that the socialist history of ideas in science and medicine had as much or more to offer than Thompson, *and* that it might be more germane conceptually to the scientized world in which we lived. In 1978 Karl Figlio's article on "the social constitution of somatic illness in a capitalist society" sat next to the one by Thompson in the same issue of the then modish journal *Social History,* but apparently that didn't mean that it deserved the same attention, or even had to be read. In retrospect it may not have mattered. In the future for them was Joan W. Scott's critique of the integrative, unifying, and teleological functioning of Thompson's celebratory notion of "experience" (1991), just as in the future for me was Ian Hacking's nerve-touching *Social Construction of What?* (1999), a Foucauldian work that might be said to have conventionalized Bruno Latour's castigation of social constructivism as "ugly idealism."[x]

The Wellcome Unit in Oxford was a more disciplined place. Although teetotal and vegetarian in orientation, it was more conducive to my subdisciplinary historical interests. It was under the direction of its founder, the enormously learned historian of science and medicine, and Labour Party activist, Charles Webster. He had set it up in 1971 shortly after Bob Young established the one in Cambridge. Webster had moved to Oxford from Leeds, where Thompson, John F. C. Harrison, Asa Briggs, and many other future doyens of social history cut their teeth (mainly in the area of adult education, continuing there the teaching that had gone on among the troops during the Second World War). Webster's work in the history of science and medicine was parallel to theirs in its anti-elitist challenging of older "top-down" stories of "great men" and "great ideas," and in its fondness for historical texts and "actors" (a word hated by Webster for its dehumanizing tones) who were unsung and hitherto un-singable in the positivist construction of the field by its practitioner-authors and historian apologists. Coffee was rung at 11 a.m. and tea at 4 p.m., and by an unspoken rule no discussion was allowed of one's own immediate intellectual headaches. Solidarity was forged not only through lively twice-weekly seminars but through a compulsory reading group in which we worked our way, line by line, through primary sources relevant to the history of medicine. All of this reinforced a belief that could never be spelled out because it was too much taken for granted—that there was a causal relationship between social contexts and the way in which individual belief and behavior was shaped. This was the central assumption of the social history of medicine movement, which the Oxford Wellcome Unit spearheaded under Webster's direction. Among its members at the Unit were Paul Weindling, Pietro Corsi, Irvine Loudon, Jonathan Barry, Emily Savage-Smith, and Maggie Pelling, the latter's whose *Cholera, Fever and English Medicine* (1978) I was to both pit against and align with Ackerknecht's essay on anticontagionism. That Pelling and I could have our methodological differences yet remain close colleagues was instructive; it proved to me that different historical orientations need not necessarily signify incommensurability. At the time we were all social historians of medicine first and foremost, marching with our assumptions in a political procession headed by Webster.

The Problem of Medical Knowledge, published in 1984, in which this essay appeared, was not intended as a counter to more conservative schools for the history and sociology of science and medicine. These were then represented by Jerry Ravetz in Leeds (though Ravetz himself was a Maoist) and by the "Edinburgh School" under the direction of Barry Barnes, David Bloor, Donald MacKenzie, and Steve Shapin. By no means were all of the contributors to the volume Marxists, but there was a good sprinkling of those of us associated with Young and the "Radical Science Collective" committed to neo-Marxist-type thinking. Karl Figlio, regarded by all (including Young) as the greatest intellectual ever to weigh in on the history of science and medicine, was a contributor, as was Edward Yoxen, a bright and highly principled Marxist then working on the politics of molecular biology and genetics, and married to one of the founders of the Radical Midwives movement. The general practitioner turned medical historian Chris Lawrence (a great admirer of Temkin and a friend of Young) was also a contributor, as was the medically trained, historically minded sociologist of medicine David Armstrong. Along with fellow liberal Nikolas Rose, Armstrong was Britain's leading Foucauldian practitioner (although never an evangelist of Foucault). His work was in the van of a theory-driven sociology of medicine that nudged forward and eventually superseded that of the more policy- and labor-orientated school of Meg Stacey and Margot Jefferys, with whom Webster and other Labour Party members tended to align.[xi] The book therefore reflected the late 1970s and early 1980s collaborative mix of scholars, all passionately interested in, and often politically involved with, issues of health and illness in contemporary society. We wore slightly different clothes, but we spoke much the same language (or thought we did) in our contributions to the same conferences and workshops.

A second edition of the book was mooted in the late 1980s, but perhaps because the editors wanted to ax some chapters and insert new ones (defying the idea of a "second edition"), nothing came of it. This was probably just as well, for its moment had passed. By 1990, social constructivism was a bygone, along with the politics that had inspired it. In the history of medicine what was left was the empty shell of social context determinism. The version of the essay presented here is that which I prepared for the "second edition." It differs from the original only by the addition of a few references and some cleaned-up prose.

The chief defect of all previous medical history ... is that the object, actuality, sensuousness, is conceived only in the form of the object or of perception, but not as subjectively.

—Karl Marx, "First Thesis on Feuerbach," 1845

It is clear that the old teleological and ethnocentric perspectives on medical history have gone into steep decline. It is also apparent that attempts to study medical ideas in purely intellectualist terms are moribund, along with attempts at mechanically and causally relating medical ideas to social, political and economic circumstances. To all three of these approaches—the "Whig," the "intellectualist" and the "reductionist"— stigmas have come to be attached. Rightly so. Yet, if historians can now do little more than situate ideas and events of medicine's past into their "full and appropriate contexts," the returns for all the recent investment in debunking would seem relatively slender. Indeed, it might be argued that with the ascent of social contextual studies there have been significant losses: Whig historians, for all their shortsightedness, at least had a strong commitment to the present; intellectualists had a deep interest in the nature of medical knowledge; and reductionists clearly recognised the primacy of economics and politics in human affairs. Such commitments and interest have not been entirely abandoned, but they can hardly be said now centrally to inform the historical agenda.

This essay attempts to redress some of these losses without abandoning the view that medicine needs to be studied as part of its social, political, economic and cultural totality. Focusing on the belief held by some early-nineteenth-century medical men in the non-contagious nature of epidemic diseases, it seeks to go beyond mere contextualization. It explores some of the ways in which "anticontagionism" as the intellectual product of a particular social context can be considered as mutually constitutive of its producers' social interests, or can be studied as part and parcel of the social relations in which it was articulated. Thus, rather than looking merely at the *belief* in anticontagionism and commenting on the ways in which debates over it *interacted* with social and ideological interests, I seek to approach the *substance* of the knowledge, to comment on the ways in which it was *interpenetrated* by those interests. The aim is to make a start on relating a particular form of knowledge to a particular form of society without falling prey to crude reductionisms. Attention is therefore confined to the writings of only a few leading spokesmen of anticontagionism in Britain. A socio-political interpretation is sought on the basis of what can be detected through the analysis of language and metaphor.

The social basis for the medical belief in anticontagionism was laid down as long ago as 1948 in a classic article by the Marxist historian of medicine Erwin Ackerknecht.[1] Initially, I conceived the present essay as a homage to Ackerknecht's insights. I would, I thought, deepen them by pushing them beyond the positivist parameters in which he could be seen to have set them. Since Ackerknecht was among the pioneers of the social history of medicine, the critique of his epistemology could usefully serve as critique of much subsequent writing in the history of medicine. Hence my liberty in replacing "materialism" by "medical history" in the famous first line from Marx's "First Thesis on Feuerbach."[2]

However, with the publication of Margaret Pelling's *Cholera, Fever, and English Medicine, 1825–1865* (1978), it becomes necessary to broaden the essay's scope. Pelling, in the course of tracing the development of epidemiology, has forcefully challenged not only Ackerknecht's interpretation of anticontagionism, but also the validity of the terms themselves, "contagionist" and "anticontagionist." Thus it becomes impossible to address this subject further without taking heed of Pelling's arguments. As it turns out, an examination of Pelling's work serves only to enhance the appropriateness of the epigraph.

In his article "Anticontagionism between 1821 and 1867" Ackerknecht sought to explain why, shortly before the formulation of germ theory by Pasteur and Koch, the idea that there were specific disease agents that could be directly transmitted from person-to-person (contagionism) experienced a deep depression and devaluation, while "anticontagionism" reached the peak "of its elaboration, acceptance, and scientific respectability." Basing his interpretation on a wide range of European, English, and American sources, Ackerknecht accounted for this apparent paradox in terms of the vitalisation in the early nineteenth century of a critical spirit in which old ideas came in for attack as irrational, antiquated and unscientific. In this I think he was right. Contemporary discussion on contagionism was not in his opinion merely theoretical or even medical, but animated by "powerful social and political factors." In particular, Ackerknecht's discussion centered on contagionism's material expression: the quarantines and their bureaucracy. Thus,

> the whole discussion was never a discussion on contagion alone, but *always on contagion and quarantines* [his emphasis]. Quarantines meant, to the rapidly growing class of merchants and industrialists, a source of losses, a limitation to expansion, a weapon of bureaucratic control that it was no longer willing to tolerate, and this class was quite naturally with its press and deputies, its material, moral, and political resources behind those who showed that the scientific foundations of quarantine were naught, and who anyhow were usually sons of this class. Contagionism would, through its associations with the old bureaucratic powers, be suspect to all liberals, trying to reduce state interference to a minimum.[3]

Ackerknecht's main point was that, in a context of ignorance on the cause of the transmission of disease, arguments for and against contagionism were evenly balanced. Hence, under these conditions,

> *the accident of personal experience and temperament, and especially economic outlook and political loyalties will determine the decision. These, being liberal and bourgeois in the majority of the physicians of the time brought about the victory of anticontagionism* [his emphasis]. It is typical that the ascendancy of anticontagionism coincides with the rise of liberalism, its decline with the victory of reaction.[4]

Until Pelling's study, the factual basis, the approach and the conclusions of Ackerknecht's article were not seriously questioned. To the contrary; as the social contextualist approach gained wide appeal, the interpretation was endorsed and the contents were much elaborated through studies of the cholera epidemics of the nineteenth century. The latter have greatly enhanced our knowledge of the basis for the faith in anticontagionism among liberals in the medical profession, while the political and social divide between them and contagionists has been illustrated and particularized in several different national contexts. Thus any challenge to Ackerknecht's interpretation signifies no trivial engagement with some antique article. It is, as Pelling fully appreciates, to contest the formidable body of scholarship that has rendered Ackerknecht's article "a standard historical account."[5]

So what gives Pelling her boldness? Simply this: that a detailed understanding of the complex development of epidemiology in Britain in the first half of the nineteenth century supports neither the assumption that anticontagionism was dominant among medical men nor the supposition that before the middle of the century medicine was distinguished more by its "human" and "political" qualities than by its "scientific" ones. Pelling undermines these assumptions by revealing the variety and subtlety of the influences on epidemiological thinking in the first half of the nineteenth century, and by detailing the many implications of its advancement. She does so in part by exposing the lack of homogeneity among the medical profession, revealing the fallacy of accepting statements such as those of Edwin Chadwick and the Board of Health as representing "official" medical doctrine in England.

Thus Pelling exposes Ackerknecht's paradox as a myth, which she accuses of grossly distorting historical reality in two important ways. In the first place it does so by purporting that the epidemiological theory that was developing in the first half of the nineteenth century was no less "scientific" than the germ theory that appeared later. (The great merit of Pelling's work is to reveal that it is only by reading pre–germ theory medical history from the perspective of the triumph of germ theory that the idea of a historical paradox emerges in the first place.) Secondly, Ackerknecht's distorts by turning the majority of English doctors by the mid-1840s into anticontagionists when they were not. Most medical practitioners, Pelling insists, were *contingent* contagionists holding that the causes of epidemic diseases were multifactoral, though related to the environment.[6] Ackerknecht wrongly supposed that all nineteenth-century adherents to disease theories involving miasma or notions of filth were at all times and in all cases anticontagionists. It is on the basis of these shortcomings that Pelling comes to her conclusion that the terms "contagionist" and "anticontagionist" are entirely inadequate, for they misleadingly summarize the contemporary concern with epidemic diseases in terms of simple opposites. Medical reality was far more complex and multifaceted.[7]

As far as it applies to England, then, Ackerknecht's thesis on anticontagionism is demolished by Pelling. Her research is impeccable and her carefully weighed statements mercilessly precise. Yet it would be wrong to conclude that anticontagionism has now been swept into the trash can of history. Pelling's argument, after all, is not that the terms "contagionist" and "anticontagionist" are useless or historically mythical, but rather, that these terms are insufficient and misleading *for understanding the history of epidemiology*. Although Pelling does not emphasize that in the second and third decades of the nineteenth century anticontagionism was a bald issue over which there were heated debates, neither does her evidence deny it.[8] Nor, furthermore, does she attempt to bury the anticontagionists' claim that "there is no subject which better deserves the serious consideration of the physician, the merchant, the statesman and the philanthropist [than anticontagionism]."[9]

More serious, if less obvious, is the way in which the force of Pelling's argument detracts from the need to think socially and critically about epidemiological thought in general and anticontagionism in particular. The empirical basis of her attack on Ackerknecht's article obscures that its major achievement was to reveal that the *medical* debate between contagionists and anticontagionists was largely a *social* debate. That Ackerknecht was misguided to the extent of medical support for anticontagionism in Britain hardly diminishes his achievement; in thinking about knowledge-claims socially, rather than solely in terms of the intellectualist history of ideas, he was a generation ahead of his time. Thus, Pelling's dispute with Ackerknecht inadvertently fouls the path to a more thorough social understanding of anticontagionism by obscuring the extent to which she actually *shares* epistemological terrain with Ackerknecht. To grasp this we need to reflect on what in fact *is* paradoxical about Ackerknecht's paradox: his accounting for the demise of anticontagionism.

On the one hand, Ackerknecht held that the belief in anticontagionism was socially informed and that it triumphed with the ascendance of the liberal bourgeoisie. On the other hand, though, he held that disbelief in anticontagionism was not in the least social, but rather was the result of the inevitable triumph of "real" truth-bearing medical knowledge which, only by an interesting coincidence, arose in more reactionary times. Ackerknecht thus relied on and perpetuated the positivist notion of a separation between "pre-scientific" knowledge which is "false" because it is socially informed, and that which is "true" or "scientific" because it is not. It is now rather better understood that this notion was purposefully advanced during the so-called "scientific revolution" of the seventeenth century and was contemporaneous with the rise of modern capitalism and its valuation of human labor. Positivism can be regarded as the metaphysic of capitalism, for the effort to separate would-be "objective facts" from "values," or the effort to separate "science" from "pseudoscience," "ideology," "scientism," and "society" transpired within, and is inseparable from, the context in which the effort was made to separate "objective" human labor from "subjective" human essences—the separation highlighted by Marx.[10] Although Ackerknecht was not concerned in his article to take "correct" science to explain "false" science—as have so many commentators on, for instance, phrenology and Lysenko's agronomy—and although he can hardly be accused of wishing to promote the metaphysics of capitalism, his thinking was nevertheless within (and thereby covertly endorsed) the positivist mould.[11] The validity of positivism's dichotomized (subjective/objective) worldview was not only implied in the article, it was reinforced by being part and parcel of the interpretation.

Once this is made apparent, it follows that any argument against Ackerknecht's interpretation of anticontagionism based solely on historical "facts" must necessarily share the positivist foundations of the argument itself—indeed, must strengthen those foundations by deflecting attention from Ackerknecht's overriding interest in the social context of the medical belief in anticontagionism. Although Pelling does not preclude a social consideration of medical knowledge, nothing is further from her own empirical-analytical agenda. It forms no part of her project to exercise what Goethe called "the highest wisdom . . . to understand that every fact is already theory" (leaving aside the romanticism inherent to his claim), or that every wisp of facticity is a priori "social" through its human construction.[12] As anticontagionists proceeded to the subject of epidemic diseases according to the principle that "it is really a question of science to be decided by facts which everyone can understand—a question of testimony, to be determined by evidence which everyone can appreciate"—so Pelling proceeds to historical "reality" on the basis of "the facts" of epidemiology's past.[13]

It would be unfair to brand Pelling an old-fashioned "internalist." Although her book swings the pendulum of medical history back in the direction of "scientific qualities" that Ackerknecht juxtaposed to "human and political" ones, she is not insensitive to the nature of the social context in which epidemiological developments took place. Nor is she unaware of the social interests of certain of the major figures involved. In the course of slamming Ackerknecht, she explicitly welcomes the inclusion of social factors into historical discussions that are otherwise regarded as purely "scientific."[14] But the reference to the *inclusion* of "social factors" is itself an insight into Pelling's positivist thinking. This is also evident through her reference to "propaganda" cluttering up "theory," to "philosophical preconceptions" being resorted to when theory seemed insufficient, and in the repeated assumption that "practical" considerations are somehow neither social nor ideological but are definitely separate from theoretical considerations.[15] She shows her colors, above all, through her thesis that in the first half of the nineteenth century a dichotomy existed between medical theory and sanitary practice—a dichotomy that is held to have retarded the *progress* of "theoretical" developments.

I could go on to point to how positivism's mechanical interactionist philosophy—its treating the world and its occupants as a set of interacting unsensuous facts—is reflected in Pelling's subscription to the infectious or soaky-sponge model for the transmission of ideas that upholds the fiction of individuality. Throughout her study we find people being "influenced" by other people and "absorbing" the thought of others, rather than telling us why certain people at specific moments in specific contexts were more receptive to the articulation of certain ideas, or how, at particular historical moments, social actors can produce the same theoretical insights independently.[16] But there is no need; whereas a great deal of recent history-writing declares its "relevance" by attachment to Ackerknecht-like social contextualism (and so makes it difficult to perceive the partial metaphysic that it underwrites), Pelling's book graphically reasserts the ideological assumptions that are culturally built into modern science and medicine and which are mystified through its historical study. Inadvertently then, the book calls for its supplement, at the very least for a counteraction to its minimizing of interest in epidemiology's social constitution.

Ackerknecht did not, as I've said, lay claim to an understanding of anticontagionism as knowledge mutually constitutive of the social. In fact his contextualist study worked against such an understanding in several ways. Most importantly, it made a case for the belief in anticontagionism as only *reflecting* certain class interests, values and preconceptions, and that the primary concern of anticontagionists was with how, through this way of understanding disease, those interests, values and preconceptions were merely socially *deployed* (as opposed to being socially made).[17] Further, it sought to make a direct causal link between anticontagionism and liberal bourgeois mercantilist interests—an illustration perhaps of how Ackerknecht in the 1940s remained very much attached to economistic thinking of the sort embraced by Boris Hessen and J. D. Bernal in the history of science. According to him (and them) knowledge was ideologically instrumental; it was simply a product of, and servant for, capitalist economics. Ackerknecht was of course justified in making the link between anticontagionism and bourgeois economic interests inasmuch as the historical record abounds with explicit calls by anticontagionists for the abolition of quarantine. But in excluding what the historical record also reveals, that commercial interests were as well, if not better, represented among contagionists as anticontagionists, and that much of the discussion on quarantine pre-dated the rise of the anticontagionist controversy, his economistic thinking is open to question.[18] In the final analysis, it is less Ackerknecht's historical oversights that are crucial, than the trivialization of the knowledge of anticontagionism consequent upon them. By conflating anticontagionism with one of its prime rationalizations (quarantine abolition), Ackerknecht denied the inherent status of the knowledge itself, rendering it merely *epiphenomenal* to socioeconomic interest.[19]

So, although Ackerknecht's article provides a better basis for thinking socially about anticontagionism than Pelling's critique of it, the article no less than the critique deflects attention from the investigation of anticontagionism as a historically contingent intellectual product. Above all this is because Ackerknecht's accounting for anticontagionism is grounded on a reified conception of knowledge: knowledge is the "thing" which merely interacts with external material and social circumstances. Like Pelling after him, Ackerknecht did not think of intellectual products and material circumstances as interpenetrated or mutually constitutive.[20] No more than Pelling later did he perceive anticontagionism as an abstract knowledge reassembly of its producers' actual and idealized conception of social reality.[21]

To approach anticontagionist theory in this way, that is, as a "mystified mediation," we need to go back to what lay at the heart of anticontagionism: the notion that disease was not transmitted by the direct passage of some chemical or physical influence from a sick person to a victim, but rather, was contracted through exposure to certain disease-inducing "miasmas" in the environment.[22] Anticontagionists held that epidemics (excluding smallpox, syphilis, gonorrhea and measles) resulted from exposure to pestilential conditions in different locations at roughly the same time. Since the object lesson of anticontagionism was the avoidance of those conditions (as well as the habits and influences which predisposed illness) anticontagionism, no less than contagionism, was an expression of concern with contamination. Hence the rise of interest in anticontagionism at the turn of the nineteenth century can be seen as a resurgence of the same concern with pollution that apparently brought the idea of *contagium animatum* to a peak of elaboration and support at the turn of the seventeenth century.[23] Arguable, in both periods, midst huge social disruption and uncertainty, there was heightened need to reorganize the understanding of "reality" in order to gain control over it. Both turn-of-the-century cases bear testimony to the observation of Mary Douglas that any concern with filth is also a concern with disorder and that an interest in eliminating dirt represents a concern with reorganizing the environment in the name of purity.[24] Since neither in the seventeenth century nor in the early nineteenth was knowledge of the bacterial transmission of diseases known, there is no basis for supposing that this interest in filth was conceptually akin to the contemporary pathogenic-inspired dirt fetish. Greater is the basis for supposing that in both periods heightened concern with filth and disease related to secularizations of social control. As the earthly punishment of disease resulted from the sin of filth, so earthly hopes of salvation depended upon conforming to notions of maintaining cleanliness.

But if both seventeenth-century contagionism and nineteenth-century anticontagionism were manifestations of similar underlying concerns with ordering the social, the question emerges why the particular anticontagionist variant should have risen to prominence when it did. A part of the answer—the most superficial part, I think—is that supplied by Ackerknecht through his identification of anticontagionists as a bourgeois counter-culture challenge to the socio-political and economic status quo. Contagionism having become by the turn of the nineteenth century a traditional-looking body of knowledge, the anticontagionist attack upon it was part and parcel of a larger critical assault on ruling ideas, institutions, and social order. Characteristically, the "determination to disbelieve every thing that has obtained pretty general credit" (as a writer in 1822 referred to the mentality of anticontagionists) was forwarded most aggressively by those with newly awakened sensibilities to their social and cultural marginality.[25] Although slightly later in the century, especially during the cholera outbreaks of 1832 and 1848, many of these same persons were involved in overt attempts to control the working class, during the first third of the nineteenth century anticontagionists were attempting first and foremost through the idea of anticontagionism to challenge the social and cultural impositions of the old order.[26] Confident in 1824 that the middle class "contain[ed], beyond all comparison, the greatest proportion of the intelligence, industry, and wealth of the state," and that the retainers of the agrarian aristocratic hegemony were decadent, leading anticontagionists such as Southwood Smith crusaded as revolutionary ideologues for intellectual, and hence social, liberation.[27] The new middle class, Smith asserted, should take "nothing upon trust," and should believe "nothing upon authority."[28] Like the social meaning of Francis Bacon's rejection of all arguments based on authority (in the *Novum Organum* of 1620), anticontagionism epitomized heretical sectarianism. It challenged reigning views of what was considered "natural," what was considered "practical good sense," and what was considered sacred (indeed, what in "primitive" society, anthropologists tell us, has often been regarded as *contagious*).[29] So fundamental was the anticontagionists' attack on the hitherto sacred order of things that language itself was called into question and in fact *was* the debate. Time and again contagionists were assailed as "wordmongers" mystifying reality, an accusation that, in turn, legitimized the anticontagionists' new discourse of medico-social explanation.[30]

Even if the vast perceptual and conceptual gulf between contagionists and anticontagionists was not explicit at the time, it would not be difficult to see how the assertion of anticontagionism was integral to the "de-naturing" and "de-sacralizing" of the old order. In claiming to be "more in unison with the state of science in the present enlightened and improving age" than contagionism because it accorded more with the observable phenomena of nature ("objective reality"), anticontagionism supplied an appropriate symbolic (secular religious) resource for discrediting the authority and worldview of the retainers of the old order.[31] In the abstract form of medical knowledge, it was a kind of sacred cross—literally with the power to defy plagues and with the potential to save mankind—worthy of repeated exaltation in the early volumes of that bible of revolutionary bourgeois consciousness, the *Westminster Review*. No wonder that Charles Maclean, the person largely responsible for introducing the idea of anticontagionism into Britain, and whose attachment to the knowledge stemmed from social and political repressions he experienced in India and England, was worthy of deification by the *Westminster* reviewer Southwood Smith as "one of those extraordinary men who is capable of concentrating all the faculties of his mind, and of devoting the best years of his life, to the accomplishment of one great and benevolent object."[32] To Smith, the Philosophical Radical, it mattered little that Maclean was actually to his Jacobin left. What was important, especially to ideologues of change rebelling against the social ideology of Calvinism (Smith was typical in having dumped Calvinism for Unitarianism) was that Maclean's theory appeared to legitimize a reconception and reconstruction of a perceived social reality. As opposed to the Calvinist universe girdled by the blunt, direct but arbitrary predetermined laws of God, anticontagionism was held to signify a manipulative, active universe of rational and uniform natural laws. Although in its own way anticontagionism was no less fatalistic than contagionism and could be interpreted as allowing God an even greater hand in the capricious spread of deadly infections than Calvinism, in its context it optimistically harkened to the reform and control of the environment and to the "freedom" of individuals to be responsible for their own destiny.[33]

But this reading of the knowledge resource of anticontagionism does not explain why it was called forth when it was; it only harkens to its social function *after* it was articulated. To approach its temporal specificity it is necessary to look to its substantive significance in relation to its context of shifting social structures and power relations. Attention needs paying in particular to the significance of anticontagionism's negation of the view of infections as spread by personal contact, and to its alternative view of this transmission as conducted through the atmosphere. The first thing to notice is that what was at issue as far as anticontagionists were concerned was not the actual physical cause of disease. On the spontaneous generation of disease-causing miasmas, anticontagionists hardly ever spoke. Between the application of cause and the appearance of effect of epidemic diseases there was, it was said, "no discernible connection." Indeed, the epidemic influence of air was often referred to as "occult."[34] The words are significant in light of what centrally concerned anticontagionists: the means or mechanism of disease exchange and communication. Fundamentally, in the midst of rapidly shifting social and economic arrangements, anticontagionists were mediating through the concept of atmosphere the idea that the relations between things could not be taken as straightforward, or could not be seen in a mechanically causal way as if in a "direct line," as in the person-to-person exchange relations in contagionism.[35] Rather, to cite the words most frequently employed by anticontagionists in their discussion of disease, relations were ambiguous, diffuse, indeterminant, not uniform, flexible, fluid and "diversified to the highest degree."[36] To the rigidity and directness of the contact infection model of contagion, the atmosphere juxtaposed a metaphor of dynamic and indirect, pervasive plasticity. "Air," said Maclean, is "of all the agents which act upon the living body, that which exercises the most diffusive influence."[37] What the stress on atmosphere signified, then, was a naturalization—indeed celebration—of dissolution. To the supra-concrete reality of the old nature of things, or to ruling conception of the way in which things and people ought to relate to each other,[38] the atmosphere provided an unfixing solvent.

But in itself the atmosphere suggested no alternative reassembly of reality. The anticontagionists' refusal of a label for the property of air that causes epidemics—lest the mere giving of a name "would cement it into a thing of substance in men's imagination"—illustrates the greater insistence on dissolution than on reconstruction.[39] Yet, as partly disclosed by the renewed importance that anticontagionists attached to the inherently hierarchical concept of miasma (those noxious exhalations emanating from low-lying areas), at the same time they manifested a need to redefine and reconstruct social reality in more precise orderly terms. The insistence on the certainties of science and the natural laws of animal economy was simultaneously a search for new order and meaning. It is to be seen in Smith's claim that although there was diversity in epidemic diseases, "yet in all countries, the periods at which they commence, decline, and cease, are determinant and exact"; and again in the "remarkably uniform" correspondence posited between epidemics and seasons. Better still, we can see it in Smith's statement that "epidemic diseases are governed by laws as precise and uniform as contagious diseases; but . . . these laws are not only not the same, but the very opposite."[40]

Thus anticontagionism was called forth when it was in order to articulate a substantially modified view of the natural order of things and the interactions between them. In this respect its function was no different from that of other naturalistic models in the history of thought.[41] But unlike, say, Bichat's physiology or Gall's phrenology, which supplied corporeal models with which to make isomorphic comparisons between real and idealized social organizations, anticontagionism was an articulation more in terms of a general *want for* new classifications and divisions, rather than a concrete illustration of them. This want is obvious in the way in which virtually every opponent of contagionism began with an attack on the way in which diseases had hitherto been classified, and then proceeded to argue the need for new rankings based on clearer distinctions between types of disease.[42] Maclean, for instance, asserted at the outset of his pamphlet of 1820 his "general conclusion":

> for, every disease, according to the laws which it obeys, must at last be ranked in the class to which it belongs, whether epidemic, contagious, or sporadic. The list of Hippocrates includes, fevers of every kind, as continual, whether mild or ardent, intermittent, whether quotidian (diurnal and nocturnal), semi tertian, perfect tertian, quartan, quintan, and nonan; chronicals, and erratics; dysentries, diarrhaeas; quinseys;

heripneumonics; palsies; erisipelas; and even ophthalmies and haemorrhages. This list I am by no means disposed to curtail.[43]

For present purposes it is less important that these re-definitions and classifications frequently only baffled lay and medical persons alike, than that they reflect the aspiration radically to reconstruct reality.

This runs to a final point: that anticontagionism was part and parcel of changes in the conception and regard of labor under the advance of industrial capitalist production. Although anticontagionists in the early nineteenth century primarily thought of themselves as opposing contagionists in the medical profession, and largely confined themselves to "factual" medical argument, their wider project was with wresting from ordinary people the limited control they had over their existence. (Indeed, this extended to legal control over the *dead* bodies of paupers, secured through the passage of the Anatomy Act in 1832, which had been framed by Smith and fellow Philosophical Radicals.)[44] Although anticontagionists (as "anticontagionists") ultimately failed to accomplish their ambition of revealing themselves as the only custodians of the truth about the vital forces of life and death (just as the medical profession as a whole failed to do this for most of the nineteenth century), they can be seen as attempting to draw such truth further into their own professional hands.[45] Commonsensical understandings of how diseases were contracted, or how, through isolating the sick, epidemics were to be managed, were explicitly referred to by anticontagionists as founded on "delusion."[46] To the specialists of science and statistics alone people should apply for mastery over their lives. Just as factory labor alienated workers by depriving them of holistic understanding of the productive process (through increased deskilling or separation of mental from manual labor), so anticontagionism was accessory to the erosion of people's synthetic understanding of everyday reality.

But this correspondence between the knowledge and its socioeconomic context, though necessary to make, is, if projected as explanation, only slightly less reductionist and mechanically causal than the economistic correspondence pursued by Ackerknecht. It similarly suggests conspiratorial thinking either on the part of anticontagionists or myself. By focusing attention exclusively onto the context of the knowledge, the correspondence forestalls the investigation of capitalist social conceptions buried within and mediated through the substance of the medical knowledge. To approach these it is necessary to return to the central organizing principle of anticontagionist thought, the idea of atmosphere. In particular, attention needs paying to the way in which the idea of air both bound people within an ethereal whole yet separated them individually from each other. At one and the same time the atmosphere of anticontagionism minimized the importance of immediate personal contact (the basis of mutual aid), while it elevated the importance of an external depersonalized environment through which social relations might be negotiated. Air can thus be seen in the thought of anticontagionists as a means of alternatively conceptualizing the emergent realities of the industrial capitalist urban environment. Harboring unknown poisons, the air took on the distinguishing hostile qualities of life in new industrial towns. Like the towns, the air required purification and, in this sense conveyed the Utilitarian-inspired ambition to moralize individuals by moralizing society's social framework.[47] Air, once the very stuff of human breath, became in this body of thought an alien "thing" amenable only to the analytical understandings and moral and political interventions of environmental manipulators. Looked at this way, the knowledge of anticontagionism can be seen not as causally linked to the economy, nor simply as a direct reflection of it, but rather as a mystified mediation of the constitutive changes in the social relations of production contingent upon the advance of urban industrial capitalism.

Conclusion

Charles Rosenberg, in a summary footnote to his study of the changes that occurred in social life and medical thought and practice in America over the cholera years between 1832 and 1866, confessed that he had "taken the perhaps ingenuous course of ignoring the causal and, to a large extent, the temporal relationships" between material developments and intellectual changes. He added that "this ancient philosophical problem, with its rigidly dichotomous statement, seemed to me one best avoided. The historical, rather than the philosophical, sensibility must conceive of this relationship not as one of cause and effect, but in terms of a 'dynamic equilibrium' between intellectual and material change (the categorical distinction between which is, in any case, artificial, and *a*historical)."[48]

Despite the slightly didactic tone and the seeming confusion of concepts, the statement is remarkable for its candor: while Rosenberg excuses himself for having not treated the relationship between material and intellectual change, and deplores the project in terms of its simple causalities, he nevertheless confesses to the problem and the ingenuousness of ignoring it.

Like Rosenberg, I have not sought to discuss concrete material realities, nor deal over an extended time period with the problem of the ideational in relation to material and social change. To do so would require a much lengthier study. The idea of the non-contagious nature of epidemic diseases (even leaving aside the wider problem of the "miasmatic theory" of disease production) remains enormously complex, diffuse and elusive, in large part because of the shifting social context of anticontagionism's use which continually re-defined its meaning.[49] I have not, therefore, sought generalization on the subject outside one particular period in Britain, nor attempted to provide a complete understanding of its nature at that particular moment. Rather, I have tried to treat the relations between material-social change and intellectual change by briefly illustrating some of the ways in which anticontagionism as a knowledge product can be seen as mutually constitutive of the historical conditions that gave rise to the social context in which the knowledge was called forth. This is precisely the point of treating the substance of knowledge as social mediation: the positivist-informed separation of history from philosophy on the one hand, and social context from knowledge on the other, are collapsed when the totality of historical constitutiveness is sought. The historian may then go on to study temporal changes in thought and material life (at both macro- and micro-sociopolitical and socioeconomic levels) not with functionalist expectations of "dynamic equilibriums," but with expectations of unraveling the interpenetrations that have occurred over time.

Although I have sought no such comprehensiveness, it should be clear that this is a fundamentally different task than that of seeking only empirical-analytical corrections to medicine's historical record, or seeking merely to socially contextualize medical thought. Valuable as these historical exercises often are, they, along with projects to determine the intellectual origins of ideas or to establish whether certain social conditions preceded or followed upon the rise of certain bodies of knowledge, must be seen in and of themselves as operating against a constitutive understanding of medical knowledge. They remain locked within a causal-interactionist framework which was itself socio-economically fashioned. Within that framework it is possible to expose and counter the historical propaganda of particular victors in history, such as germ theorists. But the real intrusion upon history—the positivist metaphysic that defines the framework itself and sets up historical procedures in terms of the metaphysic's own inherent objective/subjective, fact/value dichotomy—is not only left intact but unwittingly strengthened through its further obfuscation. Only by becoming aware of the determinant metaphysic and attempting to initiate morally and politically intended alternative means of inquiry will the study of medicine's past begin to contribute to an understanding of the socio-historical present.

3

"Framing" the End of the Social History of Medicine

THIS ESSAY WAS WRITTEN around the turn of the millennium, some two decades after the previous essay. By then there had been a sea change in Anglo-American intellectual life, with the powerful tide of the semantic or literary turn pressing all before it. Academic history was not immune, and voices such as that of Patrick Joyce (a Manchester social historian who had turned postmodern) were coming to be heard in leading journals.[i] For the most part, though, historians were resistant to the rising tide, rightly sensing the challenge it posed to their faith in objectivity, never mind its charge that their discipline was but a literary-fashioned metanarrative. There was fighting back, from both the left and the right. But most historians (myself included) pressed on almost regardless. We produced ever-more-nuanced studies of historical objects, albeit with a greater tendency to "the cultural" rather than "the social" (the latter increasingly understood as an invented historical category that needed to be handled with care). "Discourse" also came into the vocabulary, although as often as not, merely as a substitute for a generalized body of thought. Discourse analysis was nowhere to be seen, despite (or maybe because of) the earlier attentiveness to language in the Marxian tradition. There were ventures into historiography, but these were primarily descriptive accounts of the history of history-writing in different national contexts, referring mainly to approaches to the craft—Rankean, Marxist, Annales, and so on. Foucauldian historical epistemology or "historical ontology" (to use Ian Hacking's term) was largely confined to the philosophical reaches of the history and sociology of scientific knowledge. In the history of medicine social constructivist accounts of knowledge had largely fallen to the wayside. Ironically, given the earlier hostility to them from medical practitioners, they were now picked up by some of their own tribesmen.[ii] Prevailing for the most part were social contextual studies of known objects, such as sick persons, healers, or diseases. But the study of contexts remained at arm's length from the historian's own.

At what became the University of Manchester's Centre for the History of Science, Technology and Medicine, where I was based from 1984 to 1998, interpreting the past was the order of the day, rather than thinking about the politics and philosophy of its practice. I profited enormously from the company of John Pickstone, who directed the Centre and whose intellectual generosity, acuity, and curiosity were unbounded. Jon Harwood, Jon Agar, Roberta Bivins, David Cantor, David Edgerton, Mary Fissell, Bill Luckin, Jack Morrell, Stephen Jacyna, Mark Jenner, Mark Jackson, and Steve Study were among the many who came to reside at the Centre or be closely associated with it. We worked relentlessly at our craft, mostly on nineteenth- and twentieth-century projects, and our seminars and publications were nothing if not robust in terms of their empirical weight and synthetic cunning. We raised hard questions—largely informed by social history and the history and sociology of science, medicine, and technology—and sought valiantly to answer them. We came to engage in various ways with Pickstone's impressive synthesis of the history of "ways of knowing" (his unifying "big picture" of what, in 2000, appeared as *Ways of Knowing*). But we never questioned our faith in historical truth itself. On the rare occasions when one of Patrick Joyce's graduate students wandered across the street from the history department to speak at our seminars, the reception was at best puzzlement.

Perhaps because our funding from the Wellcome Trust made us relatively secure, we didn't feel any great need to question the constructedness of our own ways of knowing. We were confident in them. And we largely took for granted the importance of our place in the academy, never doubting that intellectual practices would always triumph over pragmatic managerial ones.

By the time I came to write this essay I was, in retrospect, halfway to thinking about the ethereal envelope in which history is written. If everything around one was a construct (however that might be interpreted or valued), it was hard not to regard history-writing in the same light. Such reflections were pursued further in this essay's sequel, "After Death/After-'Life': The Social History of Medicine in Post-Postmodernity," published in *Social History of Medicine* in 2007. By then I was also far more appreciative of Foucault's thinking on the historicization of culture, but not yet with the problem of the historian's self within it and, hence, with the historicity of historical critique. At the point of writing this essay I was only coming to terms with the terms for such thinking, and still believed that history-writing was something determined causally by the context for its writing. One of the purposes of this essay was to bring out that determinism in relation to the new postmodern context.

But nagging at me behind its writing was a deeper concern: where had the political bottom of history-writing disappeared? More especially, what were now my own politics in the new intellectual environment? Indeed, what was my relevance in a world that had not only moved beyond investment in the study of history but, within the field of history-writing itself and cultural studies more generally, had moved away from attempting any direct political involvement with the so-called real world? Of course Anglo-American postmodernism was political in its own way, radically challenging conventional perceptions of the nature and exercise of knowledge/power. It also did political work, which was, if largely unbeknownst to itself at the time, mainly *for* the culture and ideology of neoliberalism. Historical and sociological studies of science also did political work, by exposing the epistemological assumptions behind the production of scientific knowledge. In this connection, the landmark text for a generation was Steven Shapin and Simon Schaffer's *Leviathan and the Air Pump* (1985), which exposed how the bedrock of modern science and the foundation for all its truth claims, experimentalism, was itself socially and politically constructed. Such studies extended the contextual approach to scientific knowledge and practice begun in the 1970s, and they fanned the flames of what became the "science wars" of the 1990s initiated by those seeking to defend scientific realism against the relativist incursions of postmodern critical theorists who were challenging not only history-writing as "objective" but science, too. Yet historical and sociological studies of science, as much as anglophone postmodern literary studies, moved away from any manifest desire to change the world—away, that is, from what had shaped my idealism in the 1970s. Increasingly, it was enough to be clever at analyzing reigning truths, historically or otherwise, rather than thinking to do something ground-moving with the knowledge produced. The sails of political idealism flapped for lack of wind while, not uncoincidentally, neoliberal managerialism took hold inside the academy as much as on the outside. Writing this essay was my way of trying to grapple with these discomforts, as much as struggling to come to terms with postmodernism in history-writing. Could one, and should one, aim to reconfirm older ways of looking at history or, instead, search for new theoretical frameworks to think about it as a form of critical inquiry, as suggested by some postmodernists? This grappling was a kind of hermeneutics for myself. Looking back, it was my crisis of faith, my "Dover Beach" in a postmodern era.

Its composition was not unlike my critique of Erwin Ackerknecht's study of anticontagionism presented in the previous chapter, inasmuch as it mediated what I was wrestling with through a focus on someone else's text.

This time the text was that by the American doyen of the social history of medicine (and student of Ackerknecht), Charles Rosenberg. His concept of "frames and framers," which he first formulated in the policy journal the *Milbank Quarterly* in 1989, seemed to me important for the strategy that I took it to be advancing—preserving the social constructivist approach to the history of medicine against the "realist" storm clouds gathering around it. But what were the politics and political implications of this strategy in an intellectual world that had moved beyond social constructivism? My intervention was intended as a commentary on this move, not a defense of social constructivism, let alone an attempt to conduct a context-based social constructivist analysis of Rosenberg's own thinking. I was seeking only to set out his strategic thinking in its intellectual context, and in the neoliberal politics of its moment—all as part of an effort to further my own self-understanding.

Still hovering around the edges of the essay was the idea of situated bodies of knowledge mediating socioeconomic and political interests. But by the time I wrote it I was coming to agree with scholars in social studies of science that the social constructivist program had been far too ambitious, and far too naïve in its search for symmetries between knowledge production and discrete socioeconomic and sociopolitical contexts. Cracks in that thinking had been opened by Andy Pickering's *Mangle of Practice* (1995), with its hard-headed insistence on the indivisibility of scientific knowledge from its practice and use. And it was cracking big time through the onslaught of Bruno Latour and his colleagues investigating contemporary science. The latter were increasingly coming to challenge the idealism involved in the social constructivism program in science studies and in sociology as a whole—not to mention the asymmetries between the analysts and the analyzed in a work such as *Leviathan and the Air Pump*.[iii] At the same time came the realization, articulated by Joyce, that "social construction" presupposed a distinction between what is and what is not "the social" and, further, that "social context" becomes meaningless when the distinction between "text" and "context" is dissolved.[iv]

Like many academic productions, this one stemmed from an invitation to a conference. Frank Huisman, then at the University of Maastricht in the Netherlands, brought a group of us together there in the summer of 1999 to discuss the historiography of medicine. We did, and for entertainment we toured what for most of us were the unheard-of labyrinthine caves cut miles into the soft local marlstone. It was there that the Nazis had stocked their V2 rockets and, in the nineteenth century, Jesuits, burnishing solitary candles, had devoted a day a week to reproducing famous paintings on the walls of the caves and to carving perfect simulacrums of the Alhambra and other worldly wonders. It would be hard to say that in that cold, utterly leaden darkness the craft of the historiography of medicine had some E. M. Forrester–like moment.[v] Yet, we all felt somehow touched, aware that we were making some kind of "passage," although none of us could put a finger on it back in the sunlight of the seminar room.

There was chiaroscuro in my own career too. The previous year I had moved to the University of East Anglia in Norfolk to set up a Wellcome Unit for the History of Medicine. The department of history, where the Unit was based, was a very different environment from what I had been used to. Here was no hothouse of intellectual questing of the sort I had blithely taken for granted and upon which, I realized, I was wholly dependent. There were good scholars there, among them Geoffrey Searle, author of the excellent study of social thought *The Quest for National Efficiency*. He was regarded as the grand old man of the department. And there were some budding young scholars, such as the eighteenth-century historians Andy Woods and Mark Knights, and the postmodern-minded medievalist John Arnold. But there was no collective assumption in the department that ideas really mattered. In order to create such an environment one would have to work at it (and come to bear some wrath for doing so). It meant searching out colleagues to forge alliances, such as my old friend the historian of ideas in science Ludmilla Jordanova, who was then in the university's department for art history, or the political sociologist of contemporary biopower Brian Salter, who was then in the university's Department of Health Studies.[vi] But above all it meant the slog of writing grant applications in order to bring in companionable scholars and good graduate students, and with the success of that, the experiencing of a further degree of ostracism from the history mainstream. It also meant playing departmental politics, a game I was unused to and which I eventually lost. My only regret, as I headed a few years later to University College London (happily with my new recruits in tow), was that I never had a chance to meet the novelist W. G. Sebald, whose office had been only a floor below mine in the university's hugely successful creative writing school. Ever spinning his holocaustic *Rings of Saturn,* Sebald met his maker in a fatal car crash in December 2001.

Not that any of this was made apparent in the essay; its main result was to turn a special issue of the *Journal of the History of Medicine and Allied Sciences* into a defense of Rosenberg's reputation.[vii] Another unintended result was the arming of those who, for one reason or another, had an interest in disinvesting in the social history of medicine.

Apart from involving me in something close to writing my own obituary, framing the end of the social history of medicine is difficult for the simple reason that the subdiscipline's roll into the grave was far from obvious or straightforward. The end was more in the manner, literally, of a passing—a sluggish, uneventful meltdown, nowhere much noticed or commented on. Indeed, the walking dead are still many among us: explicitly "social" histories of medicine continue to be written, and undergraduate courses in the social history of medicine (not to mention a society and a journal by the name) continue to be subscribed to. This is odd because, in somewhat more than literal fashion, the end was heralded over a decade ago in 1989 by Charles Rosenberg. The acknowledged doyen of the field, Rosenberg proposed the banishment of "the social" in the social history of medicine and its replacement with "the frame."[1] The substitution was soon widely endorsed.[2] However, the expressed motives for it (as well as the unexpressed ones I will come to) were different from the historiographical ones that *might* have been posed. Thus, the actual ending of the social history of medicine was obscured at the same time as the literal ending was (in that other sense of the word) "framed." Such is my contention. This essay therefore explores what can be styled the framing of the framing, though I wish to imply no conspiracy.

That our concern has to do with more than merely the substitution of words first became apparent to me in the mid-1990s when a publisher approached me to write a social history of medicine. The prospect was enticing. Here would be a chance to pull together some of the many different themes that had come to comprise the field: the politics of professionalization, alternative healing, the study of patient narratives, welfare strategies, constructions of sexuality and gender, madness, deviance, diseases and disability, public health and private practice, ethics, epidemiology, experimentation and education, tropical medicine and imperialism, along with the various actual and rhetorical relations between medicine and war, medicine and technology, medicine and art, medicine and literature, and so on. More than that, the book would afford an opportunity under such headings, chapter by chapter, to review how, over the past two or three decades, historical and historiographical understandings had broadened, sharpened, and deepened. There was need for such a book and there probably still is. But when pen came to paper paralyses set in. The problem wasn't the survey of the literatures involved, formidable though they had become. Rather, it was more the nature of, and need for, the overall packaging. A *social history of medicine?* In 1992 Andrew Wear proudly declared that it had "come of age," to which Ludmilla Jordanova roundly responded that it was "still in its infancy."[3]

What caused my pen to falter, though, was the realization that somehow, somewhere along the way in the 1980s and 1990s, all the key words had lost their certainty of meaning, and some (*pace* Rosenberg) had even been threatened with excision. No longer could they be taken for granted: "history" and "medicine," no less than "the social," had become deeply problematic. The hinges of the triptych were rusting up and coming unstuck; historiographically, the whole was no longer the sum of its elemental parts. Such a realization was more than a little worrying politically, inasmuch as the implied loss of disciplinary coherence could only disable the critical, if often socialist, thrust of an enterprise that had been honed over the previous quarter-century in alliance with social medicine, medical sociology, and social history—an unusual mix of radical politics and policy, critique and theory, and empirical practice, such as represented by Thomas McKeown, Ivan Illich, and E. P. Thompson, respectively.

I have no desire to defend the use of "the social" in the social history of medicine. It may as well be admitted that the subdiscipline has had its day, done its bit, much in the manner of older forms of intellectual, political, and economic history. Given that much social history of medicine has been intellectually flat-footed and theoretically unreflective—at best revisionist, at worse dominated by empiricism and even scientism (if a vast number of demographic contributions are included)—moving on may be no bad thing. The intention here is more modest: merely to capture something of the wider intellectual context that has necessitated this moving on and ought, I believe, further inform it. This essay has been conceived mainly as a record of what has happened, a chronicle of an "ending" insofar as the business of history-writing ever has closure.[4] It seems worth doing, for while the "history wars" have been conducted over the past few decades as vigorously as the culture wars and the science wars of which they are a part, it is not at all clear what exactly has changed, or how and why. Least of all is this clear to new entrants to the history of medicine.

Admittedly, exactitude is easier called for than accomplished, given the almost ineffable, inevitably partial, and certainly very messy nature of the context within which the meaning and practice of history-writing have recently been challenged. For these reasons among others it is not possible to provide anything like a social constructivist account of the historiography of the end of the social history of medicine—an analysis, that is, in terms of a situated body of knowledge mediating socioeconomic and political interests. (The subdiscipline was never "schooled" enough for that; indeed, it was always in tension with the aspiration to de-ghettoize itself by merging into the historical mainstream.) Nor, for essentially the same reasons, is it possible to provide either an analysis of the discourse of the historiographical body, or a deconstructivist semiotic account of "the text," since neither the infrastructural "body" nor "the text" as such exist. We can, however, lay out some of the features of the subject's problematization with reference to its wider intellectual and, to some extent, sociopolitical context. This is easiest done by focusing first on the key words, starting with "medicine" itself. We can then turn to an analysis of Rosenberg's article on framing, endeavoring to locate it within the politics of historiographical change. Finally, by way of a postscript on the postmortem, as it were, some observations can be tendered on the possible place of "the social," "the political," and "the medical" in history now.

"Medicine," "History," and "the Social"

The word "medicine" has always been troublesome if one stopped to think of it. Recently described by John Pickstone as a convenient omnibus term, resembling in this respect "agriculture," or maybe "engineering" or "electronics," it invariably consists of more than merely the professional practice of licensed healers in all their economic, political, and social settings.[5] It is more, too, than just the knowledge of diseases and processes affecting the body in sickness and health and the prevailing technologies for corporeal intervention.[6] Worldviews and ideologies like humoralism, evolutionism, and environmentalism, or Christianity, communism, and fascism, have necessarily been as much as part of it as more specific sets of shifting discourses, rhetorics, and representations. "Cultures of healing" and "healing in culture" as objects of study can be as conceptually broad or narrow as the analyst chooses to make them. As both a knowledge and a practice, medicine (like religion) can be experienced at the most intimate level of being and politics, as well as, simultaneously, through precipitates from the highest reaches of global ideology and economics. And not just "experienced" in any simple sense; as Foucault appreciated, medical sites and personnel have historically been bound up with mutations of political thought into their modern forms. In the course of such processes, individuals have come to describe themselves in the languages of health and illness, and to accept the norms of "the normal" and "the pathological" as the basis for circumscribing their mortal and moral existence.[7] Obviously, then, as Jordanova has insisted, we cannot treat medicine simply as knowledge, or merely as "another form of science."[8]

At root, medicine is about power: "the power of doctors and of patients, of institutions such as churches, charities, insurance companies, or pharmaceutical manufacturers, and especially governments, in peacetime or in war."[9] Recently, however, there has been a seismic shift in the nature and exercise of the powers that constitute "medicine." Whether our gaze is on the disarray of health services in post–Cold War Eastern Europe, China, and the poorer countries of postcolonial Africa, or on the fast-changing reorganization of medical systems in the West, things look very different from what they did ten or twenty years ago. Crucially, welfare medicine has been felled. National health services now embrace the logic of private, multinational corporations.[10] In Britain's National Health Service this thinking is evident in the adoption of Private Finance Initiatives for the building of new hospitals and the turning of general practitioners into NHS "fund holders" operating according to the rules of internal markets. The shift is further made apparent and exercised through the emphasis since the 1980s on "evidence-based medicine," the mechanism enabling the state to pay only for treatments for which there is statistical evidence of benefit—a concept and practice of particular appeal to managers and accountants. Many of these changes have been pioneered in the United States, where doctors now queue to obtain degrees in business administration in order not to be irrelevant in the so-called medical reform process.[11] As this suggests, fears by the medical profession (especially in Britain and the United States) of losses of autonomy no longer stem directly from the state, as they were thought to in the past, but from the rationalizing forces of business management, encouraged by neoliberal governments. In the United States, more than one-half of all doctors are now salaried employees of medical corporations, and the consequent rhetoric of an "embattled profession" has been taken as compelling evidence of the decline of professional authority in general.[12]

Professional authority—power—in medicine has also been seriously challenged by outside "interference" in the hitherto professionally sacrosanct area of clinical decision making. Both cost-calculating managers and evidence-weighing governments now act as third parties in such decisions, even though doctors remain the principal targets of complaint in what they sometimes perceive as a "culture of blame." Through the sensational exposure of medical malpractice and incompetence, further reason is found for withdrawal of the state's former compliance in the profession's self-regulation (existing in Britain since the Medical Registration Act of 1858). Today the profession's ethical governance is increasingly given over to the courts through legal regulation.

Such redistributions in the power around medicine are reflected in, and inseparable from, equally profound changes in doctor-patient relations—to the extent that the word "patient," too, has lost much of its certainty of meaning and has become open to contestation.[13] In the face of fragmentation, de-personalization, and multiprofessionalization in the delivery of medical care, calls have been made for "narrative-based medicine," the ethics of which allege "the primacy of the patient's voice."[14] Meanwhile, increasing numbers of "health consumers" have turned to untested therapies by unregulated practitioners while at the same time demanding more evidence-based regulated medicine. Many of the same health consumers (wealthy, white, and western for the most part) partake in what Roy Porter once dubbed the "MacDonaldization of medicine."[15] They demand as a "right" the freedom to shop unfettered in the "supermarkets of life," picking and choosing reproductive technologies as readily as vaccines, kidneys, hearts, and assisted suicides. Much of the laity has also been medically reskilled and empowered through websites and illness support-groups, and now comes to act as a jury for a medicine on trial.[16] Clearly, in the pluralized medical marketplace there is not one, but multiple sources of authority, and the idea of the passive patient is noticeably passé.

This rearrangement of power is further reflected in the refutation by some medical ethicists of the very existence and operation of a universalistic morality, such as that purported for the late twentieth century by the philosopher John Rawls. The idea of a moral consensus (Rawls' "reflective equilibrium") presupposed social homogeneity. But this has become increasingly difficult to imagine and less desirable to sustain in liberal democracies where multiculturalism and multifaith are hyped as politically correct. Hence, in theory at least, different, equally "rational," interpretations of what is medically ethical have come to co-exist, based on religion, culture, and ethnicity.[17] Alongside, and not entirely separate from this development, has been the undermining of the instrumental rationality of modern medicine. Models of linear progress, and belief in rational control over the processes of medicine, have been seriously eroded by the manifestation of previously unforeseen risks and the negative side-effects of bio-medical "progress"—erosions that mark, in Ulrich Beck's terms, the shift from "simple" to "reflexive modernity."[18] Once these risks began to be registered in the minds of insurance companies, medical professionals, medical management teams, the state, and potential individual "clients," the would-be rationality of bio-medical research began to be demystified and demonopolized. Contributing to this view was heightened awareness of how multinational and monopolistic-tending pharmaceutical companies controlled much of the research as well as the allocation of products. Increasingly, the public began to decide between different plausible or probable scientific claims. In this situation, as has been pointed out in reference to the pressures for redefining brain death, political groups came to make use of scientific expertise and counter-expertise in order to push forward their own favorite practical and legal solutions.[19]

Unsurprisingly, such fundamental change in the relations, organization, consumption, and overall conception of modern medicine has had an impact on how medicine is thought about historically. Specifically, the notion of the social control of docile bodies, which was basic to the social history and historical sociology of medicine as it developed in the 1970s and 1980s, has come to seem dated as an analytical imperative. No longer is it quite so obvious to regard medicine simply as a powerful means of imposing social order though "disciplinary normalization." Perhaps in the past as in the present, the relationship between medicine and the laity entailed wider interactions between self, society, and knowledge, all according to competing priorities and the different material constraints of everyday life.[20] To arrive at this conclusion in no way necessitates discarding Foucault's insights on the crucial role of modern medical language and practice as a medium for the policing and self-policing of bodies and desires. Required, rather, is discarding crude or vulgar histories of medical surveillance, social control, and the deskilling of patients—1970s and 80s contributions to the historical narratives of professionalization and medicalization. In part, these narratives were hoisted on their own petard in the 80s and 90s when the self-serving antiprofessionalism of radical feminists and critics came to support dialectically the interests of free market ideologues and anti-abortionists, along with eco-activists, neo-fascists, ravers, "hactivists," and the others who came to make up a "Do-It-Yourself" culture.[21]

Although medicine as an epistemological and discursive concern should not be conflated with medicine in the service of professional power, it is fair to say that the very idea of "medicine" or "the medical" has been de-stabilized. As Nikolas Rose has pointed out, "What we have come to call medicine is constituted by a series of associations between events distributed along a number of different dimensions, with different histories, different conditions of possibility, different surfaces of emergence."[22] Medicine is thus no longer the self-contained entity that it once seemed; as it is technically more complex, so it is correspondingly more multifocal and multivocal in its material relations. Its boundaries are less clear and more porous than formerly thought, and it has consequently become less sharp a category for analysis. In extending everywhere, it might be seen as everything and nothing—like Foucauldian "power," somehow everywhere and nowhere. What, then, is the thing to which the analytical tool of history is to be applied? And how can medicine be an analytical tool for the history of society? Not only were all the assumptions that historians made about medicine in the 1960s and 70s with respect to the state, professional power and science called into question by the realities of the 1990s,[23] but so too, was the very object or category of study.

And "history"? While it may be true that "from the time of Herodotus and Thucydides, historians have vehemently disagreed about the purposes, methods, and epistemological foundations of the study of the past," before the 80s and 90s these debates were never so extensive or intensive.[24] Parodying Marx, Robert Putman in 1993 insisted that "history matters," because "individuals may 'choose' their institutions, but they do not choose them under circumstances of their own making, and their choices in turn influence the rules within which their successors choose."[25] Of course. What was new in the 80s and 90s was the urgency to defend this commonplace—not against an old reactionary right, nor against a perceived-to-be over-deterministic ("choice"-denying) Marxist left, but against a radically iconoclastic postmodern avant-garde of philosophers, literary theorists, and cultural critics. According to these, history was nothing more than the "invention" of historians dealing in images and representations of the past to which they could not possibly have any direct access. Further, the much-loved periodization of historians was nothing more than a strategy for narrative closure. Thus historical "truth" and historical practice were to be regarded as no less contingent and subjective ("shaped not found") than the scientific "truth" discerned by sociologists of scientific knowledge. In new and more extreme ways than in the past, the old hoary question of objectivity was back on the agenda.[26]

Professional historians were stunned and deeply threatened by such attention to their methods and assumptions. A torrent of defensive publications was issued by (to name but a few) Arthur Marwick, Bryan Palmer, Joyce Appleby, Lynn Hunt and Margaret Jacobs, Richard Evans, Geoff Eley, Eric Hobsbawm, Raphael Samuel, and Gareth Stedman Jones.[27] Despite the different political orientations of these historians, they all shared the same professional and political suspicion of "postmodern postures," while to differing degrees confessed to their own complicity in the old rationalist and increasingly demoralized Enlightenment search for objectivity.[28] As the political culture veered to the right under Reagan, Thatcher, and Bush, confusion, apathy, and uncertainty set in.

Nor was the "postmodern challenge" the only cause for concern.[29] In new and far more extensive ways than in the past, "history" (like medicine) was being "managed" in both crude and subtle ways by contending communities of opinion—a feature of the present to which, in fact, postmodern writers drew attention. In the museum in the bowels of the Statue of Liberty, as in countless other public sites, political battles for control over representations of history were fiercely fought.[30] Allied to this (again not unlike in medical practice), history was increasingly subject to naked market forces, both within the academy (for example, through assessment exercises linked to research funding) and outside it in the expanding commercial heritage and leisure industries.[31] Ideologically hand-in-hand with the "rationalization" of history departments went the global expansion of capital-intensive "Disney history," the latter providing the new discipline of museology with an unlimited supply of case studies in the manipulation of historical representations. Forces of a slightly different nature, more sinister for being less public, were also at work in the history panels of grant-giving bodies, not least in the history of medicine. Increasingly the tendency of such bodies was away from humanities-style appraisals to more science or social science models, with emphases on practical applicability, "relevance" (to short-term political interests), directed goals, and publicly accessible "outputs" and "impacts." In Britain especially, the idea of the historian as a devoted, critically-minded intellectual was increasingly derided as a relic of ivory-towered times. The fetid breath of managerialism hung heavy.

    As for "the social," by the mid-1980s it was deemed by Francophone semiologists as absorbed into "the cultural."[32] A decade later, the prospect of an "end of social history" not only had been raised, it had been realized.[33] In what superficially appears in retrospect as an intellectual parody of the thinking of the then dominant political parties, Francophone poststructuralists thoroughly underpinned Maggie Thatcher's politically expedient class consensual claim that "there is no such thing as society." Although Thatcher and her free market cronies were scarcely like Jean-François Lyotard and his disciples in their quest to deprivilege the intellectual importance of the economic and the political, the effect was much the same: "a loss of political appetite for the old frameworks of social analysis" and, in particular, for the validity and relevance of Marxism.[34] While the new political elite effected their ideological cleansing in the name of a new "end to ideology," poststructuralists put to flight all notion of structure, agency, and social determinism.[35] Operating from the aesthetic critique of modernity first elaborated by Nietzsche in the late nineteenth century, the French "new philosophers," as they were often called in the English-speaking world (notably, Jacques Derrida, Gilles Deleuze, Lyotard, and Jean Baudrillard)—or the "young conservatives," as Jürgen Habermas referred to them because of their abandonment of all hopes of social change—demanded freedom from political forms of life, and the rejection of the "tyranny of reason," technocratic rationality, and the old (Marxian) emphasis on the economic.[36]

The pre-"postmodern" Foucault—concerned with liberating the revolutionary process from ritualized and dogmatized Marxism—had only invited *consideration* of whether power was "always in a subordinate position relative to the economy."[37] Derrida and other linguistic deconstructivists, however, argued forcefully for the impossibility of reducing technologies of violence to instrumental political power, economic interests, and social control.[38] The Foucault of *Surveiller et punir: Naissance de la prison* (1975) who appeared to offer a critique of capitalism ("it is largely as a force of production that the body is invested with relations of power and domination") was increasingly re-fashioned as an avant-garde literary theorist.[39] This other Foucault, identified largely with the *Histoire de la sexualité* (volume 1, 1976, English translation 1981), doubted that power was always in the service of, and ultimately answerable to, the economy. Instead, he insisted on the grandiloquence and rules of discourse in *constructing* bodies.[40] Here, as everywhere in what became the flight into cultural studies, sociological categories were binned in favor of semiotic ones, which were often inflected psychoanalytically, as in the bogy of "narrative *fetishism.*" The nineteenth-century master narratives of modernity provided by Marx, Durkheim, and Weber were shed through the literary turn to the discursive—what the Trotskyite Bryan Palmer dubbed the "descent into discourse."[41] Although sophisticated Marxist theoreticians, such as Ernesto Laclau, were as involved as anyone else in deconstructing "society" as an intelligible and essentialist totality, it was above all apolitical literary and linguistic theorists who compelled Western intellectuals to problematize the relationship between discourse, structure, and knowledge, and to question whether "structure" and "knowledge" had status at all within the analysis of discourse.[42] It was they who compelled historians to attend to the historicity of concepts such as "class," "the people," and "the social"—that is, to see these concepts not as inevitable, natural, or god-given, but as historical products which had become naturalized over time.

Thus, while capitalism restructured itself monopolistically and globally in the wake of the collapse of socialist modes of production and social systems, and while Islamic fundamentalism arose and ethnic cleansings swept parts of the planet, and (closer to home) academic job markets shrank from their 1960s heyday, Western intellectuals were increasingly inclined to reflexivity, the play of signifiers without referents, free actors, representations, "irreducible ambiguities," and evocative explorations into the interrelations, reverberations, and tensions within and between disparate political and fictional genre. Within the professional practice of history, in the face of despair at the collapse of the old totalizing social discourses, there emerged compensatory fascination with constructions of identity, interiority, and other "imaginative geographies."[43] And, as the language of historians became more subtle and self-referential, the more the cord to political action was severed. Having come to doubt the modernist interest in politics and rationality, intellectuals were thus impaled on the spikes of their own profound sense of skepticism and powerlessness. In cultural studies, at the same time as an emancipatory politics was raised to new heights through the pursuit of *différance* or "the Other" (abetted much by the work on gender and "orientalism" of Donna Haraway and Edward Said), the difference between murder and murder fiction was not only increasingly blurred (as Robert Darnton was overheard to remark), but, thanks in part to Hayden White, was ceasing much to matter.[44] By some reckonings the abattoir and Auschwitz were not a great deal different conceptually speaking. And "the body"—resurrected not merely as a textual site of contestation and struggle, but *the* locus upon which power was seen to be inscribed (that is, out with a framework of social class and race, or even social context)—became so discursive, flexible, and fragmented as to be almost immaterial.[45]

All "reality"—not just messianic historical materialism—was in a hopeless mess, whether or not (à la Baudrillard) one saw it as having given way to virtual reality, or to a "hyper-reality" of endless copies without originals (with the video as the icon).[46] Power was no longer something found at the end of a gun or under the heel of a boot, nor even something negotiated through the exercise of social relations in a material world. It was now a monolithic object dispersed through discursive fields of knowledge production. All the world was become a text, "the author" of which was unclear and of no great concern. According to Derrida, following Barthes, the author was dead.[47] Along with the concepts of "class" and "society" and other such scientific pretensions of modernity, "ideology" (however de-vulgarized) virtually disappeared from the historical vocabulary. By 1992, when Francis Fukuyama's *The End of History* appeared, social history as the forum for the study of such concerns was an embarrassed bygone—a residue on the outer edges of a vastly expanding and hyper-inflating, pastiche-reducing, postmodern marketplace of infinite diversity, eclecticism, and discontinuities (in some eyes, anarchic, nihilistic, and narcissistic, if not entirely weightless, decorative, and commercial). Certainly social history's political anchor was weighed by it. Already by 1980 the political thrust of the social history seeded by E. P. Thompson was perceived as "marooned on a sea of increasingly diffuse cultural analysis."[48] By 1990, his own metaphor of the tide had overwhelmed it; like the "social," it had sunk into salient silence. As yet no one had the historical capacity to know why.

Not only in history generally but specifically in relation to the history of medicine there was no call for "the social." On the contrary; reviewing twenty-five years of the Society for the Social History of Medicine in 1995, Dorothy Porter dealt in understatement when she concluded that "New historiographical trends in cultural history may make a society dedicated to the social history of medicine redundant."[49] Jordanova in "The Social Construction of Medical Knowledge," published in the same special issue of *Social History of Medicine,* suggested a disciplinary relabeling to "what might best be called a cultural history of medicine."[50] Lynn Hunt's *New Cultural History* (1989) may have been more in mind than the likes of Jonathan Sawday's *The Body Emblazoned* (1995), Judith Butler's *Bodies that Matter* (1993), Elizabeth Grosz and Elspeth Probyn's *Sexy Bodies* (1995), Athena Vrettos' *Somatic Fictions* (1995), and the psychomedical studies of visual and literary representations of disease and illness pioneered and pursued above all by Sander Gilman.[51] But by then Foucault was under everyone's skin, whether they liked it or not; beyond sex and madness and far into the intricacies of gender, literary-turned-cultural excavations of the body were extending rapidly.[52] Grumbling there was, and criticism too, such as at the disconnection of the critical analysis of modernist mentalities from the historical and sociological examination of modernity itself.[53] But, without doubt, the pundits of what was rather nebulously referred to as the New Historicism in literary and cultural studies (main domicile Berkeley, main journal *Representations*) had taken the sunshine from social historians of medicine while borrowing their wares. The "somatic turning" mopped up, providing "a new organizing principle within Anglo-American intellectual activity."[54]

Body studies [the focus of Chapter 3] were in many ways the epitome of the linguistic turn in cultural studies, and they were fundamentally at odds with social histories of medicine, however constructivist, Foucauldian, and contingent-emphasizing they might aspire to be. This was not simply because they engaged with synecdoche, metaphor, analogy, "close reading," and other techniques honed in English departments. Nor was it because they privileged texts over contexts, and preferred to deal with inscriptions of modernity on the body and embodied cultural practices over narratives of modernity in medicine and health. (To a degree, the pioneering social historians of ideas in science and medicine, Owsei Temkin, George Rosen and Erwin Ackerknecht, might be praised or blamed for having done similar.) No; somatic studies challenged social histories of medicine because they undermined the reductive or determinist assumption at the latter's heart—the notion that everything is ultimately socially constitutive. Following Derrida, somatic studies asserted the contrary, that nothing is reducible to anything (whilst proceeding to reduce most everything to discourse). The thinking was not only nonpositivist and nonteleological (as in the best social history of science and medicine) but nonontological: fundamental or absolute structures, inalienable human natures, and essentialist categories like "society" were denied; everything was to be seen as emergent or immanent. "Instantiation" became a favored word.

Social historians of medicine could thus be accused (as by Patrick Joyce) of intellectual naiveté for treating the body merely as a corporeal entity and for failing entirely to recognize "the social" as itself a product of modernity.[55] Even in their most novel pursuits (such as the "understudied" experiences of patients, and the role of the laity in medical practice), social historians of medicine were left to look like mindless empiricists—a look particularly fixed on the faces of those unschooled in Foucault and the epistemology and sociology of scientific and medical knowledge, whether à la Frankfurt, Edinburgh, or Paris.[56] Retreat was on the cards. At best, market-wise social historians of medicine sought to put old wine into new bottles through books and courses re-labeled "the body."[57]

By the 1990s, moreover, there appeared no longer any *need* for the "social" in the social history of medicine. According to Rosenberg's 1989 essay enough had been written to convince all but the most moronic that "every aspect of medicine's history is necessarily 'social' whether acted out in laboratory, library, or at the bedside."[58] Since this "all-is-social" theme had been the historiographical mission of the context-celebrating subdiscipline since the mid-1970s, there was no need to reassert it. Mission accomplished. As Rosenberg submitted, the "social" in the social history of medicine and science had become tautological—"as tautological as the 'social construction of disease.'" Harking back to Erving Goffman's *Frame Analysis* (1975), Rosenberg proposed that we speak instead of "framing disease." This then became the main title of the edited volume in the *cultural* history of medicine in which his 1989 essay was reprinted. Thus social historians of medicine, far from being forced into historiographical worry over the re-theorizing of "the social," were provided with a means to ignore it whilst carrying on business as usual.

As I said, it is not my intention here to defend "the social" in the social history of medicine; I merely wish to register the synchronicity of the postmodern turn outlined above with Rosenberg's insistence on exchanging "the social" for "the frame." Arguably, the contextualization of his essay on framing disease is as important as the historiographical claims made within it, and as important as the possible impact of those claims on the writing of the history and historical sociology of medicine (so that now, amazingly, we find the writing of social constructivist accounts of disease left to members of the medical profession).[59] Rosenberg's essay deserves close attention, then, despite the fact that it was neither intended nor received as an important theoretical paper.

Rosenberg's Frame

Written in the late 1980s for the policy outreach journal the *Milbank Quarterly,* the immediate context of Rosenberg's essay was the intellectual burden imposed by AIDS. Indeed, only a few years before, Rosenberg had written specifically on this subject for the same journal.[60] To many observers, the biological realities of AIDS confounded the assertion of diseases as "mere" social constructs. As such, Rosenberg's essay was also an intervention in the much wider and more heated "science wars" between the defenders of scientific realism, rationalism, and Truth, on the one hand, and the philosophers and sociologists of science-promoting varieties of relativism, on the other.

Although Rosenberg's essay did not touch on contemporary controversy over the extreme relativism of the literary deconstructivists that challenged the sacred objectivity claims and assumptions of historians, that backdrop was widely understood. In the late 1980s discussion of relativism was extensive, especially in America, and especially among historians—in part because of a debate unleashed in 1988 by Peter Novick's magisterial history of the "objectivity question" in the practice of American history.[61] The fourth and final part of Novick's book, "Objectivity in Crisis" recorded an experience that was only too familiar to those like Rosenberg in departments of the history and sociology of science where post-positivist distinctions between fact and value, science, scientism and ideology, nature and culture, and biology and society had long been discussed, especially after the publication in 1962 of Thomas Kuhn's *Structure of Scientific Revolutions*. In such places, Lyotard was received not simply as another postmodern philosopher, but as one claiming to pronounce specifically on the nature of science. Dependent in part on the insights of Kuhn and the philosopher of science Paul Feyerabend, Lyotard's claim in *The Postmodern Condition: A Report on Knowledge* (1979; English translation, 1984) was that science was no longer in the business of truth seeking, but rather, in the manufacture of incommensurable theories.[62] As everywhere, these debates were emotional, increasingly public, and politically charged. Indeed, there was more than a whiff of McCarthyism to them. Setting the pace, Hilary Putnam in *Reason, Truth and History* (1981) saw the cutting edge of relativism as deriving explicitly from Marx, Freud, and Nietzsche, who taught that "below what we are pleased to regard as our most profound spiritual and moral insights lies a seething cauldron of power drives, economic interests, and selfish fantasies."[63] By the 90s relativism of any type or strength was tending to be equated with radical relativism and to have many of the same "damning associations as Communism, whether you're a party member of not."[64]

Rosenberg's essay was by no means simply a reactionary response to the alleged naked biological realities of the syndrome that smote Foucault. A moderate relativist himself, Rosenberg was fully aware of the provisional nature of knowledge. His essay can be read, rather, as an effort to de-privilege "radical" or "hyper"-relativism and hence spare the social history of medicine, and social constructivism in particular, from the arrows of political outrage then surrounding it. Necessarily, therefore, in this context, the renunciation of "the social" and its replacement with "the frame" was a calculated political act; the verb "to construct" was in fact being substituted by "to frame" as "a less programmatically (i.e., politically) charged metaphor."[65] While shifting attention away from the political implications of social constructivism, Rosenberg's "frame" emphasized the relations between biological events and their individual and collective experience and perception. In other words, it was as a strategy both for a compromised relativism and for a pluralistic approach to history. Through "framing," the fragments of historical evidence might be disciplined but not over-narrativized; likewise, agency could still be ascribed to social, economic, intellectual, and political forces without granting overdeterminacy to any of them, and the role of individuals could be preserved. (Rosenberg's harking back to Goffman's use of the frame should not therefore be regarded as incidental, since Goffman had held to a social constructivist view of the self.)[66] As a descriptive category, "the frame" avoids responsibility being attached to any particular interest group or set of historical actors (including historians).[67] A versatile metaphor, it has the look of structure, but commits one to no particular theoretical architecture or politics.

Yet, by these very means, and in common with a great deal of American liberalism, Rosenberg's formulation can be seen as a part of an effort to return to a "commonsense" pragmatism around which consensus might be built. In that sense it can be regarded as a part of an act to *reframe* the writing of the history of medicine within a political philosophy that, while it may not be as ideologically distinctive as Marxism, is no less ideological. One of the characteristics of this political philosophy is to make any retreat from it impossible by burning all the bridges to the old conceptual machinery while purporting (to a degree) to incorporate them. It is noteworthy how Rosenberg in his essay both praises the pioneering social history of medicine of the socialist Henry Sigerist (to illustrate how social history can be written without regard to relativist social constructivism), even as he draws attention to the antiquated positivist nature of Sigerist's regard of medical knowledge. Links between medicine (and the production of medical history) and politico-economic purpose are thereby, in effect, dismissed as vulgar (positivistic) Marxism, at the same time as (to cite Simon Schaffer dealing with the same "facetious blackguarding" in the history of science), over-robust "connections between natural knowledge and social interests are damned as sociological relativism."[68] Note, too, that Rosenberg's highly qualified endorsement of the provisionality of knowledge ("Knowledge *may be* provisional") is made in the course of staking a claim for his own historical revision (". . . *but* its successive revisions are no less important for that").[69] This maneuver conforms to what Stuart Hall has called the "discursive struggle" over the delegitimizing of opposing ideologies (or discourses), where the "older" machinery is presented as, at best, "optional."[70] In common with the postmodernists critiqued by Fredric Jameson, it's all within the spirit of pluralism: "As with so much else, it is an old 1950s acquaintance, 'the end of ideology,' which has in the postmodern returned with a new and unexpected kind of plausibility."[71]

Thus, Rosenberg's pseudostructural notion of "the frame" might appropriately be described as "within the frame" of the depoliticizing thrust of Francophone postmodernity as received in the USA. Given that "framing" was also a concept deployed by the antirealist Derrida for deconstructivist purposes, it is tempting to suggest some closer affinity.[72] This temptation should be resisted, however, for no two projects were less alike philosophically and practically. Nevertheless, Rosenberg's framing does fit to a loosely Foucauldian (skeptical-relativist) culturalist flow. Without entirely dismissing either the social constructivist approach to medical knowledge, or the left-wing political tradition behind both social medicine (from Ryle to McKeown) and the social history of medicine (from Sigerist to Charles Webster and Roy Porter), the notion of framing occludes both whilst appearing merely to encourage the historical analysis of yet more "full and appropriate contexts." If the production of historical and historiographical knowledge is but one form of the production of knowledge that accompanies the restructuring of capital (along with the restructuring and rearticulation of practices), then the substitution of "the frame" for "the social" in the history of medicine may be regarded as constitutive.[73] Perhaps this is only to state the obvious: that, to mimic Marx, Gramsci, and Putman, historians don't choose their history under circumstances of their own making. That the act of writing history creates texts and constructs knowledge seems obvious, but there is never an intellectual free market any more than there is a "free" economic one. Like Rawls' assertion of an ethical consensus for the late twentieth century, or Thatcher's "end of society," the would-be historiographically hegemonic—the invented "consensual"—must always do political work, if only through the process of occluding. The displacement of "the social" by "the frame" was one such political act.

But this is not the only way to think about Rosenberg's "frame." As germane might be a view of it as seeking to re-embody a history whose analytical categories had become destabilized—as destabilized (or generally messy) as the medicine and society of "late capitalism." Thus, the illusion of "the frame" may be less that of structure than stability, and less that of a totalizing contextual view than putative coherence. The warm embrace of the idea of the frame by historians of medicine in the 1990s might be explicable in terms of this illusion; at the very least it enables writing history of medicine without having seriously to question the terms of analysis. It places a comfort blanket over the conceptual diaspora of modern history. This partly explains why Jordanova's call to render the social history of medicine a territory for critical engagement and debate has gone largely unheeded.[74] Jordanova sought to explore the categorical messiness of the historiography of medicine as it was emerging in the 1990s. She invited us to re-examine the repertoire of categories supplied through the social history of medicine's response to the old positivist and doctor-centered, or tribal legitimating history of medicine—categories such as medicalization, professionalization, culture, representations, health, disease, hygiene, sexuality, and the family. Rosenberg's "frames," by contrast, can be seen as the effort to put some kind of lid on that messiness and its investigation—in part to contain, in part to avoid the ever-more-apparent instability of the categories. If Jordanova's mission was one of disciplining the social history of medicine and bringing it to "maturity," a part of Rosenberg's strategy was to avoid the subdiscipline having any crisis of identity or even the look of it.

Post-Frame, or the End of the Beginning of the End

History, it might almost be said, has rendered the frame as superfluous as "the social" it sought to replace. The conditions for its possibility have surrendered to others, on the whole less politically despairing and historiographically fraught. This is not to suggest that with the change of millennium the slate has been wiped clean and the epistemological agonies of the last two or three decades magically erased. On the contrary; notched-up but still grating against the old modernist certainties and categories of "social" analysis are such post-modern insights as that on the impossibility of sustaining universal categories and truths (like "the social"); the fallacy of essentialist or fundamental causes; the idea that power resides in the making of discourse, that language has the capacity to shape what it represents (be it the "orient," "sex," or "disablement"); and the inability to distinguish in principle between scientific-rationality and the stuff of religious belief. We have not heard the last of the theory wars, and the search for "the soul of history" carries on.

But just as there are signals from Francophone intelligence and elsewhere of a return to the political, so there are signs in the practice of history that we have passed beyond the notion of a postmodern declension and the attendant fear that "a radical skepticism could yet defeat us all" (as Mary Douglas put it in the mid-1980s and the philosopher Peter Sloterdijk struggled against in his *Critique of Cynical Reason*).[75] There are indications that we have superseded the would-be hegemonic neo-deterministic view of language as the root of "what it is possible for people to think and do."[76] To a considerable extent "postmodernism" has been desanctified, and there are clear signs that the "civil war" between discursivity and historicity, or more broadly between "the cultural" and "the social," has entered a more accommodating phase, less privileging of "the cultural."[77] For some historians, this largely means the recognition and reconciliation of these different categories, as in *Beyond the Cultural Turn* (1999) and *Reconstructing History* (1999), although these still attempt to rescue history from the perceived tight clutches of postmodern cultural studies.[78] For others it opens the way to de-polemicizing the debate between "discursivity" and "reality" in order to re-politicize it in terms of the political and ideological context in which it occurred.[79] For still others, however, such as Antoinette Burton, campaigns like these serve only to validate the old disciplinary demarcations, and to foster anew static, stolid, and unitary understandings of their natures. Burton demands, rather, that we inquire into the naturalizing use to which the categories "the social" and "the cultural" continue to be put in contemporary historiography.[80] This idea cannot be developed here, but as it suggests, historiographical discourse has already moved beyond the postapartheid politics of the social versus the cultural. Indeed, as early as 1995, Catherine Casey, in *Work, Self and Society After Industrialism,* provided compelling illustration of how one might "return the social to critical theory" without necessarily discarding discourse analysis, or returning to a conventional social analysis that merely privileges material social relations.[81] In many ways this is the drift of some of the most engaging recent work in the history of medicine, though to call it "history of medicine" is to force it into an increasingly anachronistic box.[82]

A review of this literature is not for here, but three general points are worth making. The first is that it owes large debts to somatic cultural studies and, in particular, to Mary Poovey's *Making a Social Body* (1995)—a work that specifically challenged the idea of "the textual" and "the social" as antithetical or mutually exclusive domains of inquiry.[83] Going beyond the earlier insights of Foucault and Jacques Donzelot on the formative role of medicine in the invention of "the social," Poovey has demonstrated how the modernist abstract of the "social body" was itself generated in the early-Victorian state in response to cultural and political anxieties about anatomy and contagion, poverty and disease.[84] Subsequent studies, such as Erin O'Connor's *Raw Material* (2000), take this further, revealing the operation of constraining somaticized metaphors not only in the Victorians' own social critiques (as in the writings of Thomas Carlyle), but also, in such contemporary cultural practices as postcolonial discourse and *its* critique and, indeed, in the practices of cultural studies as a whole.[85]

Second, while none of this literature is intended as contributing to the social history of medicine or even to the history of medicine (despite publisher's classifications), it might be said to fulfill the erstwhile ambition of the subdiscipline to join the mainstream of history. As Jim Epstein remarked in a review of Poovey and similar studies: in the face of worry over postmodernism's threat to social history, such work powerfully testifies to "the very real openings available for writing new kinds of social and cultural history."[86] Seen from this vantage, "the end" of the social history of medicine might almost be considered as that running from nemesis to omniscience.

Third, and finally, this literature brings us back to politics. For one thing, it marks a turning away from the over self-referential, hall-of-mirrors-type cultural studies of recent years, much of which was written for the sake of its own disciplinary ends, or for the sake of illustrating interesting tensions in the literary history of modernity. It is not just that the new work firmly registers Foucault's point about the body as the place where (as O'Connor puts it) "power has historically assumed its most monstrous and its most liberatory incarnation."[87] It is also that the literature amplifying this point now often passionately embraces a belief in the transformative potential of cultural theory to force thinking beyond categorical constraints. Thus, for O'Connor (drawing strength from Haraway), cultural studies are a "radical and necessary form of activisim," which is all the more driven by the realization that "genuinely searching academic work is fast becoming a vestigial structure, a useless and hence expendable appendage to a culture that neither values nor understands it."[88] In a more straightforward way, Lawrence Driscoll, for instance, maps the Victorian discourse on drugs to expose how its cultural construction continues (devastatingly) to constrain social thought and political action. The point of Driscoll's discourse analysis is directly to effect political reform.[89]

Just as this literature helps us to understand how "the social" was very largely invented and problematized within and through the vocabularies of medicine, so at a deeper level (if only by extension) it compels us to think about "the political" by encouraging the problematization of "the medical" as constitutive of "the social."[90] To be blunt, we can no longer speak of "the political" or "the medical" or "the social" because we no longer know what holds these categories together. Such categories must now be seen as hypothetical at best; like "nature," they are labels that do not explain, so much as beg explanation. Hence the complex and diverse phenomenon called medicine cannot be said to exist inside "the political" or "the social," any more than the political or the social can be said to exist within (or to structure) "the medical." It is only through the material organization of the objects and resources having to do with medicine that "the medical" and "the political" can be seen as held together or given agency; that is, through technologies, expertises, texts, architectures, and the material (actual) social relations that go with them. The latter would include not just the relations between doctors and patients, but between doctors and doctors, patients and patients, doctors and families, researchers and the state, pharmaceutical companies and the law, and so on, and so on. Admittedly, these material relations of medicine are not the only connections between people, and they can be distinguished from those around, say, education, consumerism, religion, the military, diplomacy, and the law, even if they may often overlap or be in tension with them. Whether or not they are *primary* relations in a world as "medicalized" as ours may be open to debate; what is more important to stress is the lack of any need to privilege them over discursive and epistemological considerations. The need, rather, is to understand that it is these material social relations that actually produce the political of which the emergent discourses and epistemologies of medicine (themselves capable of acting as material forces) are a part. In other words, like "the social," "the political" cannot be regarded as a transcendent category with assumed inherent force.

Happily, this leaves the territory and the practice of the history of medicine wide open, even if it collapses the old disciplinary boundaries. There is no basis for privileging "the cultural" over "the social," any more than there is reason to lord medical ethics over medical economics. And equal is the opportunity to revisit such old sites as professionalization, the idea of medical elites, the technical content of medical knowledge, and so on, whose recent loss to the history of medicine has been lamented.[91] The prospect, then, at the end of the social history of medicine is not a return to what was lost when "the social" got re-theorized and "framed," but rather, to a different kind of post-structural "political" understanding of the phenomenon that is medicine within a historiographical frame that is more critically aware of its own values, perspectives, and aims.

4

The Turn of the Body

THROUGHOUT THE HUMANITIES IN the 1980s and 1990s enormous importance came to be attached to the body. It was one of the reasons why the history of medicine suddenly looked sexy. However, by the end of the first decade of the new millennium, when much of that initial interest had dissipated, there was still no article that sought to summarize and problematize the phenomenon. This essay, written in 2008, was my attempt at it. I was then no longer so interested in the historical topics that had preoccupied me in the 1980s and 1990s. I was more intrigued by the rapidly changing sociopolitical and socioeconomic context of biomedicine and biotechnology, and by the generally unperceived connections that could be made between that and history-writing and intellectual moves more generally. Also, I had had almost a decade to register the criticism that my friend and sometimes sparring partner Paolo Palladino had sounded in his review of *Medicine in the Twentieth Century*, the edited volume that John Pickstone and I published in 2000. Paolo, far more persuaded than I was by Foucault on the intertwining of the body and modern forms of political power, noted how few of the chapters in that volume made any reference to Foucault's work, including my own three chapters, on bodies "disabled," "ethical," and "dead."[i] A contribution to the volume on "The Historiographical Body" by Mark Jenner and Bertrand Taithe gathered together various of the fashionable body bits but, in retrospect, didn't go far enough to help students understand how Foucault's anti-essentialist take on the body had fundamentally shifted the ground for writing the history of the body or, rather, the historicized body—the impact, as it were, of the postmodern history of the body on modernist history-writing. In the absence of any such sustained discussion for historians, this essay was intended to fill the breach. It was also my attempt to look at Foucault more closely for my own sake, for by this time I had also met the Foucault enthusiast Claudia Stein (see Chapter 1) and had begun investigating with her the so-called visual turn in relation to the so-called somatic one.

Crucial to it was a doctoral dissertation that I'd been privileged to examine early in 2008: Adam Bencard's "History in the Flesh: Investigating the Historicized Body." Bencard's study was unique in historicizing the historicized body, and it remains a far more detailed and nuanced investigation than that elaborated here. Although I felt that there were systemic problems with the intellectual nut that it valiantly tried to crack (namely, to move the ontology of being into the historiography of the historicized body), overall it brilliantly confirmed the worth of a history of the encounter of the historicized body in historical scholarship from before Foucault and after, or before and after the much referred to, but seldom historically analyzed, somatic turn. The latter clearly did more than render the body fashionable in historical and cultural studies; as hinted in the previous chapter, it fundamentally challenged the nature of history-writing as a form of inquiry. Confirmed was what I had sensed in writing the previous essay: that ours was not the time to reconfirm old ways of looking at history and merely apply sexy labels to it, but rather to search for fundamentally new ways to think about it.

In many ways this was the challenge that was first registered in the journal *Representations* and in the cultural history associated with it. But this itself was soon contested in a variety of different ways, all of which opened up spaces for new kinds of essentialisms, some of which were ideologically worrying. This essay reports on these moves and, in the process, again reveals the far-from-straightforward way in which historical thinking moves in general. Like new epistemologies for the practice of medicine or new epistemic virtues in science, one methodological move does not simply or easily displace another so much as obtain the permission to outdance it when conditions are appropriate.[ii] Here, admittedly, I focus more on the "dance" than on the dance floor, and in an outside observer sort of way.

When it was first drafted I had not felt the need to defend a historical perspective or to consider the strategic worth of defending the discipline of history. More in mind was the bafflement of my graduate students who, while often au fait with postmodern body talk, tended to have little idea of where it all came from, or how it had been constituted through literary and cultural theory and the sociology of contemporary social trends. If there is some repetition here with material presented in earlier and subsequent chapters (particularly in relation to positioning Foucault), it is out of an interest in keeping the essay as self-contained as possible, for student readers especially.

It was first delivered at a workshop in Madrid on "Polyphonic History" organized by Javier Moscoso in honor of the cultural historian Peter Burke. As the workshop's title suggests, this was a "postmodern" event in its unpoliticized commitment to—nay, celebration of—disconnected historical bits and pieces. Presentations ranged from straightforward social histories and feminist critiques to ludicrous, bio-reductive defenses of globalization. It was published in *Arbor Ciencia, Pensamiento y Culture* in 2010. With difficulty, I have resisted the temptation to revise it in light of my subsequent thinking on posthumanism and historiography—thoughts triggered, in fact, by the comments by Adam Bencard cited in Chapter 1. In keeping with this volume's narrative structure, retrospective reconsiderations would not be in order. I have therefore confined myself to editing for purposes of greater clarification and to supplementing some of the references.

The last decades of the twentieth century witnessed the coming into focus of the idea that our bodies have a history. No longer could this most taken-for-granted of entities continue to be taken for granted. Suddenly, an object that had been no one's particular concern became virtually everyone's preoccupation—including historians. The late Roy Porter, in an essay on the subject in the revised second edition of Peter Burke's *New Perspectives on Historical Writing,* declared that "body history" had become the "historiographical dish of the day," having proclaimed only a decade before, in the first edition of Burke's *New Perspectives,* that the topic was "in the dark" and "too often ignored or forgotten."[1]

Porter's suggestion that the body had hitherto been largely ignored by historians was not strictly true. Foucault in *Surveiller et punir* (1975) had observed that "historians long ago had begun to write the history of the body." Thinking specifically of some of the work of the French Annales School, and probably that of his teacher Georges Canguilhem, Foucault pointed out that historians had shown how the body was "a target for the attacks of germs or viruses, ... to what extent historical processes were involved in what might seem to be the purely biological base of existence; and what place should be given in the history of society to biological 'events' such as the circulation of bacilli, or the extension of the life-span."[2]

Nevertheless, Porter was right that the flow of historical scholarship on the body had greatly accelerated in recent years, even if he didn't understand how or why. He listed historical interests that had come to revolve around the body—demography, art, biology, and so on. What he failed to grasp was that "body history" had only accelerated after Foucault had drawn attention to the importance of a *non*-purely biological view of the body—a non-essentialist and politically invested view of it that undermined how historians had previously conceived it, or rather, *not conceived it*. Furthermore, it was not "body history" as such that came to excite intellectuals, but the notion of *a historicized body*—a body *as a concept* that had been made up in "modernity." Its study was not to be conducted in terms of some universalized, coherent, timeless entity but, on the contrary, as something fragmented in its different historical constructions over time and place. This was the new register of concern that by the millennium's turn had bookshelves groaning under the weight of the body "at risk," "at work," "at war"; "in question," "in theory," "in language," "in shock," "in pain," "in parts," and widely "othered" in subaltern studies and around the politics of identity. The historicized body "of the artisan," "the disabled," "the mad," "the Jew," "the erotic," "the beautiful," and "the saintly," were among the many now to be "explored," "contested," "expressed," "invaded," "imagined," "emblazoned," "engendered," "embarrassed," "experienced," "dissembled," "dismembered," "reconstructed," and "im/material"—to draw only from some of the titles of Anglophone monographs sporting "the body." Admittedly, some of this literature was little more than a cashing-in on a fashionable trope, but other of it, deeply informed by the literary turn, substantially renegotiated understandings and approaches.[3]

With a vengeance the "somatic" moment had arrived, the use of the word itself signifying a difference in approach from that which might have simply been called "body history" in conventional social history. The body was now a "problem" connected to that of postmodern subjectivity, not simply an object to be historically reified and pinned down for "objective" analysis. As such, nor was it purely an academic matter conceived in cultural isolation. Indeed, its moment was one in which, in popular culture, the preoccupation with individual identity had come to rotate around the body. Whereas in the not too distant past people might have shaped their idea of themselves in relation to their church, profession, trade union, or political party (if their culture had encouraged them to inquire into their "identity," that is), now they were thinking of it in relation to their corporeal selves—their body's health, capabilities, sexual orientation, and aesthetics. Especially in economically privileged countries, people were becoming increasingly obsessed with their bodies. Concerns over health and fitness, dieting, weight loss, obesity, personal grooming, drugs for sexual and mental "enhancement," tattoos, massaging, body piercing, tanning, cosmetic surgery, gender reassignment, organ transplantation, and so on had left the socio-political preoccupations of the 1960s and 1970s far behind. "Somatic society" had arrived.[4] AIDS, to be sure, was important in opening the floodgates to this corporeal attentiveness inasmuch as it put every body at risk. But bodies had also become big business, the focal point of an expansive internationally expanding consumer culture, linked to and much fed by the bio-tech-pharma industrial complex.[5] The relatively greater disposable income of most people in rich countries facilitated it. While bodies had long been important in human existence, not least for social ordering, by the late twentieth century in a culture of utilitarian individualism the fixation and fascination with them had become many people's primary concern. The body became "the privileged site of experiments with the self," as Rose has said, the exercise and slim-foods industries, along with Viagra,

Prozac, and "smart drinks" doing much to promote it.[6] Academics, being themselves a part of the culture of commercialization and the celebration of the individualized body became increasingly interested in "the many ways in which the body is engaged in accounts about what it does."[7] Dedicated journals sprang up, such as *Body and Society* (1995), and scores of academics committed themselves to filling their pages.

Given that these intellectual and economic investments in and institutionalizations around "the body" are still very much a part of contemporary culture, it is hardly surprising that there exists no across-the-board historical account of how Western intellectuals came to engage with the body in the late twentieth century. To explain fully the conditions of possibility for it would require nothing less than a cultural history of our recent times with none of its material moorings omitted. It would have to attend to shifts in politics and political-economic theory, commercial practices, and the status of nation states relative to global configurations, as much as to developments in biomedicine, bioethics, visual culture, and communication technology. Further, it would have to engage seriously not just with feminism and gender politics, and with the push for gay, lesbian, and disability rights, but as well with all the waves of theorization in these and related areas of intellectual concern and activism.

This essay attempts nothing so ambitious. Its purpose is simply to sketch the encounter of the body in historical scholarship since Foucault, and examine some of the devious routes it has subsequently taken. Such a task is not quite as straightforward as it sounds, however. Academic history-writing is in fact a poor Archimedean point from which to try to draw the recent history of the body for, paradoxically, it has been a territory less devoted to that particular exercise than one fundamentally challenged by it. The modern discipline of history was philosophically and methodologically assaulted by the "postmodern" literary turn in Western intellectual life that elevated the body to a privileged site. The "somatic turn" (the concern with the historicization of the body) was broadly a means to explicate and illustrate how concepts and categories like "the body" and practices like "history" served to naturalize, rationalize, and cohere an understanding of the world that was increasingly felt by many late twentieth-century intellectuals to be fragmented. The body and the discipline of history could both be described as products of (and for) "modernity"—the project whose grand narratives of progress, social contract, and Enlightenment served to disguise the fact that terms like "the individual," "the social," "nature," and even "reality" were not "objective," epistemologically autonomous entities, but "historical and normative creations, designed to handle the exigencies of political power and political order."[8] Conventional history-writing came to seem inherently modernist inasmuch as its business was to invent or apply coherent narratives of the past, and through them (and through its own narrative structure, as Hayden White made clear in his pioneering *Metahistory* of 1972) shape understandings of the present. Modern history-writing, in other words, came to be seen as serving much the same kind of sense-making as the category of the modern body in *its* capacity to cohere and constrict understanding. In fact, historically, the categories of "the body" and "history" mirrored each other: the invention of modern history as a would-be objective discipline coinciding with the invention of modern medicine as an enterprise seeking to objectify the body. In tandem, the profession of medicine sought to objectify the body, while the profession of history sought to objectify the past.[9] Both were products of the modernist (more generally, Enlightenment) project that invented the idea of such disciplines.[10]

Overall, the focus on the body extending from the postmodern linguistic-cum-semantic turn led to intense concentration on the nature of history as a form of discourse. "Body history" was not about the history of the body as a discrete object of inquiry, but about new ways of representing knowledge, including historical knowledge. Thus any attempt to explicate how the body became the "historiographical dish of the day" risks obfuscating what the focus on the body in intellectual culture actually signified—a critique of history along with other modernist constructions of the world. Any straightforward "history" of the historiography of "the body" therefore runs the risk of obliterating the politics involved in the somatic turn in history-writing. Too easily, it can end up being little more than a reification of body history, if not a reification of the body itself as an essentialized entity reducible to its biology. The danger, in short, is that of cohering a historical narrative around the very thing that came to serve above all to problematize and de-stabilize the notion of historical coherence. Although it may not be possible to avoid some of those pitfalls in a brief sketch of the historiography of the body in the academic discipline of history, in what follows it is from out of this concern that an emphasis is placed on the politics of this problematization and destabilization. As we shall see, the discipline of history has been subject to not one but several interconnected turns around the body, each of which has served to bring to the fore questions on the nature of "the past" and the place of the present in interpreting it.

Foucault

As indicated, no one was more important for drawing attention to the body in history than Michel Foucault. In a variety of important and controversial publications around corporeal themes, Foucault sought to explicate the complicated ways in which power was inscribed upon the body. For him, it was through somatic discourse, or through discursive practices operative in and upon the physical body, that modern power came to be constituted and exercised. What he came to call "biopower" refers to the somatically shaped and shaping knowledges and practices that aimed both at normatizing the health of individuals (through the defining, measuring, and categorizing of bodies), *and* at the managing and regulating of human populations.[11]

This was a notion of power that did not derive simply from social and political institutions. Since the late eighteenth century, Foucault believed, innumerable systems had come into place to encourage people to self-regulate in the interest of preserving and extending their lives.[12] At the same time as this "care of the self" came to be willingly pursued, nation-states for their part—largely for military and economic reasons—became intent on the health and welfare of their citizens in the aggregate. With the intensification of nationalistic ambitions towards the end of the nineteenth century and into the twentieth, nation states increasingly implemented disciplinary technologies around the body—practices around dress, drill, and diet, for instance—to render the body ever more amenable to productivity and notions of social order. Increasingly, nation states also implemented regulatory techniques to measure and monitor the body around "norms." Indeed, the notion of "the norm" or *normativity,* Foucault suggested, is that which links the biological and social disciplinary techniques aimed at individuals with the regulatory ones aimed at populations.[13] The "modern" body—the normatized body—was the aim and the outcome of the concerted action of both.

Foucault's concept of biopower is not to be understood as something negatively experienced, or merely acting repressively from only outside the body—it is, after all, literally *embodied*. It is a productive agency, much as language itself was regarded in the work of Jacques Derrida and the various linguistic theorists indebted to him. As such, the take up of Foucault's notion of biopower radically challenged conventional sociological understandings, not least Marxian ones, where power was seen to act as an external force for coercion and domination, whether it was exercised instrumentally in a crude mechanically-causal fashion, or mediated through ideology, "false consciousness," or other perceived capitalist machinations of "truth." Thus, whereas within medical sociology this conventional notion of power was translated into the concept of "medicalization" to mean the territorialization and exercise of knowledge/power by the medical profession, for Foucault medicalization embraced the whole of modern somaticized culture in which identity and the meaning of life were fashioned through the body, or through the notion of biological life itself. In effect, argued Foucault, it was through the body—through the various political investments of knowledge/power in and around it—that the modern subject (us) was made up. This making was held to lie outside the ambit of individual cognition and control, and outside the instrumental or mediated dominations of ruling authority. Challenged and ultimately bankrupted by this understanding, therefore, were both the instrumental *and* the analytical power attributed to "medicalization" by sociologists since the 1960s.[14]

For Foucault, the body was a referent for the discourses that he sought to analyze, and for the question of who we are. This had little purchase power, however, in the one area of historical scholarship where the body might be thought to have mattered most—the history of medicine.[15] Here the body was taken for granted as an unproblematic biological given; there was little comprehension of it *as a form of knowledge* continually being invested and re-invested in power relations, or in and of itself *constituting* politics.[16] Rather, the body was merely something acted upon by multiple manipulating forces. Often its study was in terms of "patho-" and "psycho-logics" in social normativization. More at issue, especially within the academically ascendant field of the *social* history of medicine in the 1970s and 1980s, was the role of the medical profession in their exercise of power upon and through the body. Social historians of medicine, operating in the wake of influential social critiques of the power of the "medical establishment," above all Ivan Illich's *Medical Nemesis* (1976), largely followed in the train of E. P. Thompson in elaborating a version of "history from below" around the social power of medical profession in relation to patients.[17] Here, too, "medicalization" was conceived in terms of professional knowledge and power, whether studied from the top down or the bottom up. By illuminating historically the social structuring, the exercise, and the effects of this power in the "real world," social historians of medicine could believe they were contributing to the politics of social change. They were not idly "interpreting the world," as Marx famously indicted his fellow philosophers, but in their own way were serving "to change it." Thus, to them, Foucault looked all-too-much like a philosopher providing no obvious political solutions to anything; indeed, through his non-Marxian, anti-structuralist attention to somatic discourse he could be (and often was) interpreted as a counter to prescriptive social politics, if not in fact providing merely a disguised apology for something more reactionary. In effect, Foucault's corporealization of power in general, and his de-centering of the notion of medical power in particular, robbed the social history of medicine of its "medicine," debased its political interests, and bankrupted its explanatory power. It is hardly surprising, therefore, that when Porter came to admit to body history having become "the historiographical dish of the day" it was with remarkably little attribution to Foucault.

But by then—indeed, by the 1980s—the Rubicon had been crossed in history as almost everywhere else in the social sciences and humanities. What Foucault would have identified as a new episteme was beginning to reign, even if in academic history-writing the accommodation to this corporeal "regime of truth" was slow, uneven and partial. Its signature was evident in the move away from the sociological paradigm (and Marxism in particular), and the displacement of "the social" (now understood as a historically constructed category) by "the cultural." Above all, it was signaled by heightened attention to the body within cultural studies of all sorts, and within feminist studies especially. Social politics were displaced by concerns with the politics of identity and the construction of the supposedly autonomous modern "self." The wider context was one of increasing commercialization around the individualized body and the widespread sense of the disappearance of a "genuinely democratic space under the thickening blanket of privatization and the declining welfare state."[18] As in the world of music, songs of social protest surrendered to songs of the self, or personal subjective experiences (Joni Mitchell being among the first to do so). As important was AIDS, not because it *caused* epistemic shift, but because (initially, at least, in not being easily explained) it appeared to render arbitrary the conventional distinctions between the cultural and the biological, as well as the disciplinary boundaries historically separating sociology, ecology, and biology.[19] AIDS encouraged new modes of thinking about knowledge and perceptions of power, and in this respect became a testing ground for Foucault's thinking on biopower for many cultural theorists and activists. Indeed, it was largely around AIDS that the anti-essentialist and anti-structuralist "literary turn" fused decisively with the "somatic turn" [see Chapter 5].

The Body in the New Cultural History

It was through the portal of Foucault's anti-essentialist approach to the body that "the new cultural history" came to embrace the body and, in the process, re-constitute itself in fundamental respects. To be sure, there were various prior encouragements to it, not least the famous study of the sacred body of the monarch in the middle ages by the Annales historian Marc Bloch.[20] Other social interactionist–type historical engagements with the body (perceived as "man's most available metaphor") came by way of the anthropology of Clifford Geertz and Mary Douglas.[21] These authors, not least because of their lucid prose, inspired a number of historians, including myself, to focus on corporeal matters.[22] Some of this work drew on the phenomenological canon extending from Edmund Husserl, the philosopher who believed that experience was source of all knowledge, and who sought to establish a philosophy that put essences back into existence. Here, the writings of Maurice Merleau-Ponty were particularly important, part one of his *Phenomenology of Perception* (1945; English translation, 1962) being entitled simply "The Body," with "experience," "knowledge, and "perception" regarded as fundamentally corporeal.[23] While body history—or more precisely, the historicization of body fragments—was still mostly the preserve of literary theorist and feminist scholars (often one and the same), increasingly there were meeting points and crossovers with historians. By 1989 various of these interests came together in the hefty triple-decker *Fragments for a History of the Human Body*, edited by Michael Feher, a touchstone in many ways for a new cultural history that accepted that the somatic turn was an offshoot of the literary one, and came to approach the body as entrenched in sensibility, images, illness vocabularies, and related symbolic practices. Hard structures were abandoned in favor of poststructuralist "negotiations," and language was embraced as a productive force constitutive of reality rather than simply reflective of it. Thus the social reductionism inherent to social history could be dispensed with. No longer was anything to be reducible to its social construction, and nothing was reducible to any single cause. Social categories were no longer to be seen as prefiguring consciousness or culture or language, but rather, were to be seen as dependent upon them. Such categories were to be understood as instantiated through their expression or representation. All in all, taking shape was a more constructivist view of the world and the meanings attached to it, along with a more ontological, rather than teleological, view of history.

Pioneering in these respects within historical scholarship was *The Making of the Modern Body* (1987) edited by Catherine Gallagher and Thomas Laqueur. As the book's Introduction pointed out, this was "a new historical endeavor," deriving "partly from the crossing of historical with anthropological investigations, partly from social historians' deepening interest in culture, partly from the thematization of the body in modern philosophy... , and partly from the emphasis on gender, sexuality, and women's history."[24] As the list belies, what was "new" here carried considerable baggage from the past. Indeed, "the making" in the title inhered old causal narratives and teleological undercurrents—perhaps even whiffs of nostalgia for the sociological notion of power (if not the politics) of that other famous "making," by E. P. Thompson, *The Making of the English Working Class* (1963). The same might be said of Laqueur's *Making Sex: Gender and Sex from the Greeks to Freud* (1990), the monograph that expanded his germinal essay on the construction of a two-sexed model of gender difference, which was first published in his and Gallagher's edited volume. But overall there was far more that was novel than not in *The Making of the Modern Body,* so much so, that it might be said in retrospect, that the "making" that was *most* apparent through it was that of a new corporeal regime of truth moving into history writing. Biological essentialism was routed, and constructivism embraced in the place of causality and linear narratives. Instead of the body being perceived as a naturalistic biological entity that could be taken for granted, it was regarded as something that itself had a history, and whose very construction *in* history could be reckoned a central historical problem. Within an intellectual discourse that owed much to the 1980s-born literary "New Historicism" and *its* debts to Foucault in terms of the making of modern identity, the body within the new cultural history was becoming a tool for thinking beyond categorical constraints.[25] Such thinking was also a means to out-think conventional history writing, for not only was the body that was historicized within this intellectual discourse perceived as inherently unstable and fragmented, but so too was the notion of history. History was no longer to be understood as a stable or unified body of facts, or a neutral "background" against which any object or event might be tracked. That view of history could be construed as itself *a*historical. Rather, history was coming to be seen primarily as a set of changing representations of the past—*Representations* being the journal (first issued in 1983) where Laqueur and other historians joined forces with their New Historicist colleagues at Berkeley.[26] Situating bodies historically in their appropriate "representational regimes" was part and parcel of the re-thinking of the meaning, purpose, and shape of academic history-writing. Increasingly, therefore, history (as in the history of the body) was approached as a text: authored, discursive, and malleable in every respect. It was a made-up text that became a resource for (historical) constructivist and (literary) deconstructivist analysis, neither of which enterprise was now very separable.

Thus did the new cultural history render the body and historical epistemology privileged sites for literary and cultural analysis. But it was not long before the nature of that privileging was called into question. The problem with the representational approach, it came to be seen, was its acceptance of the body *only as a representation*. "Representationalism," it was lamented, held that the body (like everything else) was *nothing more than* a discourse, or something structured entirely by language.[27] In effect, in the course of de-essentializing the biological body, the new (representationalist) cultural history of the body *essentialized and naturalized language*. Left out was the *lived experience* of the body and of history. Bodies had flesh, just as the past had substance, it was insisted by social and cultural historians hoping to hold on to a materialist basis for history and the agency of individuals and groups.[28] Among historians, as opposed to many in cultural and literary studies, "real experience" mattered, for, ultimately, there was no basis for the practice of history without it, nor little point in fighting to change oppressive remnants of the past or insidious features of the present. At most one could only struggle against the textual stranglehold of the New Historicism.

In relation to the history of the historicized body, this "tempered realist position," as we might call it, was one that had in fact been argued for as early as 1987, albeit somewhat inadvertently, by the feminist and historian Barbara Duden. Drawing on the insights of the Foucauldian medical sociologist David Armstrong, Duden's impressive and acutely self-reflective *The Woman Under the Skin* (English translation, 1991) sought to capture "the vanished reality of the 'corporeal self'" through the casebooks of an eighteenth-century German medical practitioner. At its heart was the pursuit of the experience of the self that might have existed before the "biological" realities of the modern body silenced it. Written largely independently from the work of Laqueur and the "representationists," *The Woman Under the Skin* stands, in retrospect, at the head of what was to become an ever-lengthening queue of cultural historians insisting on the need seriously to heed corporeal *experience*.

As formidable as Duden's contribution, and more critical of body (text) talk in general, was the essay by the medievalist Caroline Walker Bynum, "Why All the Fuss About the Body" (1995). Reflecting on the superabundance of works on the body, Bynum lamented that so much of it failed to acknowledge that bodies eat, work, have sex, suffer, die, and undergo emotions, thoughts, and experiences. In the bulk of the body literature, she complained, the body is dissolved into language, and refers only to "speech acts or discourse." Dissatisfied with the Foucauldian anti-essentialist epistemologies underlying such work (although by no means wishing to revert to biological essentialism), Bynum sought to save a space for real lived bodies. Without such a space there could be no place for feminist politics it seemed—a worry that was made explicit by the American poststructuralist philosopher and "third wave" feminist Judith Butler in her influential *Bodies That Matter: On the Discursive Limits of Sex* (1993).[29] This overtly political issue was also taken up explicitly by other feminists, such as Elizabeth Grosz in *Volatile Bodies: Toward a Corporeal Feminism* (1994).

As the 1990s headed to the new millennium, cultural historians with an interest in the body came increasingly to agree on the importance of the *real experience* of the body and, a priori in the face of its alleged discursive evaporation, the need to hold on to it as something more than a linguistic representation. A consensus emerged over the need to carry out the recovery of the "real" body without any return to biological essentialism. Discourse was not dismissed; instead, it was valued as a realm of historical inquiry not mutually exclusive from that of experience; both apparently had a place in practice of history. No one summed up this position more succinctly than the leading American historian of gender, Joan W. Scott. In "The Evidence of Experience" (*Critical Inquiry,* 1991), Scott convincingly argued that the conventional historical project of making experience visible precluded the "analysis of the workings of this system and of its historicity; instead, it reproduces its terms."[30] To overcome this problem, Scott proposed that "we attend to the historical processes that, through discourse, position subjects and produces their experiences. It is not *individuals* who have experience," she clarified, but rather, *subjects*

> who are constituted through experience. Experience in this definition then becomes not the origin of our explanation, not the authoritative (because seen or felt) evidence that grounds what is known, but rather that which we seek to explain, that about which knowledge is produced.

To think about experience in this way is to historicize it as well as to historicize the identities it produces. This kind of historicizing represents a reply to the many contemporary historians who have argued that an unproblematized "experience" is the foundation of their practice; it is a historicizing that implies critical scrutiny of all explanatory categories usually taken for granted, including the category of "experience."[31]

Scott did not address the subject of the body as such; that was left to her close confidant in the elaboration of her thinking on "experience," Judith Butler. Two years later, Butler argued that language is "performative" in shaping conceptions of the body.[32] It was this view, enabling human experience to be understood as embodied, that became dominant among feminist scholars on the body by the new millennium. In some ways, it harkened back to Simone de Beauvoir's *The Second Sex* (1949), which, affirming the existential principle that existence comes before essence, argued that the "reality" of the political circumscription of women's lives came *prior to* the creation of any notion of the "essence" of womanhood that served to naturalize those political conditions.[33] The work of Butler and others served effectively to re-open an intellectual space for feminist struggle in the face of its closure through the earlier over-emphasis purely on language.

Postscripts: Essentialist Re-Animations of the Body in Post-Postmodernity

In part because of these intellectual achievements, interest in the history and historicization of the body began to wane towards the end of the first decade of the twenty-first century. Increasingly, in a context of subtly shifting politico-economic circumstances, the subject had the look of exhaustion, just like the literary-driven postmodernism that had breathed life into it. For historians who had never embraced the literary-cum-somatic turn this was welcomed as a means to reassert conventional historical understandings and practices. This was widely apparent in art history, a discipline that in the 1980s and 1990s found itself besieged by the postmodern politics of "visual culture" [see Chapter 6]. Writing in 2006, the art historian Martin Kemp, for instance, railed against the "deconstructive criticism of historical culture" that treated the past "as a sour land over which to exercise present concerns and anxieties." In its place, Kemp sought to revive "an agenda for history [that] was more common in the past than it is today."[34] Postmodernism with its iconoclastic anti-essentialism, he hoped, could be written off as a mere passing fashion, not something that had radically re-shaped thinking in general, and historical thinking in particular.

But like most reactionary huffing and puffing, that by Kemp and like-minded scholars had little purchase among those still in the mindset of postmodernism and unwilling to become geriatrics, and still others (we will come to), who might be depicted as their bio-captivated descendants. For the former, the kinds of worries around the body in relation to the practice and theory of history discussed above did not suddenly disappear. They still loomed large, albeit on a different register and inclining to a different cultural politics than that of historians from Laqueur to Scott, or Duden to Butler. For the most part, the new register was without explicit address to "the body in history" or the "historicized body" as such.

In *History and Theory* in 2006, for instance, a discussion was held on the notion of "presence" that was very much a sequel to the questions raised by, but not resolved through, the somatic turn. Of central concern to its contributors was the question of how to make the lived experience of the past a part of the living present, or how to put the experiential sense of "presence" into history writing. However, unlike Duden and others in their efforts to recapture corporeal experience in history, the contributors to *History and Theory* were orientated less to epistemology and knowledge production, than to ontology, or to notions of the "temporality of being"—the kind of concerns that had preoccupied the philosopher Martin Heidegger and the phenomenologist Merleau-Ponty.[35] How, they wondered, could "authenticity" (the Heideggerian notion taken up by Sartre) be put back into the domain of historical study.[36] As with some of the scholars trading in the historicization of the body in the 1990s, this quest for authenticity was explicitly pitted against "representationalism." However, the "presentationalists" perceived representationalism as heuristically retrograde: the besetting sin of a literary turn that had overlooked that representations are not only shaped by experience but also *determined* by them. As maintained by one of the contributors to the discussion, the Groningen philosopher-psychologist turned historiographer Eelco Runia, representationalism in common with postmodern historiography effectively purged history of reality. "Presentationalism," on the other hand, or the quest for "presence" in history, was about being in touch with it.[37]

Another contributor to this forum, the anti-Derrida literary theorist Hans Gumbrecht, imagined the concept of presence being developed in opposition to a "meaning culture" (that is, a culture "preoccupied by language"):

> In a meaning culture knowledge is understood as subjective interpretation, the subject occupying an external relation to the natural world. The aim of knowledge is to transform the world [as in Marxism], and thus the temporal dimension is central to meaning-culture, along with the concepts of consciousness and processes. In a presence-culture humans are embedded in the material world characterized by its spatial, tangible relations. Knowledge tends to be understood as revelation rather than interpretation, and the idea of changing the world becomes pointless.[38]

Beyond "a reconciliation of humans with their world,"[39] what Gumbrecht was arguing for, it seems, was some kind of essentialist pre-knowledge, a position flying in the face of any historical ontological understanding of the construction of knowledge/power such as articulated by Foucault and the historical epistemologist Ian Hacking.[40] Quite why Gumbrecht desired this is not clear. Like the other advocates of "presentationalism" writing in the special issue of *History and Theory,* his efforts, he claimed, were explicitly *a*political. Presumably this meant that, unlike the phenomenological basis for the existentialism of Sartre and Simone de Beauvoir, the "presentationalist" position, à la Gumbrecht, had no interest in championing liberation and freedom. Gumbrecht had as little time for these Enlightenment-rooted social politics as he did for any postmodern tilting at dominant discourses as an armchair form of activism. For him, critique was purely the business of abstract philosophy, removed from any social or political context.

Yet it can hardly be said that presentationalism as manifested on the pages of *History and Theory* was non-political, at least in its implications. This was made plain by one of the more empirically minded contributors to the discussion, the historian Rik Peters. His essay pointed to the "striking affinities" between what his fellow contributors to *History and Theory* were involved with and what the early twentieth-century Italian "philosopher of fascism" Giovanni Gentile was up to in seeking the "cultural awakening" of his countrymen in the 1920s and 1930s:

> First, ... there is a strong resemblance between Gentile's notion of "pure experience" and the contemporary theory of sublime historical

experience; both stress direct contact with the past, the obliteration of the subject/object distinction, and the primacy of experience. Furthermore, Gentile *cum suis* perfectly understood what Runia means by presence as "being in touch" ... with people, things, events, and feelings that made you into the person you are. It is having a whisper of life breathed into what has become routine and clichéd.[41]

The similarities between fascists and presentationalists are no coincidence, Peters observed:

When we study Gentile and Fascist culture we look in a distant mirror. Looking in this mirror we see ourselves, we see our own yearning for reality, our need for presentification, and our thirst for action. We may even feel some of the enthusiasm of the hundreds of artists, architects, scientists, historians, and philosophers working together for the great common cause in the schools, universities, cultural institutes, and exhibitions. Indeed, in Mussolini's Italy, we recognize something of ourselves: we recognize a culture struggling with its own historicity.[42]

From this perspective, the presentationalists in their efforts to move beyond the literary postmodern turn and beyond the world of mere textual "meaning culture" might be seen as advocating something more than simply the return of ontology "to the center of historical theory."[43] They might be regarded as generating a space for a neo-fascist politics founded in notions of purity and cultural unity, and even reviving the possibility for the Nazi struggle for historicity around the body. Admittedly, this space—ever naively created by those who live in historical ignorance—would be but the unintended consequences of a discussion that was more metaphysical than "historical" in the sense of being tied to the analysis of a specific time and place (in the manner of Rik Peters' article). More intentional, if not claimed as such, the presentionalists were positing a new essentialism through their critique of representationalism's essentializing of language, namely, *the essentialism of the idea of the essence of experience.*[44]

The discussion in *History and Theory,* which was related to, if not specifically addressing, the historicization of the body, was but one sign of emergent post-postmodern times. Another, which was far different and was to have much greater purchase among scholars in the social sciences and humanities, was the belief that contemporary culture was increasingly fashioned around the reduction of "life" to molecular biology and evolutionary neuroscience. The uptake of this idea, however, was not for purposes of revitalizing traditional ways of conceiving "reality" past and present (corporeal or otherwise), but rather, to revolutionize fundamental understandings of cognition and the very nature of being human. Prominent in this undertaking has been the historian of the body and visualization Barbara Stafford. Although never an anti-essentialist Foucauldian, Stafford was well aware of how the deconstructivist history of the historicized body had unfolded in postmodernity, as evidenced in her *Body Criticism: Imaging the Unseen in Enlightenment Art and Medicine* (1991). In subsequent work, she acknowledged that "we live in an age of otherness, of assertive identities, of the 'diversification of diversity.' "[45] However, in a new book, *Echo Objects* (2007), written after she had attended weekly seminars on computational neuroscience, Stafford became convinced that "those of us in the humanities and social sciences" have acquired "wonderful new intellectual tools to re-imagine everything from autopoiesis [neural self-organization] to mental imagery." Converted, she was compelled "to rethink the major themes of my life's work," and, evangelically, to press this "rethink" on art, cultural, and literary historians to have them "consider seriously the *biological underpinnings* of artificial marks and built surface."[46]

> As scholars of the myriad aspects of self-fashioning we can usefully enlarge, and even alter, our humanistic understanding of culture, inflecting it with urgent discoveries in medicine, evolutionary and developmental biology, and the brain sciences. In other words, the role of culture is not just to stand outside, critiquing science, nor is science's position external, and acting on culture. Rather, we are discovering at the most profound levels that our separate investigations belong to a joint project, at last.[47]

Neuroscience, she came to believe, "enables us ... to comprehend ... reflex tendencies from the inside out." Old problems were to look anew after being "sieved through the cognitive turn," and "traditional cultural assumptions by which many of us have long lived" were to be turned "upside down."[48] Other scholars, equally prominent in anthropology, art history, and general history, came to make similar claims for the "neural turn."[49]

Superficially, neural enthusiasms in the social sciences and humanities seem the very opposite of the Kemp-like reactions to the literary-cum-somatic turn. Far from wistful hankerings after bygone methodologies and practices, they were/are proposed as anticipations of shiny new futures. However, in terms of their bearing on historical practice, and in relation to the Foucauldian anti-essentialist understanding of the body in particular, their direction is equally reactionary, if more mediated. Whatever the specific purposes might be for entertaining the neural turn, it entails a return to a positivist understanding of science (as separate from culture) as well as to a biologically essentialist (reductive and materialist) understanding of the body. Although the enthusiasts of the neural turn in the social sciences and humanities appear to do so unawares, effectively they rout the basis for the anti-essentialist cultural turn in history-writing, at the same time as they engender a new empiricist turn in its practice. Conceptually speaking, as the sociologist of science Steve Fuller notes in his assessment of the significance of one such neural production, *On Deep History and the Brain* (2008) by the Harvard medievalist Daniel Lord Smail, the book is comparable, historically speaking, to Hayden White's foundational text for postmodern studies, *Metahistory* (1972): "Whereas White situated the writing of history in the heartland of the humanities, literary and cultural studies, Smail re-positions it in the emerging evolutionary neurosciences."[50] In short, Smail's work and other similar historical productions signify the abandonment of anti-essentialist constructivist programmes in history-writing in favour of a new kind of biological essentialism or naturalism. Conventional historians might deplore the neuro-turn as mere trendy cashing-in, but the general trend to reductive bio-science for defining and explicating contemporary life remains powerful, with worrying implications for historical theory and practice [as flagged in Chapter 1], if not for Giovanni Gentile–like political awakenings.

Conclusion

In *The Culture of the Body: Genealogies of Modernity* (2001) the intellectual historian Dalia Judovitz remarks that "the fate of the body as an idea, like that of subjectivity to whose emergence it is linked, is haunted by the foreclosure of its past meanings and history. Once consolidated in the modern period, the idea of the body takes on the character of a given that renders its prior forms and modalities of existence difficult to perceive and understand."[51] Like the idea of the modern body—the historical epistemology of which so fascinated Foucault—the idea of the *postmodern body* as something constructed in postmodern epistemology remains fully to be elaborated historically. In part, that project is only possible by coming to understand postmodernism as a movement within the socio-political era of postmodernity, and not necessarily causally connected to, or deriving from it.[52] I have not pursued such a project here; I have sought merely to net some of the manifestations of the historicization of the body over the last few decades, and relate the anxieties flowing from that. Of the latter, not least for historians has been constructing a history of the postmodern body outside an *a*historical notion of "history." But another source of worry, as yet hardly glimpsed, is the new breed of essentialisms—biological, political, and historical—that now threaten to foreclose on the ontology of the present. The need better to understand these seems all the more urgent in view of the closing of the door on the modern body in modernist history, and the opening of another one—"post-postmodern," or "transhuman" or "posthuman"—seemingly destined for a historyland such as we have never known.

5

Coming into Focus
Posters, Power, and Visual Culture in the History of Medicine

THIS ESSAY WAS FIRST DRAFTED in 2005 for a session on visual culture at the Paris meeting of the European Association for the History of Medicine. Co-authored with Claudia Stein, it was published in *Medizinhistorisches Journal* in 2007. Its conception predates the essay in the previous chapter by a couple of years. However, it fits better here as a companion piece to the essay that follows. It was itself a sequel of sorts—to a chapter on the history of public health posters produced during epidemics that we wrote together a few years earlier.[i] Whereas that had concerned itself with the history of the production and consumption of these objects, here they became a means to a much broader inquiry into the historicization of "the visual" in the late twentieth century. In particular, our concern was with how and why during the 1980s and 1990s images of the kind we were looking at moved into focus in Anglo-American intellectual life. Visual studies or "visual culture" was effectively born over those decades. Thus the concern with the visual, as reflected through the interest in health posters, could not be explained simply by the fact that more of these objects came to be produced over the period, especially around AIDS. Rather, we concluded, it related to broader shifts in socio-political life and epistemology. Two shifts in particular we thought important and worth unpacking and interlinking: that on understanding the nature of power in society directed at the human body (illuminated in the previous chapter) and that on the increasing visualization of all aspects of modern life (or, more exactly, the reduction of modern life to the visual).

My interest in health posters was sparked by a short fellowship at the National Library of Medicine at Bethesda, Maryland, in 2002. Liz Fee, then its director, had invited me to make use of the Library's visual resources in order to return the following year to deliver a paper at a conference on "Public Health and Visual Culture." The subject of the conference (like the session in Paris in 2005) further testifies to how the turn to visual culture was then fanning out from cultural and media studies. In this sequel to that paper we wanted to temporally situate the visual turn in order to problematize what was increasingly coming to be accepted as a logical and inevitable intellectual development. Somehow the visual had become an object of study that everybody talked about and was increasingly solidifying in the search for its postmodern pioneers. But nobody was inquiring into the power technologies that were permitting this turn to be naturalized. Postmodernity, we began to think, had a politics, and the visual turn was crucial to it. It was these politics that we wanted to make visible. For historians to single out visual objects without comprehending the context in which they had become such shiny and powerful objects of study, it seemed to us, was a means only to compound those politics, not illuminate them. Without this, and without exploring the epistemological move inherent to the "visual turn" of the late twentieth century, historians would do more harm than good in jumping on the visual culture bandwagon. Thus, beyond wishing to introduce historians of medicine to this fast accelerating area of scholarly interest, we wanted to head them off at the pass, as it were. Unlike when we wrote the essay's sequel (Chapter 6), we were not then thinking about the politics of aesthetics. Nor had it occurred to us that in making the case for the socio-cultural relativity or subjectivity of the discourse of visual studies we might inadvertently be providing grist for the mill of an emergent neoconservative "new empiricism" eager to jump upon any kind of critique of postmodern (representationalist) thinking. Sander Gilman was subsequently to point this out.[ii]

If the visual has had a place of little significance in the Anglo-American history of medicine, it is not because its practitioners have been blind. Until recently, there was no need for them seriously to heed the visual, let alone express "anxiety" over its objects.[1] This paper reflects on why that was so, and how since the 1990s circumstances have changed so as to render the visual intellectually interesting and, for the history of medicine, something of a disciplinary challenge. If only because advocates of "visual studies" remind us that "visualizing has had its most dramatic effects in medicine, where everything from the activity of the brain to the heartbeat is now transformed into a visual pattern by complex technology," it seems an appropriate moment for historians of medicine to take stock of what it is to be interested in the visual.[2]

But while the study of everything from X-rays to post-1995 DNA microarray technology for the visual transcription of "gene expression" can prove historically productive, for the historiographical concerns of this paper a consideration of health posters serves better, specifically those dealing with sexually transmitted diseases (STDs). Because various approaches have recently been made to the study of these objects, and public battles have been conducted over some of the images they present, they more readily facilitate reflection on the reception of visuality itself, as opposed to commentary on medicine's imaging technologies. Furthermore, we can effectively contrast their relative absence from historical discussion before the mid-1980s with the attention they began to receive thereafter. This essay argues that the "age of AIDS" was crucial to their coming into focus, not because it allegedly brought health posters back into their own as one of the most relied upon means for the public transport of health information, but more so because, partly through this medium, it contributed to bringing visuality itself into intellectual focus.[3] Integral to this focusing was a fundamental shift in the understanding of power in society, with the human body becoming central to it. It is this shift that this essay explores.

The first section reviews the limited attention paid to health posters before the mid–1980s in order to comment on how the intellectual climate then prevailing did not provide the space for exercising views on the visual. We draw on the work of the American social historian of medicine Allan Brandt and the essayist and arts and literary critic Susan Sontag to elaborate this point. In the second section, we explore the reshaping of the intellectual landscape in the context of AIDS in the West; how this connects to heightened sensitivity over the visual; and how, in turn, some scholars (notably Sander Gilman and Paula Treichler) came to comment on AIDS posters within this emergent landscape. In our concluding section, we draw out the implications of this for the future of the visual in the history of medicine.

This essay is not therefore a contribution to the history of a particular medium in health education. Least of all is it concerned with the question of the possible impact of STD posters on social policy or popular behavior. Although such posters might operate by evoking a controlled form of fear and anxiety that can be channeled to the rational governance of personal and/or national life, it remains an open question how far this or any other emotional response to them can be generalized, either in terms of the intent to instill it on the part of their producers, or in terms of spectators' reactions. Nor do we endeavor to legislate on how, historically, one *ought* to interpret or use such visual objects. Instead, our attention is directed to the repertoires of discourses available at any moment in time *for* their discussion (or not), together with the wider context within which these discussions were situated, constituted, and reconstituted in the English-speaking world. For descriptive purposes, "discourse" serves us better here than "ideology." Referring to bodies of knowledge that both define and limit what can be said of anything, "discourse" avoids the over-discrete implications of "ideology" when, as W. T. Mitchell puts it, it is used to refer to "the implicit structure of political values and class interests that lie beneath a history of practices."[4] As we shall demonstrate, the history of the practice of analyzing health posters in these more structuralist terms was effectively challenged during the time of AIDS. "Discourse" is further advantageous for our essentially epistemological task in that it does not refer to a language separate from a "real" material world in which objects are located. Rather, it refers to organizing sets of signifying practices that cross the boundary between "reality" and language, and convey meaning through form as well as content.[5] We regard health posters as material objects functioning in, and constitutive of, discursive realms that change over time and place. The primary purpose of this essay is to outline that change and expose those discursive realms from the 1970s to the start of the twenty-first century.

Posters and Health Posters in History

Health posters—a variant of what are "essentially ... large announcement[s], usually with a pictorial element, usually printed on paper and usually displayed on a wall or billboard to the general public"—have been the subject of surprisingly little comment.[6] It may be that this is because they are quintessentially ephemeral objects, intended to make an impression and then disappear (as, indeed, many of them have). It might also be that their relative neglect by scholars is consequent upon health education itself being a Cinderella subject in the history of medicine when set next to the heroic tales of great doctors and great achievements. Whatever, all that exists are a few coffee-table books on health posters in which these often visually arresting mass-produced objects are further reproduced.[7] The posters reprinted are usually assumed to speak for themselves, requiring scant, if any, textual support. In fact, however, they do not speak for themselves, but rather to a narrative of public health progress, which is simply taken for granted in such books.

Serious studies of posters of any kind are also relatively scarce.[8] Until recently, most of the little that had been written on them (as propaganda for politics, for war, or for the sale of commercial goods) was orientated to the history of art with all its implicit assumptions about art as a higher form of human awareness.[9] Such studies often struggle to find a place for the poster somewhere between art for its own sake, on the one hand, and popular mass-produced products for the sake of consumerism, on the other. As one such study politely submits, posters "bridge the gap" between "high art" and "pop art."[10] Overwhelmingly, the emphasis has been on the supposed aesthetic value of these objects. When this is not explicit, it is implicit in the posters selected for reproduction. For example, a recent volume, *The Power of the Poster* (1998) produced by the Victoria and Albert Museum, invites us to investigate posters in their full "intellectual and physical context," but concentrates above all on their visual "effect" and selects images appropriate to the exercise.[11] When occasionally health posters appear in these books, a further ambivalence emerges between a supposed "vulgar" commercial world of advertising and an

Coming into Focus   117

assumed "humanitarian" world belonging to medicine and health—a legacy of the dichotomy asserted by medical humanists from the eighteenth century.[12]

It is not perhaps surprising therefore that, until recently, it was exceptional to find health posters referred to or deployed in histories of health and disease.[13] Unusual, in this respect, was Allan Brandt's *No Magic Bullet: A Social History of Venereal Disease in the United States since 1880* (1985). Some twenty posters are reproduced in this study of how American concerns over venereal disease centered on a set of social and cultural values related to sexuality, gender, ethnicity, and class. Most of the posters were issued by government departments during the two world wars as a means to combat STDs in the military and civilian population. Brandt's choice of these posters, as well as the captions he attaches to them, reveal how he thinks pictorial evidence supports his argument that at the heart of efforts to combat venereal disease there was a view of it as both a punishment for sexual misconduct and an index to social decay.

An example is Brandt's illustration 18 (Fig. 5.1), a poster that, he states, was the most widely circulated during World War II.[14] The poster—on the origins, production and circulation of which Brandt does not inform us— shows a seemingly innocent young woman smiling at the onlooker. In the left hand corner there are three small male figures, a sailor, another uniformed soldier and a male civilian who are turning their backs to the onlooker. They seem to be taking a casual stroll and their faces are turned toward the young woman. Across the poster is the warning: "she may look clean—*but*," with the "but" highlighted in red. At the bottom is the reminder that all "PICK-UPS 'GOOD-TIME' GIRLS [AND] PROSTITUTES SPREAD SYPHILIS AND GONORRHEA." In red at the very bottom of the poster the onlooker is drawn to the emergency that the poster seeks to address: "You can't beat the Axis if you get VD."

Brandt does not describe the poster or engage with its imagery; he seeks simply to summarize it with the caption, "This popular poster asserted that even the woman who appeared "pure" could pose dangers for the soldier. The choice of image and Brandt's caption replicate and exemplify the argument that he makes a few pages later, that during World War II the military changed their health educational strategy.[15] Instead of targeting prostitutes, they came to focus on the innocent-looking girl next door. The poster, Brandt submits, repeats the historically "sorry association of 'cleanliness' with chastity, [and]

Fig. 5.1. Venereal disease poster reproduced in Allan Brandt's *No Magic Bullet* (1985). Courtesy of the National Library of Medicine, USA.

Coming into Focus

impurity with disease," and reinforces the prevailing opinion about such women which (he quotes a federal committee) were " 'more dangerous to the community than a mad dog. Rabies can be recognized. Gonorrhea and syphilis ordinarily cannot.' "[16]

Brandt's caption—indeed, all the captions to the posters he uses in *No Magic Bullet*—serve to draw the reader's attention to the alleged core message of the poster, a message that, perhaps Brandt fears, his reader/viewer may not immediately understand at first viewing. The poster with its caption is thus meant to provide unmediated access to the historical narrative. It serves not merely as illustration, but as a vital part of the "truth" of Brandt's reconstruction of venereal disease in American history. What the reader/viewer is supposed to receive is visual evidence of the "real" moral concerns circulating within and (according to Brandt) dominating American society around the time the poster was produced. Invoked is the truth about social reality and medical practice that Brandt depicts. For him, the poster and social reality are one; the image serves instrumentally as evidence for the other.

There is nothing unusual in this. It is an instance of the common practice among historians of using pictures as if they were documentary evidence for textual arguments, a use identified and chastised by Ludmilla Jordanova as the "documentary fallacy."[17] It is the means, as Sander Gilman has noted, by which the historian says to the reader/viewer: "You can see the truth of my statement for yourself, as you too have this objective window into history as it really was."[18] The image and the caption also operate as a form of narrative closure, erasing any ambivalence or space for contestation in the historical interpretation. Thus the image makes the historical narrative look complete rather than partial. Closure is further effected in Brandt's case by his absence of commentary on the contemporary viewing public(s) for the posters he reproduces; conceivable differences between audiences are not his concern.

Historians can only suppose their readers' acceptance of their authority if they also assume a priori that their readers/viewers are homogeneous in their understanding and reading of such images. The reader/viewer needs further to share the historian's unspoken assumption that historical veracity can be crafted from an image, and that images in themselves speak unproblematically to the historian's construction of truth. Inherent to this construction of historical truth is a perception of power as an instrument of repression and constraint in whose service the visual is presumed to act. Just as health posters in Brandt's narrative are tools of state propaganda for national health, so in his social history of venereal disease the use of the poster is taken to be a tool of persuasion. Implicitly, Brandt assumes that his reader/viewer shares the same regard of the image as viewers of the image did a half-century or so ago. What is *a*historically understood but never stated on the operation of the power of the visual is thus assumed to be reproducible in the reader of his history.

In seeking to elaborate the history of venereal disease in terms of a "persistent tension between a rational, scientific program and a behavioral, moralistic approach," Brandt engaged with one of the most influential contemporary polemics on the power of disease, Susan Sontag's *Illness as Metaphor* (1977).[19] Brandt commends her essay for using literary techniques to focus attention on the social and cultural dynamics that contribute to the specific meanings of particular diseases. Her essay, he says, demonstrates "that disease has throughout history attracted metaphors and symbolic language that reveal implicit beliefs about the disease and its victims." But Brandt also criticizes the essay for its underlying assumption "that once a disease is [scientifically] understood and treatable, the metaphors will wither away."[20] In Brandt's view the stigma of disease can probably never be broken (only historically re-constituted), for illness is always something that society identifies and negotiates; it is not, as Sontag positivistically believed, something that value-free objective science will one day lay naked, freeing it from its baleful stigmatizing metaphors.

More interesting from the perspective of Brandt's use of the visual in *No Magic Bullet* is the attention that he does *not* pay to a 1970 essay by Sontag entitled "Posters: Advertisement, Art, Political Artefact, Commodity." Brandt had no need to engage with what was actually an introductory article for a book on Cuban political posters. For us, however, it is important for disclosing a context for thinking about posters which can be shown to be in fact shared by both Sontag and Brandt. Unique from the point of view of its substantial engagement with the history of posters and the mass media, Sontag's eloquent essay is also useful to us in revealing the same general line of argument that she was to reiterate in *Illness as Metaphor,* and in its sequel, *AIDS and Its Metaphors* (1988).

For Sontag (1933–2004), just as for Brandt (b. 1953), posters are simply footnotes to bolster a particular historical narrative, whether specific or general. Whereas for Brandt this is the narrative of public health, for Sontag it is the history of Western industrial capitalism. Posters, she writes, originate "in the effort of expanding capitalist productivity to sell surplus or luxury goods"; they could not exist before the specific historic conditions of modern capitalism. Sociologically, the advent of the poster reflects the development of an industrialized economy whose goal is ever-increasing mass consumption, and (somewhat later, when posters turned political) of the modern secular centralized nation-state, with its peculiarly diffuse conception of ideological consensus and its rhetoric of mass political participation.[21]

Posters, she maintains, serve the purpose of aggressively pushing consumption or, in politics, of selling national identities. Indeed, they presuppose "the modern concept of the public" as well as "the modern concept of public space—as a theatre of persuasion."[22] This regard of posters, as "aim[ing] to seduce, to exhort, to sell, to educate, to convince, to appeal," is predicated on a view of advertising (of any sort) as psychologically dangerous—a belief that was much strengthened in the 1960s by concerns over the manipulation of people's drives and emotions through new subliminal methods of advertising.[23] Thus Sontag sees posters as blinding people to "reality." Like the "distorting" metaphors of illness on which she would later write, they are tools for creating "false-consciousness" or corruptions of "truth" wrought through capitalism.[24] Hence, behind posters, as behind metaphors of disease, there is assumed to be an unmediated or "pure" form of reality/power; one has only to break through the constructions of reality mediated by language or by images to experience "the direct understanding of the world through one's senses and perceptions as both reliable and real."[25] Once the clouds of propaganda and illusion have been dissipated, the argument goes, people will be enlightened, led to some clearer, cleaner, more rational stage of seeing.

This view of the power of posters on "passive" spectators and consumers is therefore comparable to Brandt's view of the power of the health poster and the visual in general on the passive reader/viewer of history writing. Further, like Brandt's reader/viewer, Sontag's "public" is perceived to be homogeneous; she never considers that buying into the metaphors of disease or the messages of posters might not be a universal experience, or that different "spectators" with different backgrounds might understand differently. Inasmuch as she conceives posters only as propaganda for capitalism, she is also like Brandt in providing a closed reading, with image and text mutually reinforcing a historical (meta-)narrative. Finally, she is like him in regarding the exercise of power through the visual as something external, and only related to those who use it (the visual having no power in and of itself). Hence, images are self-contained objects merely waiting to be neatly sorted into the categories of social or semiotic analysts. For Sontag, power is a top-down force that operates more or less instrumentally on individuals and groups through "propaganda" which, as Brandt claims too, exclude and stigmatize.

From the point of view of the place of the corporeal in popular culture articulated later in the twentieth century, it is interesting that in her 1970 essay Sontag never once alluded to health posters. Putting aside the specific occasion for the writing of her article (Cuban revolutionary posters), the omission is in part because at that time health posters—especially on infectious diseases—were not the commonplace objects on buses, billboards, underground trains, and doctors' offices that they were to become twenty years later during the time of AIDS. But this is explanation only in part; for Sontag in 1970 these objects could not possibly be construed as other than adjuncts to the purpose of manipulating the human body to some political end (as by the state, for example). For her, power was produced by rational "minds" and imposed upon the minds of others so as to falsify their consciousness. Bodies in any other sense were irrelevant.

From the perspective of visual studies today (which we will discuss in more detail below), Sontag's essay is also telling in that it does not challenge a view of posters as "low art." On the contrary, for Sontag "aesthetically, the poster has always been parasitic on the respectable arts of painting, sculpture, even architecture.... As an art form, posters are rarely in the lead."[26] In this respect, as much as in her concentration on minds over bodies, Sontag reveals her indebtedness to the Frankfurt School of Marxist critical theory with its inherently elitist emphasis on the inculcation of false-consciousness among the masses through the cheap mass-produced products of industrial capitalism. Like Frankfurt School theorists, Sontag holds to a pessimistic view of the effects of the culture industry of modern capitalism on a public conceived as passive and homogeneous.

When Sontag did come to refer to health posters, in *AIDS and Its Metaphors,* her outlook on their meaning and function was largely indistinguishable from that in her essay on political posters. To indict the ideology of consumer capitalism was the whole point of her reference to public health media campaigns, such as the famous British one of 1987 on "AIDS—Don't die of ignorance"—a one-million pounds mixed-media campaign which featured posters of threatening icebergs and intimidating black tombstones.[27]

After all, Sontag believed, it was capitalism that had celebrated recreational, risk-free sexuality in the name of individual liberty. It was all a part of "the culture of capitalism" which, she noted, was, "guaranteed by medicine as well."[28] Thus she predicted that AIDS would not lead to a re-casting of the role of the state in relation to public health and individual liberties, as some commentators (such as Allan Brandt) were then coming to claim.[29] Instead, in Sontag's view, AIDS would simply strengthen the existing capitalist system of consumption and its celebration of individual freedom. Public health education campaigns, by creating anxieties over the issue of AIDS, would only encourage the production of commodities such as condoms to allay these anxieties. Moreover, this allaying process would "require the further replication of goods and services." Hence, "the culture of consumption may actually be stimulated by the warnings to consumers of all kinds of goods and services to be more cautious, more selfish."[30] For Sontag, then, talk of the "danger of AIDS" could only serve the economic interests of commodity capitalism. Here was an echo of the famous motto of the Frankfurt School, that "the whole world is made to pass through the filter of the culture industry."[31] Caught in this filter as much as everything else was our social perception of diseases such as AIDS and, hence, our stigmatizing regard of its victims, Sontag thought.

Before moving on to different perceptions, it is worth noting that Sontag's 1988 prediction seemed to be fully realized—and only a few years later. In 1992 the Italian fashion knitwear company Benetton ran its highly controversial worldwide "Shock of Reality" campaign at a cost of 70 million US dollars.[32] At the hand of Benetton's Creative Director, Oliviero Toscani, award-winning photographs on the burning social issues of the day were appropriated to the business of selling sweaters. Race, poverty, and pollution were among the images turned into billboard advertisements. And so too was AIDS. The now iconic Benetton poster-ad of the dying (or dead) American AIDS victim David Kirby in the arms of his father, surrounded by his grieving family (Fig. 5.2), appeared early in 1992. For Toscani, who tinted the photograph that had originally appeared (as a black and white photograph) in *Life* magazine in November 1990, this was the most powerful of the images in his "Shock of Reality" campaign—"a real Pietà," he declared, and "one of the most moving photographs of the whole decade."[33] As an advertisement, it demonstrated, he thought, how a global corporation was "open to the world's influences and engaged in a continuing quest for new frontiers."[34]

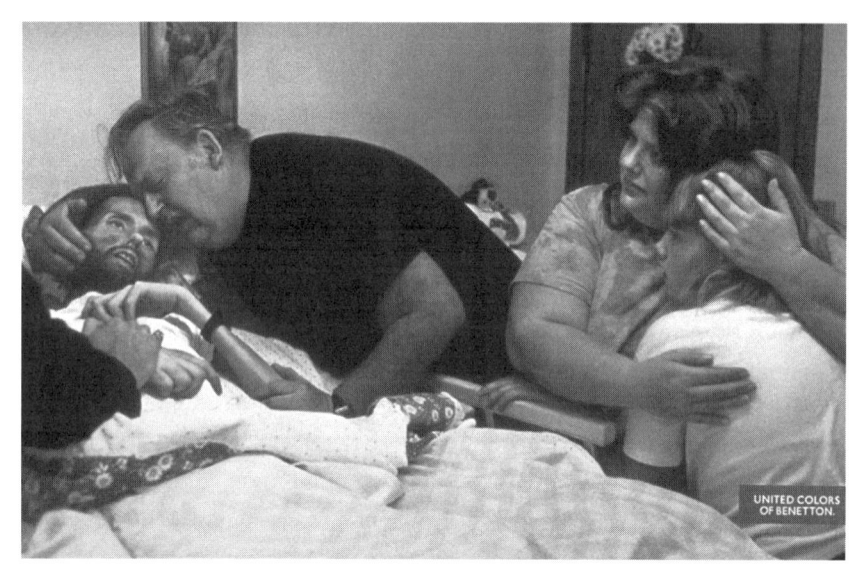

Fig. 5.2. "Dying of AIDS," a billboard advertisement designed by Oliviero Toscani (Italy, 1990) that appears to support Susan Sontag's claim of the appropriation of AIDS by the culture of capitalism. © Copyright 1998 Benetton Group S.p.A. Photo: Therese Frare. Concept: Oliviero Toscani.

Advertisements like this revealed the commitment of commercial organizations to contemporary global political and social issues. Toscani chose the David Kirby image, he said, in order explicitly to "fight against the exclusion of AIDS victims."[35]

The poignancy of the picture was indisputable, but whether a global company had the moral right to use it for selling fashion knitwear was another matter. An enormous furor was generated over it: in Germany the poster-ad was taken to court; in France billstickers refused to post it; and, in Britain, *The Guardian* (the first newspaper to run it as a full-page ad) was inundated with letters of complaint.[36] In addition to those disturbed by its apparent blasphemy, were those outraged by what was regarded as a particularly aggressive form of cultural and economic imperialism. To the latter, only too obvious seemed the truth of Sontag's indictment of the capacity of worldwide capitalism to intrude into the most private spheres of human life and suffering simply to increase the sale of meaningless consumer goods. This was exactly how the American Aids Coalition to Unleash Power (ACT UP) responded to the image. They then themselves appropriated it to further *their* campaign against passivity around AIDS. In so doing, they not only consciously revived the tradition of using art for political protest, but explicitly sought to hijack the advertisements of big brands in order to talk back to them and to re-conquer city space.[37] Thus they "piggybacked" on the image of David Kirby dying of AIDS to write beneath it: "There's only one pullover this photograph should be used to sell"—and pictured a condom. Below this they inserted their own, by then iconic logo, "silence=Death."[38] Here, so it was claimed, was the subconscious of the Benetton "reality" campaign X-rayed to uncover not only its opposite meaning, but also the deeper truth lurking behind the layers of advertisement euphemism.[39]

Intellectual Maneuvers

Between them, Benetton and ACT UP confirm Sontag's prediction that AIDS would become caught in the totalizing webs of Western capitalist culture. Yet at the same time, this example connects to a postmodern temper increasingly at odds with Sontag's outlook. For one thing, the hostility to the David Kirby poster-ad for Benetton serves as illustration of the increasingly dominant and powerful place of the visual (and the worries over it) in everyday life. It was a realization that in the 1990s would come to challenge the lingering elitist framework in the study of the visual. If only because the pictures that flowed from commercial studios like Toscani's or from giant Madison Avenue advertising agencies were highly sophisticated, graphically as well as psychologically, it was increasingly difficult to sustain the artifice between "high" and "low" art.

Secondly, in view of the reactions to the David Kirby poster-ad, and the responding visual politics of ACT UP and other AIDS activists, it was difficult to adhere to the opinion that the viewers of such pictures were either passive or homogeneous. Victims of AIDS, for example, took a far from passive interest in their visual representation in media campaigns. For them, and in particular for the emergence of a gay culture that had only recently "come out," there was reason to panic—not over world-wide capitalism, but over being forced into socio-cultural categories (such as the older medicalized one of "homosexual") transmitted through image-laden public health messages—messages, moreover, that sought to inculcate "family values."[40] For UK AIDS activist Simon Watney, AIDS presented an even greater challenge. Writing in 1987, it seemed to him that it involved "a crisis of representation itself, a crisis over the entire framing of knowledge about the human body."[41] In the same year, the American cultural theorist, feminist, and AIDS activist Paula Treichler famously declared that "the AIDS epidemic—with its genuine potential for global devastation—is simultaneously an epidemic of a transmissible lethal disease and an epidemic of meanings or significations."[42] Indeed, for Treichler, AIDS was *the*

> premier symbolic battleground of our times where war will be waged incessantly, where language and reality will continually shape each other, where women's futures—all our futures—will in part be determined, and where the health-care system in its unacceptable entirety should be challenged and transformed for good.[43]

As this suggests, the third main difference with Sontag's worldview, was the emphasis on the body in relation to power, and to ways of knowing, accessing, and constructing "reality." AIDS was not of course the cause of this epistemological shift; rather [as noted in Chapter 4], it encouraged new modes of thinking about knowledge and perceptions of power. In particular, it gave strength to the poststructuralist epistemology most closely associated with the writings of Foucault (1926–84), himself tragically an early victim of the strange disease phenomenon.

Since the 1960s, around medical knowledge and medical institutions especially, Foucault had developed a concept of power (and methods to investigate it) that centered on the micro-management of individual human bodies through various disciplinary techniques.[44] Power did not derive simply through social and political institutions, he argued, nor did it function merely by coercion. Rather, it operated through, and was inscribed upon, the body, which was "directly involved in a political field; power relations have an immediate hold upon it; they invest it, mark it, train it, torture it, force it to carry out tasks, to perform ceremonies, to emit signs."[45] As important, although less apparent in Foucault's earlier work, was his understanding of modern power as a *productive* agency, rather than (as in Marxist thinking) simply a negative or repressive force.[46] Modern power in its constitution through somatic discourse, he argued, produces corporeal knowledge that simultaneously generates new mechanisms of control.

Foucault's notion of a power/knowledge nexus around the body stimulated an enormous amount of scholarship aimed at identifying and exploring these invisible, yet most powerful technologies of "life" itself.[47] The history of mental illness, human sexuality, and the female body in particular became prized targets of scholarship.[48] Public health education, too, came under scrutiny. Its discussion moved away from structuralist-functionalist models of coercion *vs.* consent (as, for example, in the narratives of Brandt and Sontag) to a notion of citizens whose relationship to the state was "made-up" by their own biology, yet who were nonetheless capable of autonomy and a "kind of regulated freedom."[49]

All of which is to say that at the same time as ACT UP and other AIDS activists were conducting their street fights over posters and other representations of AIDS victims, many academics (often themselves activists) were trying to come to terms with "the time of AIDS." They sought to find vocabularies seemingly more appropriate to describe what they were perceiving and experiencing. To greater or lesser degrees their efforts were informed by Foucault's body-centered notion of "biopower "and "biopolitics" (the knowledge-producing "processes through which institutional practices define, measure, categorize, and construct the body" and somatically shape all experience, meaning, and understanding of life).[50] Indeed, to a considerable extent, for many academics, AIDS provided a testing ground for Foucault's ideas, with some, such as the art historian and cultural theorist Douglas Crimp, specifically targeting health posters in their Foucauldian deliberations around AIDS.[51]

Prominent among those in the history of medicine (broadly defined) working on this front was Sander Gilman (b. 1944). From the early 1980s he had written widely on the social construction of representations of race, sexuality, madness, disease, and, among other "othered" subjects, the body of the Jew. In 1995, as a sequel to his *Disease and Representation* (1988), he produced *Picturing Health and Illness: Images of Difference,* one of whose chapters offered a close reading of some forty AIDS posters drawn from the US National Library of Medicine's collection, and reproduced in black and white, full page.

Taking his cue from Simon Watney's view of a representational "crisis" of the human body, Gilman's chapter, "The Beautiful Body and AIDS," was centrally concerned, not with the effects of specific AIDS images on individual or collective behaviors, but rather with the complex and often contradictory symbols present in the posters themselves. Accepting that AIDS posters were products of advertising (and not primarily parts of educational campaigns of governments or voluntary bodies), Gilman aimed to decode them as "aestheticized" veiling of the ugly realities of death and dying. Unlike in high art, he contended, death cannot be dealt with in commercial advertising because it violates its very reason for existence: the continuous stimulation of human desire—and, ultimately, the perpetuation of the market. In order to sell any product (including health) advertisers must stimulate human appetites among erstwhile desirous humans. They need to sex-up life, not expose its liquidation. Hence, according to Gilman, instead of images or even suggestions of death and dying, AIDS posters played up the idea of "risk"—a legitimate expression of hope for life, as well as a harkening to individual choice in a culture of consumption. Beautiful and often highly eroticized bodies "with/at risk with AIDS" therefore stand in the stead of the ugly reality of AIDS and the ravaged bodies of its victims. Referring to the David Kirby poster-ad, Gilman argued that Benetton totally "misread the meaning of the representation of the body of the person dying from AIDS." They miscalculated by trespassing visually on the reality of dying, against which "even the parent's protection cannot shield."[52] This, he concluded, explained the hostile reaction to Toscani's art work. But this interpretation conveniently overlooked that the reaction was precisely what Benetton wanted; it was a part of its calculated "anti-ad" strategy aimed at sophisticated and visually well-educated consumers. Aestheticizing death and dying was the very opposite of Benetton's intentions in its "Shock of Reality" campaign.

Gilman reproduced his argument in the choice of posters for his chapter. For example, a Vancouver AIDS poster from 1986 (Fig. 5.3), revealing two half-clad "beefcake" bodies, serves apparently to illustrate "the eroticization of the act of safe sex."[53] Illness and death are absent from the image, we are told; the dangers and fears of AIDS are translated into "risk," which is symbolically represented in the form of the enormous safety pin in the background.

Gilman's interpretation aside, this was a far from instrumental ("propaganda") reading of the power of health posters. At the same time it was a rather different interpreting of the "crisis" identified by Watney. For the most part, Gilman's "close reading" of AIDS posters (like that term itself) was a project adapted from literary and art criticism. Thus it was no concern of Gilman to provide a socio-cultural background on the production and distribution of the posters he described.[54] In this respect, he was not unlike Brandt, although

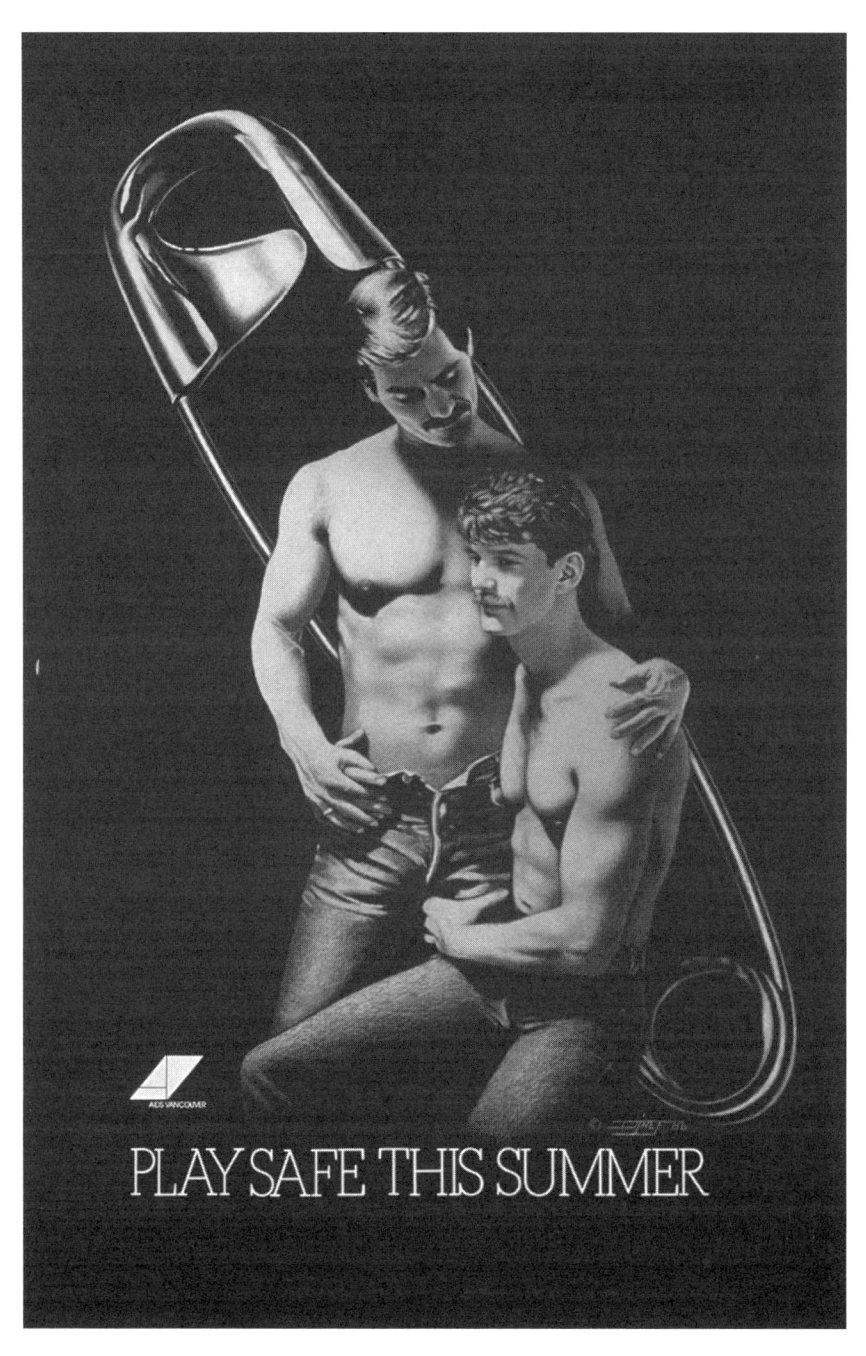

Fig. 5.3. Vancouver AIDS poster, 1986, reproduced in Sander Gilman's *Health and Illness* (1995) to illustrate the culturally acceptable eroticization of the act of safe sex in commercial and public health advertising and the "complex and often contradictory vocabulary of images" present in AIDS posters. Courtesy of the National Library of Medicine, USA.

for totally different reasons. And there is a further overlap between him and Brandt, inasmuch as Gilman also regards posters as serving the purpose of normativity (though, again, it was not his concern to demonstrate this). With Sontag, too, there are connections, which we will touch on below. Overall, however, the differences between Gilman and Brandt and Sontag far outweigh any similarities in their regard of posters. Above all, this is testified through Gilman's centre staging of visual representations in the history of Western medicine and culture, and by the concluding remark in his chapter on "Beautiful Bodies" that AIDS posters "*form the material for a new history of medicine rooted in the study of the visual image*."[55] The visual that for Brandt and Sontag could be neglected because it did not in their view constitute any independent dynamic or "forming" force on the onlooker, was for Gilman— as for many other academics in the 1990s—a welcome means to decode the "chimerical world of picturing beauty and health at the close of the twentieth century," and for unmasking the hidden processes of normatization, exclusion, and "othering."[56]

Whether this permits us to label Gilman a "Foucauldian" is a moot point; a careful reading of his chapter reveals the input of many thinkers. Nevertheless, like so many other academics then and since, Gilman subscribes to Foucault's suggestive ideas on the functioning of modern power in Western culture—that is, everyone's implicit participation in it, and the centrality of the body in it and in its representations. However, in categorizing health posters as products of advertisement and pop art, and in sharply distinguishing them (and their visual semiotics) from representations of AIDS produced by "serious artists," Gilman belies an attachment as strong as Sontag's to a conventional analysis of art and literature. Indeed, in maintaining that AIDS posters have "audiences, visual vocabularies and intended contexts other than the high art representations of AIDS during the same period," Gilman is not unlike the authors of coffee-table books on posters.[57] This is odd given that by the time *Picturing Health and Illness* was published in 1995, the distinction between "high" and "low" art had already become a much disputed practice. Since the 1980s there had emerged a new interdisciplinary field of study, "visual culture" (also known as "visual studies"), which sought to distinguish itself from traditional art history on precisely these grounds.[58] Scholars working in this new field regarded *all* visual imagery as their remit, and justified their emergent discipline on that basis. It was through it all, they argued, that meaning was made in any cultural context. In the wake of anthropological and literary redefinitions of "culture" itself as "a whole way of life" (not just the life of elites), scholars such as Nicholas Mirzoeff and W. T. Mitchell leveled high art.[59]

Some commentators in this new field of study, besides voicing dissent at the habit of treating images in terms of semiotic notions of representation—as if images had fixed meanings that could be decoded—with enough effort—were also beginning to perceive that much of what they construed as "meaning-making" in the modern world was negotiated through multiple emotional and intellectual responses to simultaneous, rather than single, visual impressions.[60] As remarked in criticism of the tendency in visual studies itself to focus on discrete sites (such as the cinema, TV, art galleries, and popular magazines), neither the eye nor the psyche actually operated within such tidy categories. "In the area of visual culture the scrap of an image connects with a sequence of a film and the corner of a billboard or the window display of a shop we have passed by, to produce a new narrative formed out of both our experienced journey and our unconscious."[61]

The implication of this non-elitist, non-hierarchical, and non-static approach to the visual was that objects such as health posters could no longer be considered (as in Gilman's work) in isolation from the rest of the visual, material, and ideational world.[62] Instead, the ubiquity of objects like AIDS posters in public spaces in modern cities, and their use in conjunction with other media, such as TV or newspaper advertisements in health education campaigns, made them central, but not the sole foci in the "struggle for signification and meaning" around AIDS in Western culture. It was precisely this approach that Paula Treichler (b. 1943) adopted in a number of searching essays written from the mid-1980s (gathered together in 1999 in *How to Have Theory in an Epidemic*). In these, AIDS posters are discussed and reproduced, but they are deliberately scattered within the text alongside reproductions of newspaper advertisements, comic strips, pictures of TV commercials, and magazine covers, as well as pictures and charts from medical journals.[63]

Treichler was among those in linguistic theory and cultural studies in the 1980s whose sensitivity to words and language increasingly reached to the visual. For such scholars, this sensitivity was nowhere rendered more acute than in relation to visual representations of AIDS, which were perceived as exercising an enormous role in the construction of the syndrome and the politics around it.[64] For Treichler, as for Crimp and other AIDS activists, the interpretation of the literal picturing of representations in AIDS posters was part and parcel of the engagement with the "crisis" around knowledge of the human body and its representation as identified by Simon Watney. Thus, very much around AIDS, the "cultural turn" of the 1980s was imported into what was fast becoming the "visual" or "pictorial turn."[65] And largely through the syndrome came convergence around the notion of "the visual as a focal point in the processes though which meaning is made in a cultural context."[66]

But Treichler's particular take on the visualization of AIDS, including its posters, continued to reflect her roots in linguistic theory, with its long-standing denial of the notion of access to reality through language. For her, unlike for Gilman, there was not and never could be a single way to comprehend the visualization of AIDS. Any visual engagement, moreover, could not be separated from language.[67] AIDS, she insisted, must always be understood "in multiple, fragmentary, and often contradictory ways."

> [W]e struggle to achieve some sort of understanding of AIDS, a reality that is frightening, widely publicized, yet finally neither directly nor fully knowable. AIDS is no different in this respect from other linguistic constructions that, in the commonsense view of language, are thought to transmit pre-existing ideas and represent real-world entities yet in fact do neither. . . . Rather the very nature of AIDS is constructed through language and in particular through the discourses of medicine and science; this construction is "true" or "real" only in certain specific ways—for example, insofar as it successfully guides research or facilitates clinical control over the illness. We cannot therefore look "through" language to determine what AIDS "really" is.[68]

In her analysis of various media campaigns around AIDS, Treichler was not interested in distinguishing "true" representations from "false" ones. This was a futile exercise in her view. Posters, like other representations, function very much like "truth regimes," she argued, borrowing a term from Foucault to describe the circular relation between truth, the systems of power that produce and sustain it, and the effects that power induces and, in turn, reconfigures. Truth, in this sense, is always power, Foucault had claimed. For Treichler, media accounts of AIDS conform to such regimes; they come to seem familiar, or true, because they "simultaneously reinforce prior representations and prepare us for similar representations to come."[69]

Treichler, then, in the wake of Foucault's interrogation of the rules at work in a society that distinguishes "true" representations from "false" ones (an alternative to the Sontag-like pursuit of fighting for or against a particular truth), aimed at finding an "epistemology of signification" for AIDS. She wanted "a comprehensible mapping and analysis of AIDS' multiple meanings—to form the basis for official definitions that will in turn constitute the policies, regulations, rules and practices that will govern our behavior for some time to come."[70] Her concern was with linking the postmodern epistemology of disease to the social practices around it—in fact to turn her "epistemology of signification" into a manifesto for all future social action. This agenda appears to stem from her realization as an activist that there exists an uneasy relationship between the postmodern celebration of fragmentation and applied social policies which tend to build on rather fixed, stereotypical images of AIDS.

Treichler argued that an epistemological rearrangement of social practice is urgently called for, especially in relation to AIDS in Africa because of the sheer scope of the problem there. In her article "AIDS, Africa and Culture Theory" she analyzed the multiple meanings of AIDS through various visual materials, such as posters, brochures, pamphlets, and other printed items issued by statutory and voluntary organizations in different African countries. Very much like in the West, she observed, fierce "battle[s] of signification" were involved Hence she urged that the analysis of visual and printed materials be rooted in their specific local contexts of production and presentation, that is, within local traditions (including healing ones) and local practices of viewing. Ultimately, she concluded, it is only such material conditions that give meaning to the various conflicting codes that are perceptible in any representation of the body in health and disease, although she herself never left the discursive domain around representations.[71]

Closing the gap between postmodern epistemology and social practices was an ongoing concern in Treichler's work. For example, in an earlier article, "Beyond *Cosmo:* AIDS, Identity, and Inscription of Gender," she concurred with a colleague writing in 1992 that

> postmodernist fragmentation and dispersion do sometimes deflect attention from realities that *should be* brutally (rather than strategically) essentialized. . . . [T]o embrace fragmentation uncritically runs the risk of duplicating the move to a market-driven consumerist model of human populations in which the fragmentation of conventional identities is a fine art.[72]

In this we can hear an echo of Susan Sontag, but it is transmuted by a generation whose experience (and lament) was of a world dominated by commercialized visual media, and whose intellectual coping with that world was heavily informed by Foucault's body/knowledge notion of power.

Conclusion

While it might be assumed that health posters came into intellectual focus in the 1980s and '90s simply because there were more of them (especially around AIDS), this essay has argued otherwise. In explanation it has singled out two important shifts in the wider intellectual and cultural context of Anglo-American society during those decades. The first was a re-conceptualization of power. Older notions, dominant among academics from the 1960s, based on Marxist theoretical agendas, were increasingly rivaled and rendered unfashionable through the elaboration of new concepts emerging mainly from Foucault.[73] Particularly important in fueling new discussion (not least on normative "regimes of truth" around public health) was his body-centered notion of power as a positive, creative force.

The other shift, broader still, was the dramatic visualization of virtually every aspect of human life. As a result of ever-more, and increasingly sophisticated media technologies, contemporaries came to perceive modern life as "tak[ing] place onscreen."[74] Warranted was a new discipline ("visual culture") to deal with the overkill that stemmed largely from high-powered global advertising companies, unshackled since the 1980s through the ideological promotions of the "open market." Arguably, the visual marketing skills of these companies invaded, appropriated, and shaped all aspects of cultural life, including medicine and public health, the two areas long perceived as outside the world of commerce or regarded as "humanitarian" and reserved for state and voluntary bodies alone. Public uneasiness over this is reflected in the reaction to Toscani's "Dying on AIDS" poster-ad of 1992 to promote the sale of commercial goods.

It is this dual shift—that of power concepts and their focus on the human body, and the increasing visualization of all areas of Western culture—that moved health posters into focus. They suddenly seemed a challenge, and around them intellectuals developed new vocabularies to bind them to discourses which simply did not exist in the 1970s when Susan Sontag wrote her essay on posters, and which were hardly countenanced among historians in 1985 when Allan Brandt published his poster-illustrated history of venereal disease. By contrast, in the work of Sander Gilman and Paula Treichler in the

Coming into Focus   135

1990s—but two representatives of the trend we have outlined—that discourse had become all but hegemonic, despite the substantial differences between these authors' particular approaches to the visual.

Today, postmodern anxieties and theorizations over contemporary life no longer have the potency they had in the 1980s and 1990s when first put forward by Treichler and others—not unlike anxieties over AIDS itself in the West. As an intellectual movement, postmodernity has suffered much the same sterilizing fate as other such movements before it.[75] This does not mean, however, that we can go back on, or erase, the understandings gained through it. Nor does there seem any reason why we should, since many of the key features that postmodernists identified (not least in relation to public health and commercial advertising) are all the more prevalent today. For example, the recently formed Joint United Nations Programme on HIV/AIDS (UNAIDS) is largely constituted by major international media and entertainment companies whose function is to disseminate information on AIDS and reap company credibility thereby.[76] Furthermore, where AIDS and STD prevention has not been commercially franchised, it has been devolved by governments to charities whose campaign publicity now often becomes indistinguishable from the slickest commercial advertisements (including the play with well-informed consumers). For example, in 2006, to raise new awareness to STDs, Europe's largest such charity, Britain's Terrence Higgins Trust, ran a mix-media campaign which was conceptualized and produced by a professional advertisement company.[77] The campaign included posters, of which Figure 5.4 is an example. According to the Trust's chief executive, the campaign was its most successful to date.[78]

What changed, then, was not the drift of "late-modernity" identified by postmodern commentators, but only the urgency of worry over it. We may have become complacent with our lot, or may simply have moved on to other pressing matters—"climate change," "terrorism," "globalization," and so on. Where then does this leave the visual in the history of medicine? This paper has sought to indicate that the use of such material must be guarded, but not because, as conventionally claimed, "looking practices" are different from "reading practices."[79] Indeed, it should be clear from our discussion of Gilman's choosing to unveil symbolic messages in AIDS posters, that the very notion of "reading" pictures is itself a construct of a particular (semiotic) agenda, which is riddled with all sorts of background (largely structuralist) assumptions about the nature of "reality" and the means for its "decoding."

Fig. 5.4. Poster used in promotion of STD health by the Terrence Higgins Trust illustrating the seeming collapse of commercial and medical humanitarian aspirations. By permission of the Terrence Higgins Trust.

Rather, historians seeking to use visual materials need to be aware that any instruction as to their use is a priori discourse laden. The coming into focus of health posters in the 1990s and visual culture in general as something of seeming great importance, something for serious critical engagement, is but a perception of one particular socio-cultural moment (precisely that which we have sought to outline in this paper). It is a discursive regime, not a universal truth. Historians of medicine should by all means be encouraged to pursue the abundant visual objects in their field, and treat such objects in terms appropriate to the context of their production. A multiplicity of approaches is also to be encouraged. But they should do so with awareness of, and open candor towards, the discourses around the visual from which their approaches derive. That is, they should be attentive to how their mindedness to the visual has been informed. Not to do so is to be in danger of practicing the equivalent of retrospective disease diagnosis—a reading backwards from present medical knowledge, blithe to the positivist epistemology embedded in the exercise [see Chapter 7]. In other words, the discourses around the visual need to be seen as just that, and not as objective tools for historical analysis that can be grasped uncritically. In short, the purposes to which images might be put in the history of medicine is less relevant than cognizance of the epistemology of perception embedded in the claims *for* visuality. Without bringing the latter into focus, the visual in the history of medicine seems destined to remain out of focus.

6

Visual Objects and Universal Meanings
AIDS Posters, "Globalization," and History

THIS ESSAY, ALSO CO-AUTHORED with Claudia Stein, furthers the focus on health posters of the previous chapter, although here dealing exclusively with those on AIDS. Again the concern is with the "visual turn" in relation to history-writing, but added on is the problematic of the spatial turn, which is explored in terms of both the conceptualization of "the global" and that of the physical aesthetic arrangement of objects in museum spaces. Supplementing the question of how visual objects came into being as historical objects is that of how change in their meaning accords with the spaces such objects come to occupy. Our thinking was triggered by the trend in museum display during the first decade of the new millennium from national to global perspectives. This rearrangement of objects did not eradicate national pasts so much as celebrate the new neoliberal economic world of connections, trade, "exchange," and universalism. Objects arranged to fit this ideology came to confirm its rightness historically. The fulcrum of this essay, therefore, is more political than epistemological. In particular, two kinds of politics concern it: those around globalization, and those around aesthetics. Neither could be said to be analytically self-evident, and in truth, the latter came into view only in the course of our thinking about the former.

The takeup among historians of the theme of globalization had long irritated us for the fig leaf it provided for global capitalism, whether it was intended to do so or not. As in the speak of Western governments, globalism in history-writing tended to be projected as if (in and of itself) it was some kind of benign force for progress, a discourse that seemed to us always to sidestep the issue of the exploitation and destruction of peoples in the ideology's train, as well as avoid vexing questions over the postcolonial insistence on difference.[i] For historians to take it up as if it were unproblematic struck us as an abnegation of critical responsibility. Yet it was largely as such that it was promoted as the agenda of the Wellcome Trust Centre for the History of Medicine at University College London, where my patrons, the Wellcome Trust, relocated me in 2002 (and where Claudia was an associate). Its director, Cook, posed global perspectives on medicine and its material culture as an alternative to the social history of medicine, the heyday of which was associated with the earlier reign at the same institution of the late Roy Porter. But exactly what kind of an alternative the global was supposed to be was never spelled out. Robust problematization of the concept was shied away from. It was only toward the end of Cook's tenure, driven in part by the need to renew the bid for the Centre's continued funding, that it came in for concerted questioning. This was done first through an in-house reading group, and then through an effort to compose a volume of essays on the subject by the Centre's staff. This essay was originally written for that volume. But the volume never materialized; the funding bid was barred, and the Centre was dissolved. The essay was then revised and published in *Medical History* in January 2011.

The other politics, those of the aesthetics, or, more precisely, those of museum display and archival storage, came into focus largely as a result of the particular illustrative means deployed in this essay: a globally themed exhibition of AIDS posters held at an arts and crafts museum in Hamburg in

2006. It dawned on us that the politics of aesthetics, no less than those around the global, needed attention from historians who were thinking to write on, or innocently to use, material objects displayed or stored in museums and archives. What the exhibition in Hamburg alerted us to was the political afterlife of material objects (such as our health posters) once they had departed their first life in other public domains. We became aware that not only do institutional spaces have their own politics embedded in aesthetic traditions as much as in policy decisions, but, simultaneously, they entertain the cultural politics of their times—local, national, and international. Moreover, the construction of such afterlives was much more complex and less obvious than we had assumed. As we came to realize, contrary to postmodernist expectations, today's application of aesthetic display for the purposes of making global connections does not necessarily break radically with the virtues and morals attached to the visual in institutions that were shaped in the nineteenth century for local and nationalistic purposes. Hence attempts to write historically on visual objects deposited and/or displayed in such places needed to take into account the complicated mix of changes and continuities in aesthetic concepts and political inscriptions. If historians didn't, then they could easily fall prey to seductive aesthetics, a slide akin to what Max Horkheimer once decried as the collapse of theory into art.[ii] Sites for viewing are never just sites for viewing, in other words; they are also political and politicizing spaces made and remade. Moreover, older interpretations continue to exist and shape newer meanings.

Entering the door of a nineteenth-century museum in the twenty-first century was also a further step to thinking about late twentieth-century critical theory in relation to the politics of history-writing. In seeking to draw out some of the political features of the early twenty-first century it was virtually impossible for us not to reflect historically on the construction of postmodernism's own episteme, as we had begun to do in the essay in the previous chapter. In turn, the path was opened to concern over the emergence of a new post-postmodern episteme, which was both building upon and substantially deviating from postmodernism. Within this new episteme, for reasons that transcend mere "political" ones, historical critique becomes increasingly absented. In retrospect, the door was opening to this volume's last chapter, on another means to the dissolution of historical critique.

Among striking developments in history-writing towards the end of the twentieth century were moves to the spatial and the visual in the reconstruction of historical consciousness. "Global history," accelerated through American hype on "globalization," and "visual culture studies" propelled by society's increasing reliance on visual communications, became fashionable projects.[1] "Global history," grounded in contemporary economics and partly pitched in reaction to Western parochialism, came seriously to challenge nationalistic and structuralist approaches to history.[2] At the same time, the "pictorial turn," with its epistemological claim for vision as the prime sense in knowledge production, came to defy not only conventional history of art, but the status and value of the discipline of history itself.[3] Wrestling with these turns proved enormously productive. In the history of medicine it led to foregrounding disease in its global dimension, revivifying, at the same time as challenging, older narratives on the "worldwide" spread of disease.[4] In the history of science it led to heightened attention to visual representations in struggles over the production of scientific knowledge and authority.[5] Both turnings did more than merely provide historians with exciting new conceptual frameworks for comprehending the past. As illustrated through the contemporaneous growth of a literature on the history of material objects in all their global distribution, they also helped broaden the range of objects deemed worthy of historical attention.[6]

This essay draws on these moves in relation to one particular material object that is both "global" and visual: the AIDS poster.[7] Their worldwide production from the mid-1980s hugely reinvigorated the whole genre of the health poster and, according to one expert, restored its original function as a communications medium.[8] Certainly, as never before was so much money, aesthetic effort, and psychological marketing put into this particular media on the part of voluntary bodies, national governments, and international health agencies.[9] Our concern, however, is with the wider conceptual frameworks that were mobilized to make these objects meaningful. In the 1980s and 1990s [as outlined in Chapter 5], a cohort of Western intellectuals concerned themselves with them along with other representations of AIDS in order to talk about the politics of identity. Those concerns, in turn, were linked to broader ones emerging at the time over the rights of citizens to equal access to health care, the privatization of medicine, and the role of the international pharmaceutical industry in the commercialization of health care. Here we focus on another aspect of these ephemeral mass-produced objects: not their "active life" on the streets and in the corridors of learning, but their "afterlife" when they were turned into items to be collected, exchanged, and stored in museums and archives. It is well known that the social life of material objects in such places is not the same as that of their initial culture of production, circulation, and consumption.[10] Museums and archives, like other depositories for images and artifacts, have particular collecting agendas and particular institutional and intellectual traditions into which new acquisitions are fitted. They also inhabit the present, embracing wider conceptual contexts that serve further to shape the organization and meaning of their artifacts.

We want to explore one such "afterlife" for AIDS posters: an exhibition entitled "Against AIDS: Posters from Around the World," which was held at the Museum für Kunst und Gewerbe in Hamburg between February and April 2006. We do not wish to make causal claims for the "importance" or "impact" of the exhibition; our interest in it is, rather, as an illustration of the more general trend towards the "global" rearrangement of material in museums by the twenty-first century. The Museum für Kunst und Gewerbe was founded in 1877 in a spirit of aesthetic modernism and German nationalism, and, as it happens, by the 1890s was host to one of Germany's largest and most prestigious poster collections. It was intended as a place to celebrate the people's arts and crafts, much like the South Kensington Museum in London, established in 1852 and renamed the Victoria and Albert Museum in 1899 at the height of British jingoism. As at the Hamburg Museum für Kunst und Gewerbe, so at the Victoria and Albert Museum over the past few decades, objects have been reorganized for exhibitions accentuating "the global."

The representation of AIDS posters at the Hamburg exhibition provides us with a means to discuss the politics of such "global assemblages."[11] On the one hand, it permits us to draw out the inherent contradictions and tensions that can be involved in any such institutional mobilization of the concept of

Visual Objects, Universal Meanings 143

"the unity of the globe."[12] On the other, it allows us to underline important continuities hidden under the more apparent or insisted upon "discontinuities" between national and global discourses, as well as between modern and postmodern politics of aesthetics—continuities rooted, we argue, in *shared* aesthetics values.[13] As important, the example permits us to reflect on how the discourse of the global affects the work of historians using material objects in their constructions of historical consciousness. As these aims and objectives should suggest, we are not concerned here with how viewers might have responded to the images or to the exhibition as a whole (an almost impossible task in any case given the uniqueness of individual psychology and experience).[14] Nor are we interested in providing a walk-through critique of the exhibition. Our main interest is in the historical context of the Museum and how this bears on the politics of aesthetics implicated in its exhibition of AIDS posters.

The Hamburg Exhibition

"Against AIDS: Posters from Around the World" was a modest, low-budget affair. It was staged mainly in order to exploit the Museum's recent acquisition of over a thousand AIDS posters from a private dealer, a purchase that enabled it to join the club of institutions harboring such collections.[15] The organizers of the exhibition selected only a hundred of the posters to display, choosing those that were most visually arresting, and others that, even after three decades in some cases, still had the power to shock, titillate, and/or amuse. In part to enhance these effects, realist anti-homophobic posters (e.g., Fig. 6.2) were mixed with erotic "body-beautiful" ones, such as "Semen Kit" (Fig. 6.1). Posters were also connected through novel imagistic and ironic associations. "Semen Kit," for example, joined nautical space with a poster of a condom disguised as a life-saving ring, which, in turn, was hung alongside a Russian poster-advertisement for rubber tires.[16] These striking juxtapositions were intended to demonstrate the variety of aesthetic choices that governments, charities, commercial bodies, and private artists employed in their efforts to inform the public of the threat of HIV/AIDS and incite onlookers to ethical behavior (safer sex). Dramatically, at the entrance to the exhibition, the visitor was confronted with a full billboard-size reproduction of Oliviero Toscani's iconic 1992 image of the death of the American AIDS activist David Kirby turned into an advertisement for the United Colors of Benetton [see Chapter 5].

Other less dramatic images played on popular solidarities around AIDS, as prefigured in the socially integrative "Against AIDS" in the title of the exhibition. These images could serve to counter any charge that might be leveled at the Museum for its use of the more erotic and more humorous and ironic ones, namely, that it was denying the pain and suffering of AIDS victims, or trivializing the world's most devastating disease. "Against AIDS" was also literalized in the predominance of posters promoting the use of condoms. It was through the display of these in particular that the exhibition sought to exemplify regional variety and similarities in aesthetic styles. Condoms, the viewer might come to see, were a wholly unambiguous global symbol for, and *the* warning against, unsafe sex.[17]

Aesthetic Framing

Although a flyer, but no catalogue, was produced for the show, the aesthetic and intellectual motives behind it are discernible through the catalogue that accompanied a much larger exhibition of posters held at the Museum in 1996 (which was curated by the same person, Jürgen Döring, and displayed some of the same AIDS posters).[18] The 1996 exhibition marked the centenary of the Museum's first-ever exhibition of posters—a late nineteenth-century entertainment that was in fact the first of its kind in Germany and one of the first in Europe.[19] In many ways the 1996 exhibition was faithful to that of 1896, its agenda more or less the same as that articulated by the main advocate for and co-founder of the Museum für Kunst und Gewerbe, the Hamburg lawyer and art critic Justus Brinckmann (1843–1915).[20] In his catalogue for the 1896 exhibition, Brinckmann proclaimed that posters and their display in museums and galleries perform the "ethical task" of elevating the masses to aesthetic appreciation.

> Art should be accessible to everyone. It should bring elevation and joy to all; not only to those who buy it or have the time to visit art galleries. In order to fulfill this purpose art has to go on the street, and has to cross— as if by accident—the path of the many thousands going to work who have neither time nor money to devote to it otherwise.[21]

These were not the only politics in the 1896 exhibition with resonance for the one in 1996, despite the significant contextual differences. The wider context at the end of the nineteenth century was that of flourishing national rivalries as well as attitudes to competition as a virtue in itself—be it between regional institutions, or between the military might and/or levels of "civilization" of different nation states. The posters in the 1896 exhibition, from Italy, France, Russia, the United States, and elsewhere, were organized accordingly and (typical of most exhibitions at the time displaying products from different countries) were competitively judged by an international panel. The exhibition thus served to cohere national identities at the same time as sell the ideology of competition along with the virtues of the products of mass production for mass (visual) consumption. More than this, while on the one hand the 1896 exhibition promoted the promise of democracy embedded in the notion of arts and crafts "by and for the people," uniting people through aesthetic education, on the other, it sold an elitist aesthetic ethics that simultaneously challenged traditional elitist views of what constitutes "art" and who "properly" can access and comment on it.[22]

A century later many of Brinckmann's views were dusted off. Jürgen Döring, the curator of the 1996 exhibition (and subsequently the one of 2006), similarly reveled in the non-elitist engagement of poster art, at the same time as celebrating the particular aesthetic distinction that Brinckmann drew between mundane commercial advertising and sophisticated poster design.[23] The 1996 exhibition visually illustrated this difference by juxtaposing examples of each. Above all, Döring celebrated and reproduced Brinckmann's ethical mission to teach people how to appreciate and enjoy the visual world through a better understanding of its underlying aesthetic principles and techniques of production.[24] This ethical educational work of the poster is all the more important today, he argued in the catalogue of 1996, because the fine arts have lost the main task they had when Brinckmann was living, namely, to interpret the world in its visual parts and provide the onlooker with an edifying understanding of it.[25]

The 2006 exhibition of AIDS posters was conceived in the same intellectual framework, according to Döring.[26] It, too, was intended to elicit an emotional response from the viewer and, from that, heighten their sensitivity to art.[27] But while Brinkmann in his day had looked forward to a brighter future for the poster and its onlookers in museums such as the Kunst und Gewerbe in Hamburg, Döring and his colleagues found it hard to be quite so optimistic. In their view—as maintained in the catalogue for the 1996 exhibition—the general quality of the visual language of posters had dramatically decreased due to the increasing flood of pictures in everyday life. Discriminating aesthetic appreciation, they felt, was less and less in evidence in contemporary culture because modern education failed to teach the classics and its iconography. Consequently, despite the fact that posters were intimately a part of the revolution in graphic design and design technology of the 1980s and 1990s, they had become impoverished—superficial, because they could no longer be "properly" designed or read.[28] Hence these objects reflected a *lack* of social and moral responsibility. The apparent proof of this lay in the popular media's focus on the human body or, more precisely, on superficial beauty and "feelings" around the human body. According to Döring and his colleagues this was the *Zeitgeist* of our times that "exceeds rational understanding."[29] Today's fashionable heroes hardly transmit any moral values, they lamented; indeed, they did the opposite—dissipating, through preoccupations with individual self-fulfillment and superficial gratification, the canon of Christian ethics and traditions of virtue behind "good art."[30]

AIDS posters apparently shared this fate in having behind them no shared aesthetic appreciation. According to the 1996 catalogue, they could communicate no meaningful visual language for the deadly disease they referred to because the health educationalists who issued them were themselves unable to formulate one. Hence, the iconography of AIDS posters appealed predominantly only to the *feelings* of onlookers. All that could really be said in their favor was that they sponsored a positive moral practice, namely caution against the spread of HIV: "the onlooker feels attracted by the picture and decides to use a condom."[31] Although it seems odd that the curators of the Museum für Kunst und Gewerbe disdained the very objects they were celebrating, they justified themselves on the grounds that such posters were expressive "of their times" and therefore wholly within the remit of an institution whose agenda was to archive and exhibit the art of the streets. By the 2006 exhibition "the times" were global, and hence a new justification was established through the parade of moral values within that discourse.

In maintaining that contemporary culture and advertising were saturating the world with meaningless, morally deprived, corporeally fixated images, the curators of the Museum für Kunst und Gewerbe were maintaining a distinctly "modernist"—as opposed to postmodernist—view of "art." In fact, in believing that the mechanics of perception operated along the lines of emotional attraction, followed by rational thought, and then enlightened responsible behavior, they were following nineteenth-century notions of the physiology of seeing and sense perception.[32] They were also assuming that onlookers were simply passive vessels for ethical education in visual good taste and, moreover, that "good taste" was the highest form of human awareness. Interestingly, in many of these respects—not least in connection with the understanding of disease—they were sharing an outlook with one of the few intellectuals in the twentieth century ever to provide sustained commentary on posters, the late Susan Sontag (1933–2004), despite that Sontag, as a theorist, was to their left in embracing the Marxism of the Frankfurt School of Critical Theory.[33] Like her [whose views on posters and on disease are discussed in Chapter 5], they too believed AIDS to be "real"—that is, to be a universal "medical" problem that was simply something for science yet to solve. As if to underline this understanding of AIDS as medical problem and nothing more, the 2006 exhibition was accompanied by health education information issued by various private and national AIDS agencies. Significantly, in both the flyer for the 2006 exhibition and in the section on AIDS in the 1996 catalogue, a medical discussion of AIDS preceded that on the aesthetics of AIDS posters. Thus a distinctly modernist view of art was accompanied by a distinctly modernist view of science and medicine as unquestionably superior forms of consciousness and practice, even to art.

Re-Picturing AIDS

How people interacted with AIDS posters during these images' "active life" on streets, buses, billboards, underground trains, and so on, and how the power and fear of AIDS operated in relation to identity, were simply not a part of the 2006 exhibition. Through the literal framing of the posters—their hanging according to the conventions of art galleries, their arrangement (three or four to a single wall), and the choice of them in terms of the quality of their visual language (*Bildersprache*)—the organizers denuded them of the local contexts in which they were created and in which they were engaged with politically, intellectually, and emotionally. This absenting can probably be illustrated by paying close attention to the history of any one of the posters in the exhibition. For present purposes, to make the point, we focus on only a few of them: "Semen Kit" (Fig. 6.1), "Stand Up Against Homophobia" (Fig. 6.2), and Toscani's advertisement for Benetton.

The first two of these were among the five British examples of AIDS posters in the exhibition. Both were issued by AIDS charities (as were the other three British examples in the show), and were relatively recent. "Semen

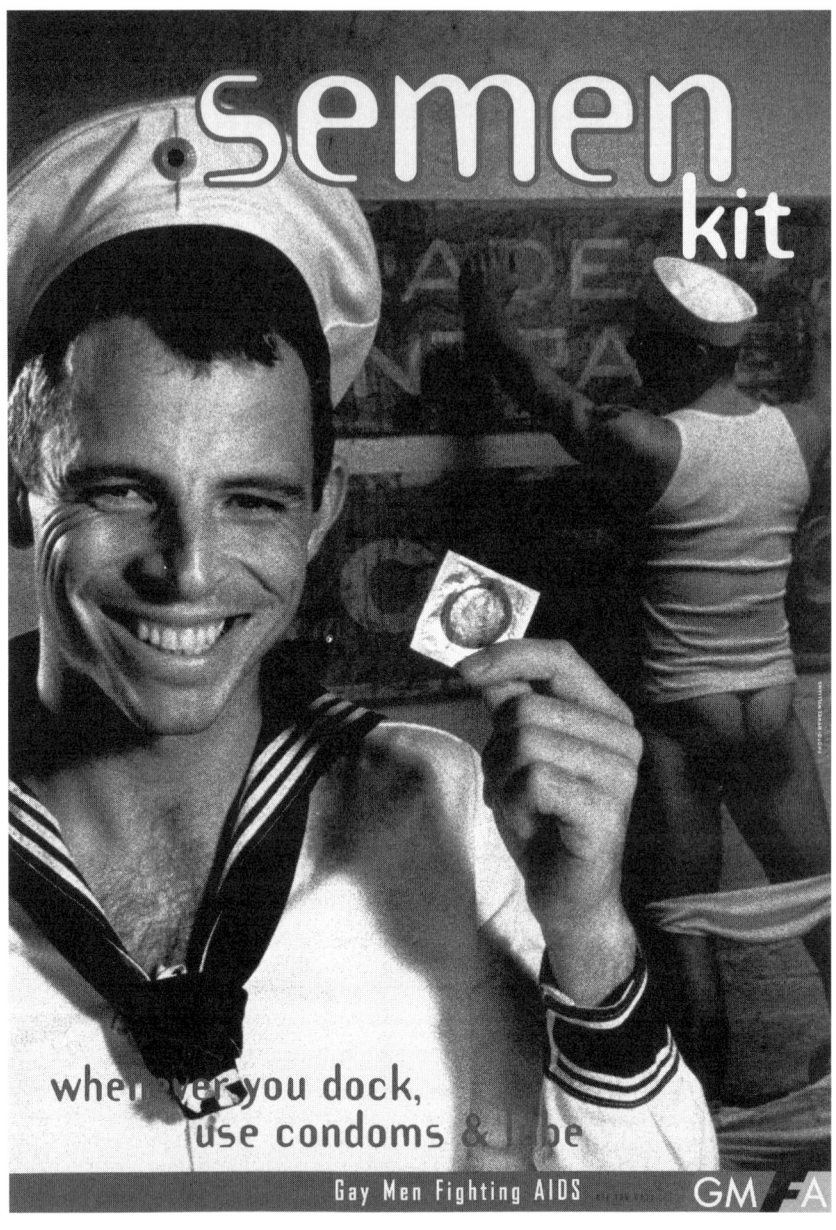

Fig. 6.1. Promotion by Gay Men Fighting AIDS, London, 1994, Hywel Williams photographer. The image was reproduced on a 15 cm card and handed out at the gay pride festival in July 1994. The text on the card mentions the UK government's poor funding of AIDS campaigns and Parliament's homophobic reaction to the proposal of an equal age of consent of sixteen for both hetero- and homosexuals. By permission of the Wellcome Library.

Visual Objects, Universal Meanings 149

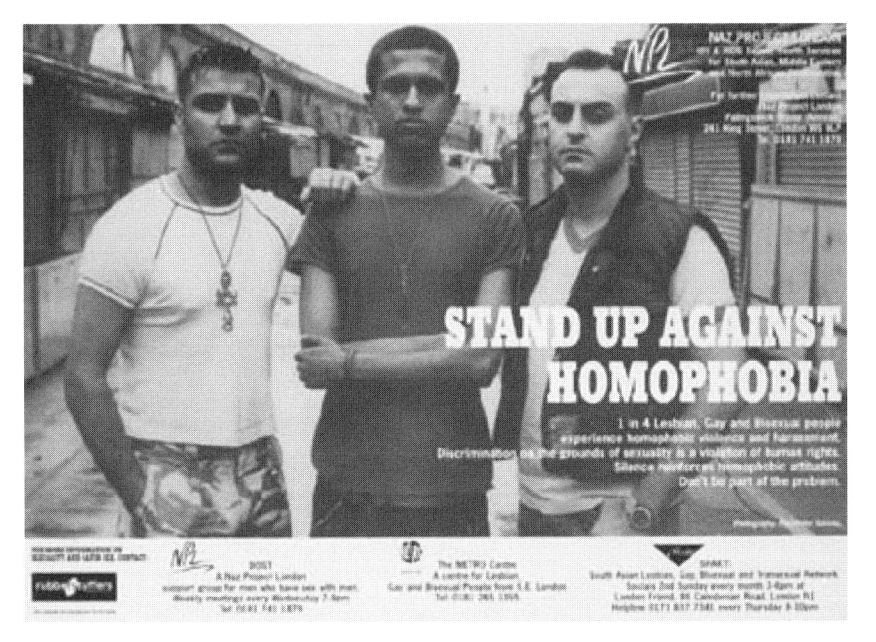

Fig. 6.2. Poster used in promotion of safe sex by the Terrence Higgins Trust, 1999. By permission of the Terrence Higgins Trust.

Kit" (1994) was produced by GMFA (Gay Men Fighting AIDS) established in 1992, while "Stand Up Against Homophobia" (1999) was distributed by the Terrence Higgins Trust (established in 1982, and by the 1990s, Britain's leading HIV and sexual health charity). That they were not issued by the British government is important, for they were in fact conveying messages alternative to it. In 1986, when Thatcher's government announced its intention to spend £20 million over the next twelve months on HIV/AIDS health "information," and commissioned the advertising agency TBWA to undertake it, the resulting "Iceberg" and "Tombstone" images were crafted to sell a health message to a general audience and to meet the government's inflated interest in family values, heterosexual sex, and nationhood.[34] Condoms were not then in the frame, nor were gays, and nor was "the global." The story of how the "patriotic heterosexual imaging" began to be turned around in the 1990s does not concern us here.[35] As in most Western countries, it was gay men—the first victims of the disease—who initially confronted the benign, sexually prudish and denying images of the establishment, and then inverted the anti-liberal homophobic rhetoric generated by "the gay plague" in celebration of their own collective identity ("the gay community"). It only needs observing that in the UK this development was significantly different from elsewhere; bearing on it were the particularities of social and cultural traditions as well as prevailing political, medical, and legal discourses. Pre-existing affirmative representations of "gays" in the public media were far less pronounced than in some other places, while the ability of the tabloid press and the Conservative government of the day to stir homophobia was greater.[36] Prime Minister Margaret Thatcher's notorious homophobic legislation of 1988 (Section 28 of the Local Government Act forbidding the promotion of homosexuality), in particular, did much for heightening alterity, struggle, and confrontation.

"Semen Kit" and "Stand Up Against Homophobia" exemplify and reflect this particular legacy. Yet it was precisely these features that the Hamburg exhibition eclipsed by having the images hung alongside foreign advertisements and alongside AIDS posters from China and elsewhere. The exhibition thereby performed not for the struggle for gay identity within a national context, but for world solidarity. "Stand Up Against Homophobia," both by virtue of being a product of 1999, and by presenting AIDS as a part of a wider social issue, served in itself to mask the historical significance of these images. "Semen Kit" performed similarly through its play to a would-be unproblematic history of overt sexual behavior among men (iconized in the image of the sailor), and by its appropriation from some American AIDS posters of the by then more-than-a-decade-old image of the gay "body beautiful."[37] That gay men in Britain, well into the 1990s, were still struggling for a viable visual public identity could not be guessed from the hanging of this poster at Hamburg. Nor could it be known from either of these images that the changes in gay identity that occurred in Britain in the 1990s were largely attributable to the uptake of a slick visual language borrowed from the culture of Madison Avenue–driven international advertising—the same visual language, from the same source, that was simultaneously and enthusiastically taken up by Thatcher and other politicians around the world in their electioneering.

But it was not only governments and gays in the 1980s and 1990s who struggled to capture the meaning of AIDS; so too did Western medicine. It was its lack of success at turning AIDS into a meaningful scientific category that, in fact, opened the medical profession to pointed confrontation, and

Visual Objects, Universal Meanings 151

opened up AIDS to the visual politics of identity. In effect, through AIDS, and in particular through the early debate over whether HIV caused AIDS, the medical profession was pulled off its pulpit as the authoritative arbiter of modern secular identity.[38] Arguably, Toscani's poster advertisement for Benetton reflects this by blurring the normative boundaries between public art and private anatomy/medicine, as well as by confusing the conventional distinction between commercial marketing and medical humanitarianism.[39] For Toscani, who regarded the David Kirby image as the most significant of those he designed for Benetton's "Shock of Reality" advertising campaign of 1992, it was supposed to show how an international corporation was "open to the world's influences and engaged in a continuing quest for new frontiers."[40] For him, this meant commitment to global social issues, above all, to the problems of poverty, race, and disease.

However, this was far from how the image was regarded by others, including gays [as noted in Chapter 5]. Here, then, was opened up yet another field of combat over the meaning of AIDS centered on its imagery, but which again there was no hint of at the show in Hamburg. Indeed, while Toscani's image still had the power to shock, it is doubtful if anyone outside of living memory who viewed it at the exhibition (such as the school children bussed in) could have guessed what it meant for many visual theorists and AIDS activists worrying over images of AIDS since the late 1980s, namely, that it was further testimony to the ongoing "crisis over the entire framing of knowledge about the human body," with AIDS, not just a "medical" problem, but "an epidemic of meanings or significations."[41] The exhibition in Hamburg gave no clue to how images like this had been regarded as testifying to a self-consciously postmodern culture, ethics, aesthetics, disease representation, and politics of identity. Nor was the reproduction of Toscani's image at the Hamburg exhibition intended to illustrate how the lines between "the commercial" and "the medical humanitarian" had become so obscured that there was now little to mark the difference between a "health poster" and an advertisement for the sale of fashion knitwear. Instead, despite that many of the images in the Hamburg show had been designed simply to sell condoms, and others, such as that on homophobia, to "challenge social injustice, prejudice and exclusion" rather than caution against AIDS itself, the exhibition cohered them all into a would-be historically uniform and medically mediated message about the struggle against AIDS.[42] Thus—although not with conscious intent—the show flew in the face of the preoccupations of the visual theorists and AIDS activists of the 1980s and

1990s. Where they had seen in representations of AIDS the postmodern play of signifiers, had argued for a plurality of subjectivities involved in visual engagements, and had construed visual perceptions of the human body in general as involving an onlooker's unconscious construction of his or her own body through the immediate act of viewing, the curators of the Hamburg exhibition saw only medicine and art in modernity.

This was not the only way in which the exhibition displaced the specific and general historical meanings attached to these objects by investing them with others. The title alone of the exhibition, "Against AIDS: Posters from Around the World," did as much. First and foremost, it constructed a particular framework for their perception, one that above all suggested that aesthetic form can travel the world regardless of local geographies and local histories of ethnicity, religion, race, rights, sexuality, and gender—not to mention alternative aesthetic traditions. This global aesthetic spin in effect harmonized a modernist Western transcendent notion of "art" with the late twentieth-century notion of a spatially transcendent capitalism—an economic system supposedly unfettered by place or national boundary.[43] The aesthetic spin and the exhibition as a whole thus further performed for notions of homogeneity and universality—attributes long associated with modernity and perceived to be at odds with the pluralities and fragmentations associated with postmodernity during the "time of AIDS" in the West.[44]

The exhibition's title also suggested that the history of AIDS was about everyone the world over being uniformly against AIDS. But that, too, was hardly the case at the point of production of many of these posters. In the 1980s, Christian fundamentalists and other religionists took a rather different line, and gay men and lesbians did not always see government campaigns against AIDS as being against AIDS so much as against themselves.[45] As noted above, in Thatcher's Britain the AIDS campaign was an occasion for the moral high-grounding of heterosexual values.[46] To the extent that people (and international pharmaceutical companies) were allegedly "against AIDS" in the 1980s and 1990s, their concerns emerged from a multitude of different and often conflicting social and economic interests. Moreover, the relative power of those interests was hierarchically organized, and differently so over time, as Virginia Berridge has made clear for the history of AIDS in the UK.[47]

Also, implicit to the entitling of the Hamburg exhibition was the idea that nations around the world were homogeneous in their fight against HIV/AIDS. This not only collapsed separate national encounters with HIV/AIDS,

Visual Objects, Universal Meanings 153

such as its very denial by South Africa's President Thabo Mbeki, but effaced the differences between the kinds of media campaigns used in different countries—including the often bitter struggles between local, national, and international agencies.[48] As important, this political gutting of AIDS posters through their aestheticization erased national rivalries and pressures involved in medically treating AIDS victims (or not, as was often the case). Through the mixing of posters from different countries, the exhibition dissolved the conventional boundaries between nation-states, while the multitude of images of condoms that it presented served visually to re-unite them around a commercial product. The images of condoms promoted the idea that campaigns for their use had actually united the countries of the world, a message curiously at odds with the Bush administration's contemporaneous funding of medical missionaries advocating sexual abstinence instead of the use of condoms.[49] In its own small way, therefore, and for its own particular didactic reasons (as well as, perhaps, discomfort over Germany's nationalistic past), the Hamburg exhibition engaged in the same kind of historical effacement and rewrite that the major international media companies were also entertaining by 2006 through *their* takeup of HIV/AIDS for purposes of reaping public corporate credibility. By then HIV/AIDS-funding was a fashionable cause, a benign branding resource for various Western philanthropic organizations.

At local, national, and transnational levels, then, "Against AIDS: Posters from Around the World" obliterated the individual history of the objects on display through a particular universalizing and seemingly neutral kind of aestheticization. At a closer look, however, it both appropriated them into an older script (a local and fondly held modernist epistemology of viewing and aesthetics) *and* a new one—globalization. By collapsing two decades of national histories into a singular and would-be unified world fight against HIV/AIDS, the history of HIV/AIDS was visually construed in terms of this new global subjectivity. Not only were particular constructions of the recent past left out (such as the local struggles around these objects) but, so too, the construction of the present—the global media industry's selling of itself through the attack on HIV/AIDS as a "global problem." Thereby, "globalization" was not only made "real" or made "true" through aesthetic representation of an ostensibly international struggle against HIV/AIDS, but by this same conceit medically re-appropriated and humanized—no matter that over the meaning of "globalization" there has been little agreement, let alone consensus on it as a "good thing."[50]

This is not to suggest that HIV/AIDS was not recognized as a global problem almost from its start. In 1987 the American AIDS activist, feminist, and visual theorist Paula Treichler, for instance, referred to AIDS as having "a potential for global devastation."[51] But for her and others in the 1980s "the global" was only a background problem (and rhetoric), and the term was yet without particular epistemic load. As we have indicated, Western nations and their intellectuals and AIDS activists were gripped more by their own campaigns, interests, and ideologies than by "global" concerns. In fact, the idea in the West that HIV/AIDS was a "global problem" was a viewpoint that itself had to be fought for through a worldwide media campaign brokered by organizations with internationalist interests. The campaign can be dated precisely to 27 May 1987 when the World Health Organization issued a press release proclaiming that "AIDS is a global epidemic that demands a global attack."[52] The WHO then produced a poster to sell the message (Fig. 6.3)—to compete, that is, with other struggles for the meaning of AIDS.[53]

The Hamburg exhibition abetted that project through a visual rhetoric of shared international struggle against HIV/AIDS, just as it unwittingly abetted the subsequent takeover of AIDS programs by the media multinationals.[54] Thus it eclipsed the history that would enable anyone to believe that AIDS had ever been anything *other* than "global," or for that matter, anything other than mainly a struggle for economic resources for better medical provision for victims of the condition. Images intended for, and often produced by, local sub-groups became artifacts for making up global citizens, with "the global" and "global citizenship" presumed good. Thereby, the exhibition did far more than merely reinforce what had become the "semantic hegemony" of "the global," as iterated through the banalities of "global warming" and "global terrorism."[55] Through aesthetics alone it rendered tangible the universalizing concept of the global. "The global" became something to experience, identify with, and embrace. Although this was not the exhibition's primary intention, it could not help but perform it, and in so doing contribute to a reconstruction of history and consciousness. There was no conspiracy in this. Far from it; the globally spun institution-serving celebration of the aesthetics of AIDS posters merely reflected and reinforced much of what everybody in the West had already come to "feel" about HIV/AIDS by 2006—that it was a serious worldwide problem about which everyone needed to be continually reminded.

Fig. 6.3. World Health Organization poster selling AIDS as a global phenomenon, 1997. This poster was translated into many languages. Courtesy of the National Library of Medicine, USA.

Reflections

We have been concerned with a particular moment in the history of AIDS posters: not that of their initial "public" life on the streets, or in pubs, gay clubs, doctors' offices, and so on, but their "afterlife" in places where they might be displayed or stored.[56] Through the analysis of an exhibition at one such afterlife location our intention has not been that of negative dismissal, nor critique for the sake of it. Nor has it been merely to expose how these often strikingly visual objects that aimed at protecting individual health had their meaning changed through appropriation into a conceptual framework different from that of their initial contexts of production and consumption. More interesting to us are the intrinsic and unremitting links between these visual objects and wider politics, or how the visual is inherently a part of the latter.

While it might have been supposed that AIDS posters came to political rest once they were retired, categorized, catalogued, and stored according to the principles of collecting institutions, the Hamburg exhibition proves otherwise. In fact, as collector's items they entered a space that was no less political than when they were on the streets in the 1980s and 1990s, and when they were appropriated to Western discourses on postmodern identity, and the role of the visual in the cultural negotiation of the self. It could hardly be otherwise, for simply by entering "retirement homes" such as the Museum für Kunst und Gewerbe they necessarily became a part of institutional agendas. In effect, here, as elsewhere, they were "framed" in agenda-serving classificatory narratives embedded in bricks and mortar. Indeed, from the moment such objects become collectors' items and are stored and/or displayed as artifacts they become epistemologically loaded through that very process of objectification. There is never an escape from the historical a priori.

Hamburg's Museum für Kunst und Gewerbe demonstrates that collecting agendas and accompanying aesthetic guidelines often have deep historical roots. But what the analysis of its 2006 exhibition of AIDS posters also shows is that old and seemingly *a*political agendas (invented to express specific national political interests) are neither lost nor rendered innocuous in the contemporary world. Rather, they come to serve new political frameworks linked to the world of today's visitors—a world in which aesthetics is the dominant means to a politics constituted on little more than the idea that "if it looks good go with it" (an outlook now as pervasive in the practice of science as in the arts of government).[57] Crucially, this new politics is sustained through,

Visual Objects, Universal Meanings   157

and for, the absenting of critique; not today the critical outlook entertained by Sontag and other pre-postmodern intellectuals, that visual representations (and popular posters in particular) covered up or cloaked lurking ideologies. Postmodernists, who for the most part were unconcerned with that view, in effect opened the space for the new politics of aesthetics that masks something different: the idea of aesthetics as void of political intention. The Hamburg exhibition of AIDS posters was, in fact, an early example of the coming-to-reign of these particular politics, with the visual alone being the vehicle for understanding and creating a ("global") community without distinction. Whereas for Brinkmann in the nineteenth century, aesthetics (in art) could be consciously *used* for the political purpose of populist democracy, with aesthetics and politics in clearly separate spheres, for the inheritors of his institution in Hamburg aesthetics (unbeknownst to them) *became* the politics, not simply a means to it. Ironically, their arrival at those politics—their unknowing performance of them through the exhibition of 2006—was through adherence to Brinkmann's legacy. Through that, Brinkmann's original political agenda was emptied of its original political purpose. Installed in its place were the politics of the *appearance* of political *un*-intentionality. An aesthetic concept born in the nineteenth century to serve nationalistic purposes thus came to operate for the political work of educating national citizens to global citizenship.

What can this mean for historians working with material objects stored in global-tending museums and archives and who are themselves now operating within a global framework? Since material objects have no meaning without a framework, and are framed in being collected, the simple answer is that historians have to take into account the afterlives of such objects as much as the object's original lives. Would that it were quite that simple, though. Harder is the problem of the historian's own place in "the global framework," which in many respects is not unlike that of material objects in the global-aspiring museum (and very much like the position of the curators of such exhibitions). "The global," whether avowed explicitly in "global history," or embraced implicitly in the practice of history-writing in contemporary culture, operates both politically and epistemologically. Just as global history's predecessor "world history" is now recognized by some of its originators as having been a product of, and agent for, its Cold War moment,[58] so for our own times dominated as they are by multi-national corporations and abiding politicians, the takeup of global themes in history-writing is widely recognized as providing, at the very least, legitimacy to a globalization discourse, even if, as often the case, the historian's immediate object has been the far from reactionary one of provincializing the West and critiquing its hegemony.[59] Of course, for some historians "the global" does politically more, overtly serving as a rhetorical strategy for the re-coherence of the discipline of history itself after its pummeling by poststructuralists, deconstructionists, and other fragmenting postmodern forces over the past thirty years or more.[60] To this end, what could serve better as a reunifying device than the holizing connective metaphor of the globe? Thus the global provides a new grand narrative—a universalizing tool—with which to reimpose the meta-narrativity of history. Although seemingly mindless of one of postmodernism's cautions, that totalizing worldviews can lead to totalitarianism,[61] these historians seek more or less intentionally what the curators of the Museum für Kunst und Gewerbe performed innocently through their exhibition of AIDS posters. In doing so, they also share company with certain art historians anxious to revive older agendas.[62] Yet neither global history nor visual culture studies need *necessarily* lead in this direction. To see the global as a discourse tied politically to institutionally specific agendas (such as the aesthetics entertained at the Museum für Kunst und Gewerbe in Hamburg in 2006) offers the possibility to take an alternative political position, or at least to liberate its historical study *from* such agendas.

Nevertheless, in terms of the less-visible shaping of historical knowledge production, the global framework can never be other than "world-making" as opposed to being simply historically descriptive.[63] No matter how hard we try to stick to the recovery of some truth of how the world came to be globally conceived (past or present), our knowledge productions implicitly reproduce and foster the unifying construct. Just as with material objects (or words or images) there is no meaning to historical events outside the conceptual frameworks we apply to them, wittingly or unwittingly. In short, there exists no resting place for history-writing; it is always already, at one and same time, fashioned *and fashioning*. Like creating a museum exhibition of posters to aestheticize the "global" nature of HIV/AIDS, the business of contributing to a global history of anything entails, by that very act, the politics of constructing a necessarily partial representation of the past.[64] As a product of its time, history-writing, too, in other words, cannot avoid making up historical consciousness.

This article cannot escape the charge that it too contributes to this process merely by discussing an event that was conceptualized in "global" terms. It might even be seen to compound the problem by drawing attention to spaces

Visual Objects, Universal Meanings

where, a priori, the historian is already politically and epistemologically implicated: the museum and the archive. However, in doing so it has sought to move the discussion beyond the tired call for attending merely to historical contexts, especially of material objects globally attributed. Our purpose has been to encourage historians to an awareness of their own immediate entanglements in history's constructedness—the constructedness of the present mediated in history-writing as much as through aesthetic assemblages of "the global."

The historian Aruf Dirlik, critically inquiring into the point of writing world history, has observed that "an awareness of the variety of world histories that have been constructed at different times and in different places . . . [must cause] any world historian worthy of the name [to] . . . be uncommonly aware of the constructedness of the past."[65] Similarly, we submit, all historians need to reflect on their contributions to the present—that is, to a culture given to re-enchantment through "the global." Quite how we should historicize material objects of the sort raised in this essay may be open to debate; what is not is the necessity to historicize our own historical projects. Otherwise we move perilously close to becoming blind participants in the historically fashioned spaces where memory is increasingly naturalized and neutralized through universalized and universalizing concepts mediated aesthetically. We end up, as it were, naïve viewers of the exhibition at Hamburg: as blind to the nature of the new post-postmodern politics of aesthetics, as to the modernist would-be universal humanity that "the global" unwittingly espouses through those politics.

7

The Biography of Disease

TAKING ON DIFFERENT VIEWS of history-writing doesn't mean abandoning ones that have gone before. This essay and the following one were not undertaken in the spirit of the previous ones, as quests in self-understanding in relation to the politics of history-writing. In terms of this book's narrative they mark ongoing sideways moves. Both were driven by a sense of responsibility to uphold critical standards in history-writing against unperceived tendencies to its backsliding. Conventionally historiographical, they scrutinize the politics of its writing and conduct some policing of its disciplinary boundaries. Both were also tasked with communicating to audiences outside academic history, this one for readers of a medical journal, the following one for bioethicists.

Published as a brief "Perspective" in the *Lancet* in 2010, its main purpose was to confront another form of essentialism in history-writing, the biographizing of disease, and the allied practice of retrospective diagnosis. Its warrant was a review of the first four volumes of a new monograph series on "the biography of disease."[1] At least three of the authors in the series were distinguished social and cultural historians of medicine, and their volumes were testimonies to it. Essentially they were committed to "framing" disease within descriptions of historical contexts, especially the social contexts of the reactions to the diseases in question. But the authors were inattentive to the politics of doing so in the context of their own times. It wasn't just that they were furthering bio-creep by adopting a literary conceit anthropomorphizing

The Biography of Disease

bacteria and disease categories. Through the implicit endorsement of these categories as historically transcendent, and through commitment to historical explanation through social causes, they were also working against a generation of scholarship in critical theory and in the history and historical epistemology of science that had successfully out-thought and historicized such positivist essentialisms and causal thinking. Historical epistemology's unveiling of fundamental concepts organizing and structuring the knowledge of different historical periods was directed to the present, and it extended to history-writing itself. But the contributors to the "biography of disease" series were oblivious to this; epistemologically, their monographs had more in common with positivist bio-science, and in this they were sharing the company of prize-winning journalists similarly enthralled by the cuteness of the idea of the biography of disease.[ii] The idea of the potential of history-writing for contemporary critique was simply not entertained.

I had half expected that some medical readers might take umbrage at my perspective. Ever since taking over the editorship of *Medical History* in 2009, I'd become acutely aware of how popular was the pastime of retrospectively diagnosing disease. I was surprised, therefore (though I shouldn't have been), when perturbed e-mails began arriving not from them, but from practitioners of history. One of them, a prominent social historian of medicine who was also the author of one of the volumes in the series under review, denounced me for performing a "disservice" to history—giving it a "bad name." This intrigued me; what was "good" history now presumed to be, and had this been universally agreed upon? Surely in the wake of the "death of the social" in postmodern discourse, there was room for some doubt. Naïvely, I had assumed that historians, even those disinclined to historiography and who might be ignorant of the scholarship on the place of disease categorization in historical epistemology, would have sided with the general drift of the piece: on the need to be sensitive to the epistemic warrants embodied in history-writing as much as to those embodied in the practices of science and medicine themselves. In retrospect, that was stupid of me; having satisfied myself as to the merits of postmodern critical thinking and historical epistemology, I had blithely assumed that my reasoning was endorsable by all. But clearly there were historians who did not agree with me and who had little inclination to question the aims and objects of their own enterprise. Many, perhaps the majority, were still thinking that history-writing was simply an exercise in the objective recovery of the facts of the past, and that the best means to this was through the study of social contexts. The historians' reaction was a wake-up call; I had to realize that not all of them were in the same boat as I was, or even on the same river. Indeed, from my newfound perspective, many in the social history of medicine appeared to be paddling backward.

I'd also forgotten something: that the most significant history wars of the past few decades between clinicians and historians, sociologists and historians, and historians and historians had been fought around disease and its representation. It was precisely here where theory was politicized, whether it was entertained as such or not. From Foucault's weighty intervention on the epistemological re-understanding of "disease" in the late eighteenth century (in *The Birth of the Clinic,* 1963), to the rise of the idea of the social construction of disease in the 1970s, to Charles Rosenberg's "representational framing" of disease and disease categories in the 1980s, and on through to Bruno Latour's *Pasteurization of France* (1988), the meaning and regard of "disease" had been at the center of heated debate.[iii] To think that I might write on this subject without re-stoking old fires was simply naïve. The science wars may have cooled in the light of the cultural victory of biology and neurobiology, but the history wars were not over, and it was now social historians who were on the side of the defenders of realism and empiricism. Clinicians, for their part, might take it all in their stride. As, in light of my article, general practitioner Martin Edwards was to remark tongue-in-cheek on the "incursions" of clinicians into the intellectual arena of historians: "We physicians offer hagiographic biographies of obscure nineteenth-century medical figures, triumphalist narratives of medical progress and—the most heinous offence—retrospective diagnosis of ailments afflicting historical characters."[iv] At least Edwards's characterization of the professionals in both fields bore some acknowledgment of the recent history of significant intellectual moves.

After the essay's publication I encountered a few further examples of the practice of retrospective diagnosis of disease on the part of professional historians. I have added these to the original text, since it entails no retrospective analysis or reconceptualizations on my part. I have also reinserted some of the original text that was deleted by the *Lancet* for reasons of space.

Ralph Waldo Emerson famously wrote "there is properly no history; only biography." What kind of history, then, is the biography of a disease? Or, for that matter, what kind of biography? Do diseases share the birth-to-death contours of an ordinary life, or the emotional and intellectual aspects of extraordinary ones? Epidemics might possibly fill the bill, and perhaps, too, the reactions to them. But it is hard similarly to think of diseases themselves in this way, even if we can point to an occasional "birth," such as Legionnaires' disease, or to a singular "death," such as smallpox. Like cholera (origins unknown) most diseases seem to live on forever, skulking around in the shadows of populations and occasionally rising up to be shot down by magic bullets where they exist.

So what kind of a metaphor is "biography" in relation to disease? The idea is not new; the bacteriologist Hans Zinsser famously pioneered it in 1935 in his "biography of typhus": *Rats, Lice and History*. But it remains a strange notion, at once appealing and deceiving. On the one hand, it cuts a disease down to human size, domesticating it. If diseases have lives like ours their comprehension is surely within our grasp. But on the other hand, what if diseases don't have lives like ours, neither private nor public? What if diseases are nothing like human life? What if, in fact, they only exist *as* concepts; or like hysteria and associated "psychosomatic" disorders, are merely social constructs? What then is the function of biographizing them? What is the point of this anthropomorphicism?

Words, John Stuart Mill observed, are the custodians of ideas. "Biography" and "disease" are certainly so, and like other words and concepts, they serve at one and the same time to open up and close down understanding. The notion of a biography of disease is interesting in that it simultaneously privileges a historical method (biography) and a particular intellectual construction of disease (as a life). Method and construction go hand in hand foreclosing on other modes of historical approach and other types of comprehension. Among the methodological alternatives to the biography of disease might, indeed, be that suggested by Mill—an approach that foregrounds the intellectual custodianship involved in the invention, or in the particular apprehension of certain disease concepts. "Hysteria," in its nineteenth-century assignment to gender inequalities, is a good example, even if to call hysteria a "disease" begs a slew of other questions. "Asthma" (which also begs many of the same questions) may exemplify for the late twentieth century how the idea of "risk" in relation to environmental pollutions is shored up. Such an approach stands in contrast to the biographical one in which diseases are treated as biological entities. By the latter reckoning, the signs and symptoms of asthma are the same across time and place; all that differs are changing theories of causation and treatment, and different contexts for experiencing the "disease." Thus the biographer's task is that of locating in the past the evidence of the disease as *we* know it, whether it was named as such or not—in other words, with the business of retrospective discovery and diagnosis. A person described in an ancient Chinese manuscript as suffering from the symptoms that *we* would identify as asthma is considered without doubt to be suffering from "asthma." Whatever the context of the disease, the disease itself is historically transcendent and loaded with agency.

Another example of the same kind of logic applies to the history of "obesity," although this is slightly trickier in that, unlike asthma or anorexia nervosa, the word can actually be found in medical texts as early as the fifteenth century.[1] Indeed, often deceptively modern-looking were the "medical," "social," and "aesthetic" concerns over it. Because of this, the conclusion is all-too-easily reached that "obesity" in the past was much the same as it is today, or that today's social concern with fatness is universal and historically transcendent. Consequently, the historian feels entitled to construct a linear history that runs to the present—a connected narrative that can even accommodate vast periods of no mention of "obesity" at all by confessing to "discontinuities" (a rhetorical device in history-writing that inherently strengthens the notion of "continuity"). Overlooked is that concerns with obesity in the past were shaped within wholly different moral regimes than those surrounding "overweight" today. These were often constituted around biblically interpreted "sins" such as gluttony, overindulgence, and sloth, or around historically shaped concerns over sexuality, fertility, and population. One cannot, therefore, use the available historical material on obesity to ask it contemporary questions. To do so is to press the past into the historian's own analytical frameworks. It is comparable to regarding concerns with sex (or the construct "sexuality") as ever-present and constant through time. No academic historian would do that today; it would be regarded as ahistorical, Whiggish, and presentist. Yet, when it comes to medically framed diseases or specific health conditions, this is exactly how many historians think. Born out of what might be called epistemic false consciousness, the practice constitutes an exercise in, and for, the essentialist regard of diseases and disease categories.[2]

The first thing to note about the biographizing of disease in history, then, is that it is foremost an engagement with the medical present, not the past. The illusion is that a history of a disease is being related, whereas in fact, *the present understanding of the disease is being confirmed*. It is *our* biological understanding of the disease that is running the show, not the pursuit of any historical understanding. Exactly the same occurs when historians seek to tell us something about the past through the medium of contemporary neurobiology.[3] Inherent to all such exercises is the underlying assumption that our culturally prized notions of scientific "objectivity" and "facticity" are historically transcendent, not historically fashioned. It is as if *our* notion of "empiricism" and *our* valuation of a scientific fact have always been around. In other words, the approach forgets that the enterprise of retrospective diagnosis is itself made up *in* modern science, as well as driven *by* modern aetiological knowledge. It is also propelled by the idea that the tools for retrospective diagnoses are ideologically and epistemologically neutral. The technique of DNA analysis of the bones of long dead lepers, for example, is never questioned in the course of directing attention to the historical "truth" of the death from leprosy that the technique is held to provide. The truths themselves are perceived as value-free bytes of information, not artefacts predetermined by the science and technologies involved (and often put to cultural and ideological use). The same transpires when the latest bio-archaeological, microbiological, and epidemiological test-kits are wheeled out to discover if black death in the fourteenth century really was the same as modern plague. In fact, all that is witnessed through such exercises is debate over the production of scientific knowledge and the prowess and politics of the technicians involved. Finding leprosy, or whatever disease in the past, merely proves the worth of the science through its would-be "historical" application. Exemplified is what the sociologist of science Andy Pickering refers to as "the scientist's account" in which "accepted scientific knowledge functions as an interpretive yardstick in reconstructing the history of its own production."[4] Far from being a "genuine" engagement with the past, it is wholly an exogenous indulgence with the science of the present, and an affirmation of the esteem of its reductive method. Behind it all is the fiction that it serves: that "epidemical evidence" is real, and just waiting to be discovered. Confirmed along the way, furthermore, is the fundamentally anti-historical notion that "reality" is not culturally constructed after all. Historians can therefore be dispensed with; only scientists are required to reveal the truth of the past. Far then from leading historians to think that science is a helpmate to understanding the past, the business of retrospective diagnosis within the biographizing of disease should lead them to the opposite conclusion. Ideally, it ought to lead them to the injunction that it is not only legitimate but *necessary* to pay close attention to the values that underlie all practices, be they scientific or historical.

This is not to say that we should banish the biographizing of disease; only that we should see it for what it is rather than the "history" that its pundits purport it to be. Nor should we necessarily dismiss the "evidence" that the practice turns up. In and of itself, this can often be fascinating—although, again, only in terms of our own cultural appetites. That said, the presentation of the evidence from retrospective diagnosis is always inherently condescending. It can't help but be. After all, *we* know what the guys in the past didn't, and since we have the biological "truth" of the disease, we can only laugh or cry at their ignorance. What other purpose can be served by the revelation, for example, that Aretaeus in the second century believed that "diabetes" was "a melting down of the flesh and limbs into urine" and that it was caused by "coldness and humidity, as in dropsy"?[5] How can it be other than scientific smugness to inform on our forbearers' belief that smoking tobacco was good for asthma, or that cholera was caused by miasmas, and so on and so forth? In stringing together such anecdotes, the biographical history of a disease is not the history that the inhabitants of the past understood; it is ours. Sadly, in the course of transporting our scientific mentality, authority, and morality into such histories, understandings peculiar to the past are often misappropriated and historically short changed. The meaning of the "signs" and "symptoms" of a disease in the early modern period, for example, is not the same as ours, although biographies of disease routinely assume it to be. Left out or ignored is that these sets of signifiers were embedded in, and integral to, other kinds of cultural discourse, usually religious and astrological. It is to them, not to later positivist medical discourses, that the historian needs to be alert.

In fact, the only actual history of a disease left to its modern biographer is some version of its social, political and economic impact. Ever since Charles Rosenberg's *The Cholera Years* (1962) this has been a popular pursuit, involving as it has the challenging of allegedly doctor-driven histories of medicine's "great men," discoveries, and medical institutions. This is even the case in one 1966 history of cholera, subtitled "a biography of disease."[6] Besides cholera, the most frequented resorts for social historians of disease have been plague in the middle ages, venereal disease in the early modern period, tuberculosis in the nineteenth century and early twentieth, and AIDS in the late twentieth century. Most such studies refer sensitively to the different contexts in which these diseases appeared. We discover how communities, families, governments, religious bodies, and orthodox and unorthodox physicians reacted, ideologically, commercially, and otherwise. Noticeable about all such studies, though, is the absence of discussion on the construction of the disease concepts themselves. This is the tricky bit that so easily defies historical imagination because its modern scientific basis is so readily taken for granted. Just as it is unappreciated that a "fact" has no point in existing outside the narrative and/or epistemological framework that requires it, so it is overlooked that this also applies to concepts of disease; as bodies of theory they are never outside the wider knowledge-generating contexts in which they are produced. It is when this isn't overlooked that the absurdity of the biographical approach to disease becomes apparent. As the French ethnographer of science Bruno Latour observed in connection with palaeontologists announcing that Ramses II had definitely died of "tuberculosis": how is it possible for a person to die of a disease that was not discovered for a thousand years or more? For Latour, this extreme example of "transplanting into the past the hidden or potential existence of the future" was comparable to accepting that a person two thousand years ago could have been killed by "a Marxist upheaval, or a machine gun, or a Wall Street crash."[7] But what is preposterous to think of with regard to some agencies is apparently all too easy when it comes to microbes. To them, the biographer of disease is only too happy to grant unlimited transhistorical liberty.

This freedom is all the more odd—and shameful for any trained historian engaging in its practice—in view of the fact that the kind of criticism of retrospective diagnosis levelled by Latour has been in existence among medical historians for over a century. It goes back at least as far as the early 1900s when the German historian of medicine Karl Sudhoff engaged in debate with the author of *The Origin of Syphilis* (vol. 1, 1901; vol. 2, 1911) over whether the then new science of paleopathology could be employed instrumentally to settle historical issues.[8] For Sudhoff it could not; indeed, it could serve only to blur what was for him the crucial distinction to be maintained between the reductive methods of science and those involved in history-writing.[9] Similarly, for the Polish physician and microbiologist Ludwik Fleck in the 1930s (writing on syphilis and drawing all his historical information from Sudhoff), it was obvious that a contemporary concept of disease could not be pushed back into a period of time when it did not exist, or be used as if it were a neutral tool for understanding the past. Fleck perceived that concepts in science and medicine are human and social inventions that develop and change in different ways specific to their times, a relativist view that was directly to inspire the "revolutionary" work of Thomas Kuhn.[10] The point has been elaborated in more recent historical scholarship on disease concepts, such as by Andrew Cunningham on plague, Adrian Wilson on pleurisy, and Claudia Stein on the "French Pox."[11] Yet the message is forgotten again and again, even by those who claim some awareness of the "representational frame" in which disease concepts are made and exercised.[12]

If sustaining an essentialist view of disease is the second unspoken feature of the biographical historical method, the third is that the exercise constitutes a rearguard move in the intellectual world we now inhabit. It isn't just that social histories of disease have had their day among professional historians; more important is that the tidy taken-for-granted categories of the past that held the idea of disease together (and cohered its history) have been challenged. Not only in literary theory, but in the lab, we live in post- and post-postmodern times. A recent message from the bench of cognitive neurology, for instance, is that the days of searching for neural patterns and structures is over; posited, instead, is a "poststructuralist" brain in which thought is conceived as the outcome of "disorderly" impulses that trigger "avalanches" of neural activity.[13] A parallel is in genomics, where the narrative "book of life" approach to DNA of the 1980s has given way to an a manifestly anti-narrativizing emphasis on the proteins inside genes, perceived as plural, porous, and not at all open to simple narrative literary text-like reductions.[14]

Biography is an inherently structuring and ordering device, the appeal of which would seem to be all the greater in the face of contemporary impressions of fragmentation and the collapse of universal meanings. Swept under the carpet by it are the pressing issues of identity and meaning raised by today's experience of the world. Its purpose is to offer reassurance that the past, as construed through a few hundred years of scientific ways of knowing, has *always* been there and always will. The idea of the biography of disease thus serves to tame the past in the interest of shoring up our culture's hitherto dominant, but now threatened, way of knowing. But in so doing, it revives an agenda for a conception of medicine and its history that is fast fading—indeed, is now passé, if not quite past. It has been overtaken by the accumulation of new ways of thinking, which perceive science as presenting but one mode of constructed thought, which is itself now fairly fragmented. Thus the biography of disease, as Emerson might have said, is properly no history at all, only biographical contrivance.

8

Inside the Whale
Bioethics in History and Discourse

MEDICAL ETHICS, LIKE THE retrospective diagnosis of disease, is a soft target for conventional critique, especially when historicized by its own professionals (bioethicists) for careerist and marketing purposes. Nevertheless, it seems criminal not to, given its international standing and authority in contemporary society and biomedicine. How Western values and virtues have been internationalized through biomedicine is a timely question, and one that historians ought effectively to comment on. But for the most part they don't. When prompted to write on the subject, they tend to fall into the thin moral gruel of contemporary bioethics itself, going all pious and politically correct. Moreover, their "facts from the past" often serve instrumentally to prop up the enterprise and its pretenses. As this essay indicates, historians over the space of the past three or four decades have provided quite specific rationales for bioethics, although with little realization of their complicity in the enterprise. One unfortunate consequence is that historical questions worth asking have been obscured, such as how ever did medicine and biomedicine come to be the principal means for the arbitration of "the ethical" in human behavior, and how did this come to get sold around the world?

More germane to this book's narrative is the question of the cost to history-writing for jumping on the bioethical bandwagon. It is clear that since the 1980s bioethics has been a threat to the funding of medical history; certain historians therefore hitched up to it as a part of their survival strategy. I talk about this opportunism in the essay, paralleling it to the bioethicists' own. But what I don't raise for discussion is how this further contributes to the general de-historicization of our times and the devaluation of academic history in particular. It is yet another case of those outside the profession calling the shots and thereby defining the enterprise. Empirical historical qualification of what doctors and medical researchers have done to patients, for example, becomes the historian's business, thus only confirming the world according to biomedicine as mediated by bioethicists. Moreover, all of this inquiry only further closes off historians from thinking critically about their own involvements in the history-writing in question. Even historical critique, such as that presented here, remains just a description from the inside, and made in the smug confidence of the superiority of one's own moral categories (those at large in our culture and shared by bioethicists). The analyzing self remains on the outside; the question of what, ultimately, is being questioned remains apart.

For the most part (speaking from that outside), historians of medical ethics and bioethics have remained locked in medicine's own project of medical humanism, unable to see the whale they are inside of. Hence, they have been unable to ask on what the project turns, and what it permits, ideologically and epistemically, and what it does to themselves as potential upholders of the investigation of how our present is shaped. It is usually enough for them to regard ethical injunctions in medicine as historically contingent, locating them in time and context, all the while casting them within a narrative of inherently ever-better moral enlightenment.

This essay, published in the journal *Social History of Medicine* in 2011, was meant to provoke them. It was also intended as an update on some of the literature on bioethics published since the mid-1990s when I wrote "The Resistible Rise of Medical Ethics" for the same journal. In that essay, as in this sequel, socio-historical critique is the name of the game, although here with additional reference to discourse analysis, as merited (in irritation) by the pretensions to it in the volume upon which this essay turns: *The Cambridge World History of Medical Ethics* (2009), edited by Robert Baker and Laurence McCullough. The volume's other pretension, to be "global," also called out for comment. Concerns with historical critique and self-understanding therefore make no manifest appearance in this essay, which is perhaps why the response to it—from historians at least—was largely consensual. Even one of the contributors to *The Cambridge World History of Medical Ethics* confessed to agreeing with one of my overarching points, that the culture-bound enterprise of medical ethics has remained remarkably uncritical of the medical practice it was supposed to be upgrading. Nothing was to be heard on the main point, however, on the complicity of historians in that same business.

> *The whale's belly is simply a womb big enough for an adult. There you are, in the dark, cushioned space that exactly fits you, with yards of blubber between yourself and reality, able to keep up an attitude of the completest indifference, no matter what happens. . . . Short of being dead, it is the final, unsurpassable stage of irresponsibility.*
> —George Orwell, "Inside the Whale," 1940

Like medical ethics before it, bioethics has a history of opportunism. Long before revelations in the 1990s that some of its institutions in the USA were funded by the pharmaceutical industry, the enterprise was hailed for saving clapped-out philosophy departments and their professors.[1] It got bums on seats, the philosopher Stephen Toulmin observed in 1981.[2] Less cynical, if more sinister, was the observation by Alasdair MacIntyre in 1984, that here was a domain where the ideological function of dominant conceptions masked professional power and authority, shielding it from general moral scrutiny.[3] Many other scholars soon came to an equally jaundiced view, so much so that by the 1990s charges of "ethnocentrism, medicocentrism, [and] psychocentrism" had become cliché among those seeking to reinvigorate what had become a bureaucratized managerial set of practices driven by an empiricist methodology and mindset.[4] For example, the editors of *Bioethics and Society* (1998) decried the preoccupation of bioethicists with socially irrelevant issues. It was a view echoed by the contributors to *Bioethics in Social Context* (2001), and by Wesley J. Smith in his broadside *Culture of Death: The Assault on Medical Ethics in America* (2000).[5] Arrogance and truculence along with a dehumanized utilitarianism had all too evidently married with exclusion and elitism in the effort of Anglo-American bioethicists to colonize the discourse around biomedical morality.[6] While some insiders themselves admitted to a certain "sway-bellied middle age," outsiders characterized bioethical erudition as nothing short of a "philosophical disease."[7] But neither appeals to a *Brave New Bioethics* (2002) nor to an ethnographically orientated bioethics more sensitive to human emotions could alleviate the problem.[8] What is remarkable, however, is not that the enterprise should have become a target for negative illness metaphors, but that many of those dishing up the critique had their snouts in the same trough, and their heads in the same dominant bioethical discourse. In the academic chase for funding, bioethics was a relatively available milch-cow, and to some extent still is in the world of fashionable research. Far from seeking to question the dominant discourse, therefore, there was every reason not to.

    Today, the space for bioethical opportunism has shrunk. In part this is a result of the tarnishing of the image of bioethics, but also, conversely, because of the success with which its practice has been institutionalized in health care services and in the biosciences more generally. The empiricist turn in bioethics has helped facilitate this institutionalization by rendering the enterprise, if not more social scientific, then more pragmatic and practical. Empiricism, naively supposed to be outside of historical invention, would produce an ethics free of ideology and idealism—in effect, an ethics free of ethics! By these and other means by the 1990s bioethics had taken up a residence in the belly of the whale of biomedicine, as Charles Rosenberg put it in 1999. Rosenberg added that for many critics of bioethics it seemed much like "graphite sprayed into the relentless gears of bureaucratic medicine so as to quiet the offending sounds of human pain."[9] The graphite was sprayed not only in America and in the West generally, although elsewhere it was for reasons un-prompted by Judeo-Christian notions of pain and its quiet. The development and international marketing of biomedical products necessitated the adoption of the rules and rhetoric of western bioethics, whether the site of production was Singapore, South Korea, India, or Cuba. Thus by the late 1990s it could be confidently stated that bioethics had become "among the most visible and influential fields of our globalized world."[10] Efforts by UNESCO furthered that process, and furthered the suggestion inherent to it, that bioethics was a value-neutral, ahistorical and universal good around which the world could and ought to unite. Social scientists joined the chorus, contributing to a literature fitting American bioethics to global scripts.[11]

But as its discourse became hegemonic and its practices routinized, bioethics could no longer retain the popular socio-political appeal that it had in the 1970s and 1980s when it appeared to be fighting the power excesses of the medical establishment. Both on and off medical school campuses, it lost its shine, giving way to courses such as those on medicine and literature—a move itself not free from opportunism.[12] In sub-departments of medical history, too (at least in the UK and USA), courses on medical ethics fell out of fashion, to be replaced by those such as on non-western traditions and various other subscriptions to global narratives.[13] Today, as a historically new version of medical humanities moves into place partly on the back of those narratives, bioethics has become further de-centralized, its opportunities for knocking less and less. Over the past few years its authority has waned still further as a result of the neuro-turn in both the natural sciences and the humanities. Neuroethics, while on the one hand continuing the bioethical business of monitoring the good and the bad in brain research, on the other purports neo-phrenologically to operate within the "neurological foundations of moral cognition," rendering parts of the brain responsible for ethical thinking tout court.[14] In theory at least, the latter exercise renders the former superfluous. Yet, although neuroethics seems a cerebral celebration of personhood gone materially mad, it nevertheless remains a fillip to the cardinal plank and birthright of bioethics, that of individual autonomy. At the same time, of course, it closes off the would-be democratic narratives of the sort that helped to impel and propel bioethics in the 1970s and 1980s.

Should we be surprised that none of this finds mention in the would-be definitive *Cambridge World History of Medical Ethics?* From the perspective of opportunism, not at all. It is hardly in the volume's interest to admit to the tarnishing of the image of bioethics or to the waning of its academic and popular appeal as a movement (even if 70 per cent of the volume is concerned with medical ethics *before* bioethics). Nor would its credibility be enhanced by confessing that a so-called world history is entirely a species of western history conceived and writ globally. After all, it is upon the back of western bioethics that historical interest in medical ethics has been sustained; before bioethics no one thought to historicize medical ethics.[15] While all history writing is to some extent opportunistic—sights always readjusting to the surrounding ideological and market-forces driven by granting bodies and publishers—the writing of the history of medical ethics is overtly so. Given that bioethics today services the medical profession and the biosciences, its history writing is predatory on them as well.

The 876-page *Cambridge World History of Medical Ethics* is by far the grandest looking statement of this opportunism to date—eight inches by twelve inches by two, with dark green covers and gold embossing, it is a perfect medical textbook simulacrum, replete even with an "advisory editorial board." Its majesty pales into insignificance the first efforts by historians to cash in on the vogue for bioethics: the two edited volumes on the history of medical ethics that appeared in 1993.[16] Those publications, in turn, it is worth noting, served to displace two previous ones, Jeffrey Berlant's *Profession and Monopoly* (1975) and Ivan Waddington's *The Medical Profession in the Industrial Revolution* (1984), which were far more politically robust than the 1993 productions, if somewhat historically misguided in their sociological enthusiasm.[17] Writing from within the then new and socially thrusting bioethical paradigm, Berlant and Waddington sought to critique the self-interested doctor-driven medical ethics that existed before the advent of an allegedly lay-driven bioethics. As such, like David Rothman's *Strangers at the Bedside* written some years later (1991), Waddington and Berlant cleaved a strong separation between the so-called "old" medical ethics and the "new," heaping historically mediated abuse on the former for its paternalism and apparent lack of respect for the newly made sacred notion of "patient autonomy" in western medicine and culture. Rothman's book, as I observed in my review of it in 1995, also served for his appointment at a centre for bioethics at Columbia University.[18] Although a solid piece of history delivered with considerable punch, it was nonetheless a version of the bioethicists' tale and, as such, a strategy in the historical legitimizing of bioethics and bioethicists. In this respect it was scarcely different from the shamelessly self-serving historical stories of origin produced by bioethical practitioners themselves, such as Albert Jonsen's *The Birth of Bioethics* (1998) and his *A Short History of Medical Ethics* (2000). In these, as in dozens of similar bioethical productions, timeless medical ethical truths (panting to be called such) troop ahistorically forward decked in Enlightenment garb.[19] Such progress narratives for the enterprise of bioethics fail to admit, however, let alone historicise, what the medical sociologist David Armstrong has referred to as "the remarkable shifts in [the] focus" of medical ethics, especially over the past two centuries (which, interestingly, he parallels to re-conceptualisations in public health).[20]

Of course, when it comes to detailing the actual birth of bioethics, tunes vary somewhat. As Renée Fox and Judith Swazey have pointed out in their ethnographic study *Observing Bioethics* (2008), bioethicists parade many different stories of the origins for bioethics, each tale staking a claim for pioneer status at the same time as implicitly legitimizing the enterprise as a whole (like *Observing Bioethics* itself, inadvertently). The multiplicity of stories is hardly surprising given that bioethics did not emerge from only one stable. As bioethicists are keen to remind critics of their enterprise, interdisciplinary (maintained as an inherent good) is the name of their game, with policy makers, social scientists, lawyers, scientists, clinicians, philosophers, and theologians all involved—albeit mostly in reverse order of socio-political significance it seems.[21] Enrolled in the "birth" were not just ordained Roman Catholic priests, such as Jonsen, and Roman Catholic doctors and their wives concerned with abortion in the 1960s, along with the blighted Kennedy family (who financed the "Joseph and Rose Kennedy Institute for the Study of Human Reproduction and Bioethics" in 1971), and Jesuit institutions such as Georgetown University, but also secular and more social policy oriented outfits, such as the Hastings Center, established in 1969, with "ethics" inserted into its title a year later.[22]

While some bioethicists persist in the belief that their institutionalization emerged "like Athene from the head of Zeus, nearly full grown," others in the trade hold to the view that bioethics was a "natural" response to moral dilemmas raised mainly through the advance of bio-technology.[23] This conclusion has been especially so where medical ethics has intervened around the boundaries of life and death. Almost all such stories, however, play to philosophical ideals, rather than social, political, economic, and religious (indeed, denominational) circumstances. As such, these romantic tales of origins, while pointing to contingencies in the birth of bioethics, serve to reinforce would-be universal, non-culturally relativist ethical beliefs—scripts modern rather than postmodern we could say. More important, they omit to tell how bioethics was appropriated by, and then congealed within, the very institutions of medicine whose social relations the bioethical "movement" initially set out to reform. As the historian Tina Stevens has argued in her study of the cultural politics of *Bioethics in America* (2000), bioethics won legitimacy within the medical establishment because it "proved far less threatening to existing social arrangements than the changes demanded by more radical, and more popular, social critics of the sixties."[24] The efforts of bioethicists effectively diffused those challenges, she submits—a phenomenon that Michael Whong-Barr has also shown to have been the case in Britain.[25] Thus the enterprise did much more than merely save the life of philosophy departments; it did as much for the medical establishment as a whole. Contrary to the tales of bioethicists reproduced in one form or another in social histories of medicine, bioethicists in fact collaborated with biomedical researchers to assist them in managing public opinion, whilst appearing to the public to be a profession for the scrutiny of medical behaviour.[26] Problems perceived to be generated by exotic technologies were transformed into problems manageable by bioethicists who, unsurprisingly, tended to treat the technologies as inherently value-neutral, at the same time as disconnecting them from the politico-economics of private enterprise bio-technology and the soaring costs of medical care and health insurance. To be publicly credible the bioethical enterprise needed to maintain the appearance of pristine neutrality in relation to both the technologies involved and the bioethical technicians. Whether by omission or commission, the historicization of bioethics by bioethicists and their apologists played to this agenda.

The Cambridge World History of Medical Ethics follows suit by registering critiques of bioethics without ever overtly testifying to them. Its extensive bibliography reveals its editors' awareness of the critical literature, while internal evidence reveals their dissent from the grossly ahistorical Enlightenment narratives for medical ethics uttered by bioethicists themselves. But instead of engaging with this literature, or with some of the worse-case examples of historical Whiggery, the volume performs a side-step by adopting an alternative strategy for the presentation of the history of medical ethics, that of "discourse."[27] Around this the volume is built. The "Discourse of Medical Ethics Through the Life Cycle" (with chapters on Hindu India, Buddhist India, the Islamic Middle East, and so on) starts it off, followed by the

"Discourse of Religion on Medical Ethics" (with the same run of chapters), "The Discourse of Philosophy on Medical Ethics" (with only one chapter), "The Discourse of Practitioners on Medical Ethics" (19 chapters geo-chronologically arranged up to and including "Contemporary Islamic Middle East"), followed, finally, by the discourses "of Bioethics" around the world, and "of Medical Ethics and Society" (covering, in separate sections, the themes of animal and human experimentation and regulation, colonialism, Nazism, and apartheid, through to contemporary "health policy").

The most that can be said for this use of "discourse" is that it unifies an otherwise unwieldy bundle of subjects (much as the "global" serves in other contexts). As such, discourse here strings together subjects that conceivably fit to other scripts, and which, by being moulded into this particular one, obscure more than they reveal about a patently fabricated *world history* of medical ethics." In other words, while there might be a strategic purpose to the bundling, it nevertheless serves political purpose, as all strategies do, by the way it configures its subject and thereby eclipses other kinds of scripting. It needs bearing in mind that medical ethics, the deciding of what is good or bad in medical and biomedical practice, has no obvious boundaries. Publications such as this invent them by restricting the subject's remit to books specifically on medical ethics, or to narrow moral debates over the biological boundaries of life, and to those around animal and human experimentation—Tuskegee, for example, where, in fact, the content of medical ethics was decided through interactions between activists, politicians, and the media, not by nerds in bioethics.[28]

"Discourse" here is a handy device, and even possibly a political resource for avoiding direct engagement with the politics of the subject's critique. It is decidedly *not* a tool for discourse analysis, nor even for opening out the many different conditions of possibility (or impossibility) for medical ethics and bioethics past and present—spaces such as Europe between the Renaissance and nineteenth century that Mary Fissell in her contribution to the volume informs us were bereft of any particular ethical governance of patient/practitioner relationships. Or spaces such as the Soviet Union, where, in the 1970s, as Boleslav Lichterman relates, the official line was that "the corporate spirit of professional ethics is characteristic of bourgeois society," whereas the Socialist society "does not need petty regulation of people's behaviour"—a thought worth savouring. In fact, all of the chapters in *The Cambridge World History of Medical Ethics* are wholly conventional exercises in history writing.

They are framed almost entirely in terms of simple causalities that speak relentlessly to "impacts" and "influences," or to "transmissions" and "exchanges" of ideas—extending to styles of reality's apprehension unsuccessfully exchanged between historians and medical ethicists, as tellingly explicated by Martin Pernick in his introductory chapter to the volume. At most, some of the chapters treat "discourse" as a frame for the study of language, or rather for "tracking the path" of it.[29] This is the case in the longest chapter, that by the editors themselves on "The Discourses of Philosophical Medical Ethics," which treats concepts and metaphors "found in speech and writing" on medical ethics. Admittedly, the authors see "limits to this method," but they are not the ones that critical theorists informed by discourse analysis would recognise.[30] Rather, they are those of "appropriation" and "misappropriation"—methodologies once common in studies of the popularization of natural knowledge, but long since abandoned for their naïve regard of that knowledge as inherently passive.[31] Here, however, appropriation and misappropriation are held to be processes that can lead historians "to overlook forms of [philosophical] influence that are non-linguistic."[32] The main example provided is Jeremy Bentham's "influence by example" in having himself publicly dissected, and so effecting "British law and medical ethics around the globe," the authors believe. Thus attention to language, they conclude (all too literally), "can both overestimate and underestimate the nature and strength of philosophical influence."

Not here, then, compulsions to deconstructionism, discourse analysis, and discursive formations. To such postmodern conventions the editors seem blithely unaware; unaware even, it seems, that these conventions are invoked simply by the use of the word "discourse," as if it could now be deployed in innocence (rather like "paradigm" post-Kuhn). Leaving aside that the literary turn has itself been under siege these last dozen years for its reduction of experience and identity to language alone [see Chapter 4], *The Cambridge World History of Medical Ethics* seems doubly hoist by its own "discourse" petard: first, by performing no analysis of the discourses it points to; and second, by nowhere recognising in the slightest that the whole of its massive undertaking is bound entirely within bioethical discourse—the western discourse without which it would have no reason to exist. Discourse immune, as it were, it is wholly inside the whale of bioethics and biomedicine.[33]

Nothing better reveals this immunity than the volume's failure to transcend the conventional historical narrative of bioethics: that it was born of social (lay) opposition to doctor-driven (profession-serving) medical ethics.

The very existence of *The Cambridge World History of Medical Ethics* bears testimony to this since, implicitly, the historicization it undertakes is only important because of the trumpeted socio-moral "revolution" staged by bioethicists. But in thus testifying, the volume conceals more than it reveals, for by reproducing the idea that "the new" medical ethics fundamentally superseded "the old," the focus is kept entirely on the plane of medicine, instead of on the more systemic shift involved in creating the possibility *for* bioethics. At root, the history of bioethics is not about medicine and morality at all, nor about the activity of bioethicists, but rather, is a reconfiguration of what it is to be human. Crucial to this reconfiguration within and without bioethics is the concept of "informed consent," which the Harvard anaesthetist Henry Beecher in the 1960s famously raised to the status of the central issue upon which must hang most of the ethical problems of human experimentation.[34] Predicated on a historically constructed notion of human nature in which persons/patients are defined primarily in terms of their ability to act autonomously or to make choices and take risks, "informed consent" per se spoke not to a revised medical ethics socially widened to the laity and broadened to a "bio"-techno remit, but rather, epistemically, to a prioritization and celebration of personhood within a particular politico-economic context.[35] "Personhood" displaced alternative more communal discourses—be they those of doctors speaking protectively as a corporate body, or communities as a whole in concerns over health care and its provision.[36] Fundamental here is not that the authority of doctors was displaced by the would-be authority of (laity minded) bioethicists (a fallacy in any case), but rather, as Foucault would have it, the replacement of one "truth regime" by another—namely, an ethics based on "the social subject" to one grounded on "the self." It is a shift that can be read ideologically, as in the defence of "the social" mounted via the "gift relationship" elaborated in 1970 by Richard Titmuss explicitly against American neoliberals arguing for a "free" commercial market in the sale of body parts and fluids.[37] But it is more appropriate to speak of it as discursive, inasmuch as it was based on a new psychological way of making up people (essentially as greedy and competitive). This, in turn, both opened up and closed down other ways of thinking and acting that were more than merely political, as well as much more than merely medical.[38]

The Cambridge World History of Medical Ethics is written wholly within this discursive regime, blinded by its surface appearances of truth. Thus the editors regard it as enough merely to provide a definition of autonomy as "a

Inside the Whale 181

philosophical concept describing the scope of individual liberty of thought and action," all the while thinking it historically sufficient to trace such concepts in philosophy's past, as if philosophy was not itself constitutive of dominant discourses.[39] How the concept of autonomy relates to notions of being human, and how these in turn relate to wider political economic formations which have become mediated through medical ethics, constitutes no part of the editors' task. As such, ironically, the volume misses the one real opportunity before it: to illuminate the historical construction of the present truth of which bioethics and its historicization are constitutive. By failing to do so—failing, that is, to historicize how people (including historians) "govern themselves and others by the production of truth"—the volume operates *a*historically.[40] It assumes the present *is* the truth, which can simply be acted upon historically by narrativizing the causes and consequences of individuals' actions and ideas. Of this *a*historicity, the editors and contributors seem mindless. Moreover, by virtue of being wed to an empiricism comparable to the bioethicists' own, they remain complicit with a practice of history-writing as a science-like enterprise that operates from some external would-be amoral position of objective viewing, not appreciating that "objectivity" is itself a historical construct.[41]

While a generation ago this position might have been excusable, today it is not. In its political naivety it helps consign to the dustbin of history, not merely the history of medical ethics (which may be no bad thing), but the discipline of history itself. In the face of the neurological reduction of all human activity, including now history-writing itself, *The Cambridge World History of Medical Ethics* serves only further to squander one of the most valuable tools in the defence of the humanities.[42] It not only contributes nothing to the understanding of ourselves and the westernized world in which most people now live, but keeps its readers blind to the urgent need for such understanding. Irresponsibly, through its disservice to historical understanding and analysis, it drives a nail in the coffin of the professional practice of history.

But if the bad news is that *The Cambridge World History of Medical Ethics* is wholly within the whale of western bioethics, the good news is that the whale has been beached. At least as far as doctor/patient medical ethics are concerned (the sort that largely shape *The Cambridge World History of Medical Ethics*), these are increasingly of only marginal interest among academics today. Western medical ethics is now more likely to refer to concerns over distributive justice in the allocation of health resources, national and global; the justifiability of risk imposition; the role and limits of "consent"; the nature and desirability of evidence-based policy making; health charity spending priorities; the ethics of screening, and so on (all of which of course also ought to invite critical scrutiny). Its practitioners are more likely to be discussing normative theorizing, public policy, law, human rights, and social environments than doctor/patient relations. These new concerns reflect that many more authorities than doctors and health promoters are now involved in articulating the rules for living.[43] "Somatic experts," as Nikolas Rose refers to them, now include genetic counsellors, support groups, projects for the public understanding of genetics and bioethics, among many others. Moreover, contemporary biomedicine interlinks the laboratory, the factory, and the stock market as never before—becoming, as Rose observes, "saturated with issues of financial value" and "intrinsically linked to the spirit of biocapital."[44] The need to distinguish the idea and practice of bioethics as a legitimating device for biomedicine and the pharma-industry from that of an embodied "somatic ethics" has therefore become compelling.[45] *The Cambridge World History of Medical Ethics* by failing to seize this opportunity to look anew at the history of Western medical ethics as a discourse constructed in its own historical present thus stands as a monument to anachronism. Although it is hard to imagine that there would ever be a market for the study of the decline of the historicization of medical ethics (a publishing market, that is, for an irrelevant subject) were that moment to arrive, this volume would afford abundant grist to its mill.

9

Cracking Biopower

I BEGAN TO THINK SERIOUSLY about the subject of this essay when I was asked to review Majia Holmer Nadesan's *Governmentality, Biopower, and Everyday Life* (2008).[i] It was one of a number of works then emerging on the topic by sociologists, anthropologists, political philosophers, and others. But I elected to review it only because, for some time before this, Claudia Stein and I had been working on a study of the politics of the visual in the making of "biopublics" in Germany and Britain circa 1880–1920—a study exploring the possibilities of drawing on Foucault's notion of biopower for history-writing. Claudia was also thinking around the concept in relation to her new work on eighteenth-century Germany.[ii]

At the forefront of those then explicating the concept were Paul Rabinow and Nikolas Rose, both long-standing interlocutors of Foucault in the English-speaking world. Their "Biopower Today," which appeared in the first volume of their new journal, *Biosocieties*, in 2006, had been available on the Internet since 2003 and was widely taken up by analysts of contemporary culture, politics, and economics, such as Nadesan. Among historians, however, the concept was barely legible and, indeed, was hardly heard of. The primary purpose of this essay, which was written with Claudia in the summer of 2009, was to make it explicable to them, as well as to help us with our own understanding of it amid the proliferation of territories for its application that scholars were articulating. What, we wondered, did "biopower" and "biopolitics" actually mean for them, and what were the nuances in interpretation in the different disciplinary contexts of its use? Why was it now so attractive? Above all, how might historians be able to use it as a tool for the analysis of past and present? The opportunity to take up such questions came after an invitation from the journal *History of Human Sciences* to review Rose's *The Politics of Life Itself* (2007) and *Bíos: Biopolitics and Philosophy* (2008) by the Italian political philosopher and theorist Roberto Esposito.

These were odd books to consider from a historian's perspective, and they differed considerably from each other in their take on biopower. Both, however, sought to expand on Foucault's use of the concept, especially in their effort to get beyond its reach in the work of some Italian neo-Marxists for whom the Holocaust was perceived as biopower's apogee. Both Rose and Esposito wanted to rescue the concept for the better analysis of contemporary politics, culture, and society. In doing so, however, it seemed to us that they overlooked a key aspect of Foucault's way of using of the concept, the working from the contemporary situation backward historically. Foucault's use was neither just a sociological describing nor a philosophical meandering, but rather a historical and political positioning. Our review sought to bring this out, at the same time revealing how the efforts of Rose and Esposito to disturb the habits of writing within older Foucauldian formulaics could be seen as intellectual maneuvers constitutive of our present times. To take this view was to adopt a quintessentially Foucauldian position, since Foucault's interest had always been with the manner in which different moments in history posed different problems for themselves and found solutions to them—indeed, with how such solutions were made to seem inevitable and necessary. Today, for certain intellectuals at any rate, biopower, if not exactly a final solution to understanding the contemporary world, is a highly prized tool for its analysis, and in this respect it does political work inasmuch as it circumvents the use of other kinds of tools. As suggested in this book's first chapter, "the politics of life" are more than what they innocently purport to be or understand themselves as being. In the end, "cracking biopower" meant reopening politics, those of Foucault's new interlocutors as much as Foucault's own.

That Rose, understandably, should have expressed some bafflement at our "eccentric, perhaps tendentious" reading of his work delighted us.[iii] It indicated that, to some measure at least, we had read across his text, rather than subscribed to its logic and politics and moved nothing on. *The Oxford English Dictionary* defines "tendentious" as "expressing or intending to promote a particular cause or point of view, especially a controversial one" (as in "a tendentious reading of history"). Tendentiousness is exactly our point, and in our opinion, there should be no evading it in history-writing. Indeed, it is arguably in the relentless efforts of historians to avoid being tendentious that lie many of the discipline's current problems. Taking a point of view, as Foucault did, may no longer be understood or wanted in our neoliberal world, but that is no reason for historians to avoid it. In fact, it is all the more reason for it.

Biopower," a decade ago hardly on any scholar's lips, is today on almost everyone's. In part, this is because Foucault's take up, and take, on biopower only relatively recently came to general attention, and even more recently got translated into English.[1] It is also due to the fact that hitherto Foucault's many interlocutors were less disposed to this particular technology of power than to his conceptualization and elaboration of the knowledge/power nexus, the archive, the medical gaze, genealogy, discipline, discourse, epistemic ruptures, and the technologies of the self. In view of over thirty years of publishing on these fronts, many academics had come to the conclusion that Foucault had been "done to death," or that his thinking, if not exactly passé, had become so thoroughly imbibed through literary and somatic turns that it could be safely left behind. As far as biopower is concerned, however, this judgment is grossly premature, since the digestion of what Foucault actually wrote on the subject has only just begun—with something of a rush.[2] Certainly, as a result of the posthumous publication of his *Dits et écrits, 1954–1988* (1994), and its recent four-volume translation into German, there has been a change of focus on Foucault's oeuvre, particularly around his "analytics of power."[3] As Thomas Lemke points out in *Gouvernmentalität und Biopolitik* (2008), until the 1990s this was of only marginal importance.[4] Now, however, it has moved to the centre of sociological and political study. In particular, two of Foucault's concepts within the analytics of power, "biopower" and "governmentality," have come in for extensive critical scrutiny.

"Governmentality" does not concern us here. Suffice it to say that it was central to Foucault's lectures at the Collège de France in the late 1970s as one of the several tools he devised to critique current understandings of "power" by pointing to historically alternative forms and fields of its practice.[5] "Governmentality" was directed to the different forms of administration and guidance of individual subjects and collectives in history.[6] The term "biopower" likewise emerged out of Foucault's attempt to identify different paradigms, foras, and practices of power. It came into particular focus after the publication of his *Surveiller et punir* in 1975. There, and again in the *Histoire de la sexualité* (1976)—volume I of *La volonté de savoir* (The will to knowledge)—he made clear that his interest was not in the great events of human history and human thought, but in the subtle, slow, and often invisible dispositions, maneuvers, and tactics that occurred in the domain of knowledge and society over time, and which traversed and linked every kind of institution.[7] These subtle changes, he argued, are reflected in specific techniques and functions of power—to be perceived less as properties than as strategies—which continuously shift their constellations over time. "Biopower," it seemed to him, was one of these different forms of power.[8]

Like "governmentality" the actual word itself was not of his own invention.[9] As Roberto Esposito reminds us in *Bíos*, its usage stretches at least as far back as 1905 when the Swedish political theorist and politician Rudolph Kjellé introduced it in *Stormakterna*, a work on "geopolitics" (a word also coined by Kjellé). It was elaborated in Kjellé's *The State as a Form of Life* (1916) where geopolitical demand was seen as "existing in close relation to an organismic conception that is irreducible to constitutional theories of a liberal framework."[10] Much the same kind of naturalization of politics or biological reconfiguration of the State was explicated through the term "biopower" elsewhere in the interwar period.[11] This was the case not only in Germany, but also in Britain, notably in Morley Roberts's *Bio-Politics: An Essay in the Physiology, Pathology and Politics of the Social* (1938).[12] After the fall of the Nazi regime this approach was discredited, only to enjoy an interesting renaissance in the mid-1960s with the emergence of a new area of research in Anglo-American political science: "biopolitics." The main assumption of this programme was that political behavior was ultimately based on biological laws.[13] Another, different, neo-humanist version of "biopolitics" emerged in political philosophy in Foucault's France in the 1960s.

Foucault's use of "biopower," however, marked a decisive break with all former attempts to reduce the nature of politics to fixed biological determinants. His was the investigation of the historical process in which biological life (of the individual and the collective) ultimately emerged as the central object of political strategies. In contrast to the idea that politics follows natural and timeless laws, "biopower" in Foucault's formulation represents a break or discontinuity in the practice of power. From his perspective, power has not always been "biopower"; rather, biopower presents a specific modern form of power dating from the eighteenth century and maturing in the nineteenth and twentieth. It was distinguished, first, by its concern with the preservation and fostering of individual life and, second, by its interest in the lives of populations. As such it was defined in opposition to an earlier "sovereign power," which operated repressively over life.

Nevertheless, Foucault's definition of biopower left much to be explained and accounted for. Not least, it left dangling the question of the historical emergence of biopolitics and "modernity" (no less slippery a term), along with the question of "life" itself. Of the two books under consideration here, that by Roberto Esposito provides a major contribution to the first question, while that by Nikolas Rose addresses the second, at the same time as demonstrating the need for conceptual clarification.

Both books shimmer with insights and acuity, Rose's *Politics of Life Itself*, through close observation and analysis of contemporary biomedical discourse and practice, and Esposito's *Bíos*, by scrutinizing the logic of the analytical framework. Although odd bedfellows in many ways—indeed, incommensurable in certain respects—they serve wonderfully to illuminate and problematize different aspects of Foucault's conceptual thinking on biopower. Both authors share his view that biopolitics is not about impositions *on* life of the sort epitomized in the Third Reich, as usually understood in political philosophy. "Politics," they submit, is no longer something that can be disarticulated from "life." Rather, they follow Foucault in his suggestion that from the late eighteenth century the nature of European politics began to change and that those changes correlated with emerging knowledges about "life." It was from that point on that that "life" and "politics" began to fold the one into the other. As Foucault put it, "for millennia, man remained what he was for Aristotle: a living animal with the additional capacity for a political existence; modern man is an animal *whose politics places his existence as a living being in question.*"[14]

To put this otherwise, "life" in its double philosophical meaning of *zoë* (simple "biological" life, or "naked life" or "bare life") and *bios* (specific human life, or qualified forms of life, including political forms of it) began to be linked, so as to refer to each other through an ever-increasing number of normative mechanisms. Today, both Rose and Esposito agree, we are much further along the road than when Foucault wrote; increasingly the politics of life emerge as life itself—the whole experience of being. Where the sociologist Rose differs from the philosopher Esposito is in how to conduct the analysis of these politics. For Rose, the concept of biopower as derived from Foucault is a valuable tool for prizing apart our increasingly biotechnologized and somaticized world, and for exploring how subjectivities and ethics are reconstituted thereby. It is a tool applied to an empirical analysis guided by the conviction that to arrive at an understanding of "what is life?" in contemporary culture it is necessary to investigate the ways of acting and thinking of those involved in this particular kind of politics. For Esposito, the analysis of biopower is not to be done empirically, but through the study of the semantic logic of the thing called "biopower." His object is to reconfigure the way we think about individual and collective "being" within a biopolitical world.

Both authors oppose the negative casting of biopower of Giorgio Agamben, Michael Hardt, Antonio Negri, and other (mostly Italian) neo-Marxists who conceptualized it as a sovereign instrument in the exercise of biological power *over* life. For these thinkers, biopower could only lead to a repeat of the Nazis' thanatopolitics of population purification. Indeed, for Agamben in his concern with the juridical and political spaces for the exercise of power (in *Homo Sacer: Sovereign Power and Bare Life*, among other publications), the "first principle" of biopolitics is the politics of death. Although Esposito avoids explicit reference to Agamben and Hardt and Negri, his text is animated by the urge to counter their negative view of biopower and biopolitics and, as importantly (through close attention to language and syntax), to counter their structuralist framing. Affirmative biopolitics, or what he calls the "vitalization of politics," is the name of the game, as Timothy Campbell clarifies in his extensive preface to *Bíos* introducing Esposito to the English-speaking world.[15] Rose, similarly, has no interest pursuing doomsday scenarios, except as public utterances that provide grist for the analysis of contemporary fears, hopes, evaluations, and judgments around our emergent form of "life." Interested less in Foucault's move to destabilize the present by pointing to its contingency than in destabilizing the future "by recognizing its openness," he is insistent that today "biology is not destiny [as the Nazis configured it] but opportunity." "[N]o single future is written in our present," he seeks to illustrate against critics and cynics who would argue otherwise.[16] Specifically rejecting Agamben's pronouncement that "the [death] camp is the diagram of the biopolitics of the present"—"the hidden dark truth of biopower," as he and Paul Rabinow elsewhere put it—Rose is unequivocal in his assertion that within advanced liberal democracies "our somatic, corporeal neurochemical individuality has become opened up to choice, prudence, and responsibility, to experimentation, to contestation, and so to a politics of life itself."[17] "Vital politics," as the politics of life are otherwise known, is thus a fairly rosy affair, or might be, so long as one does not entertain the idea that some form of Nazi biopolitics could indeed be one of the choices or opportunities, or that that opportunity has already been seized. Such views are not countenanced by Rose.

But it is not simply by this affirmativeness that the philosopher and the sociologist move forward our thinking on biopower and biopolitics. Nor (at all) is it because Esposito depends on Rose, or Rose on Esposito-like philosophizing around the politics of life. On the contrary, Esposito's problematization of biopower from a philosophical and semantic point of view is effectively justified by what Rose's empirical investigation and depiction of contemporary strategies and practices of biopower and biopolitics fail to elucidate—what Esposito refers to as the "black box" of biopower. Taken together, both books help us better to unpack that box.

From Rose's perspective, biopower is a valuable analytical resource, or simply a useful way of viewing. It is, he submits, "more a perspective than a concept" that "brings into view a whole range of more or less rationalized attempts by different authorities to intervene upon the vital characteristics of human existence."[18] "Biopolitics," in turn, are the "specific strategies brought into view from this perspective, strategies involving contestations over the ways in which human vitality, morbidity, and mortality should be problematized, over the desirable level and form of the interventions required, over the knowledge, regimes of authority, and practices of intervention that are desirable, legitimate, and efficacious."[19] Rose elaborates these strategies in their various guises in *The Politics of Life Itself*, a collection of addresses and lightly reworked previously published articles dating from the late 1990s to c. 2005. Health and biomedicine are his focus; race, eugenics, somatic ethics, the neurochemical self, and biocitizenship his subjects. Consistently, he has his finger on the pulse of modern biomedicine, providing acute, jargon-free and often gripping analysis of its ramifications and implications for contemporary consciousness and thought.

That the politics of life have been refreshed in contemporary society is a part of the argument. For it is really only *since* the end of the Second World War, and even more recently, Rose contends, that a whole bio-complex has come into being interlinking ethical and technological aspects in dramatically new ways. Thus medical agents of one sort or another now routinely exercise determination over who is to be let die or begin life, and their decisions are enhanced and regulated by sophisticated medical technologies. There is also the rise of new types of patients' groups and individuals who define themselves largely in biological terms. Others now articulate their citizenship not in terms of legal rights and social duties bound to nation-states, but as active and informed biological consumers. Perhaps most crucial of all are what he and Rabinow refer to as the new circuits of *bioeconomics* that have taken shape over the past few decades. Thus we now have "large scale *capitalization of bioscience* and *mobilization* of its elements into new exchange relations: the new molecular knowledges of life and health are being mapped out, developed and exploited by a range of commercial enterprises, sometimes in alliance with States, sometimes autonomously, establishing constitutive links between life, truth and value."[20]

This emphasis on the novelty of the bio complex, of its owing nothing to earlier historical periods, is not, however, a commitment to epistemic rupture. Rose's basic claim is that biological assumptions, prejudices, and conceptions have always influenced political categories (such as citizenship) because, implicitly or explicitly, they order membership, participation, and access to political activities. On the basis of this general assumption, Rose conceptualizes a break between the eugenic projects of the past and current biomedical practices. Biological citizenship, for example (and, hence, biopolitics in general) stands for a new regime in the twenty-first century, which breaks radically with previous eugenic ambitions. What we see is a shift of political rationalities, which no longer aim at controlling risk at the level of populations, but now instead target individuals, especially through the management of genetic risk. This thinking is in line with that of other scholars, such as Deborah Heath, Rayna Rapp, and Karen-Sue Taussig, who similarly tend to underline the novel implications of this biopolitical move and, like Rose, do not argue for an epistemic shift so much as a historical seizure predicated upon today's fragmentation and pluralisation of political spaces.[21] It all goes back to Rabinow's idea of biosociability as "the emergence of new forms of representation, new forms of community and identity politics in the context of the new genetic knowledge, and which understands human nature as culturally open and technically changeable."[22] According to Rabinow, and also Rose, there is now no simple translation of social projects into biomedical termini—like social Darwinism of old—but rather, a new configuration of social conditions via biological categories.[23]

Even so, Rose believes, the strategies of biopower are not universal or totalizing. Biopolitics is far from being the only show in town. In the course of analyzing the mutations of personhood through the concept of "genetic risk," for example, he submits that "ideas about biological, biomedical, and genetic identity will certainly infuse, interact, combine, and contest with other identity claims; I doubt they will supplant them."[24] Such reassurances lend weight to his view of the present as a place still deeply connected to the past, and which for that reason, apparently, is ostensibly less threatening. Commitment to historical continuity permits him to be wary of "breathless epochalization," or to the "overstating [of] novelty" in contemporary biomedicine and biotechnology, while at the same time, on the one hand, maintaining it and, on the other, remaining blind to significant continuities.[25] Moreover, such reassurances serve not just as evidence of Rose's commitment to "incremental rather than epochal" change; they are also the means to spare him being tarred with the brushes of reductionism and over-determinism—the sins of which he accuses other social commentators on our biological times and futures, and around which commentary he typically structures his scholarly interventions.

This tactic, and the gradualist view of history that it entertains, has less to do with the politics of life that Foucault engaged with in the 1970s and 1980s than with the politics of Rose—a liberal pragmatism peculiar especially to Anglo-American academics of the postmodern era who, accepting the failure of the Enlightened project of humanity, see virtue in avoiding moral judgments. Rose's "always open" position stands in stark contrast to, say, the anarchist democracy extolled in Chris Hables Gray's *Cyborg Citizens: Politics in the Posthuman Age* (2000). For Rose the question of how to create a common project of humanity in today's bioscientific globalised world smacks of old-fashioned humanism; it is too much of the view that humans have a responsibility to create and shape the world in which they live. (Like the sociologist Steve Fuller, Rose is convinced that the old social humanist paradigm is bankrupt, a position that makes it easy to invest naively in the rhetorical power of novelty.)[26] A product of our times, Rose shares company with many others in the social sciences who now fashionably subscribe to a view of "the social" which is not "social" at all in the sense of being filled with human agency and ideals of organization. The "social" is but an environment of strategies, mechanisms, and relationships ceaselessly re-created and reshaped (whether by humans, animals, or machines) through which people simply pass. Adopting this view [elaborated in Chapter 10], and overlooking that contemporary society continues to be structured and stuck in older ideas, practices, institutions, and mechanisms of power, Rose feels no compulsion to suggest how a society of biologically ever-enhancing individuals and proliferating identities could or should organize its collective togetherness. He only vaguely alludes to such concerns when he urges us to pursue those "practices of intervention that are desirable, legitimate, and efficacious," or when he makes the claim for an analysis that would "intervene in that present, and so . . . shape something of the future that we might inhabit."[27] But there are no concrete proposals; no suggestions as to what might come after, or out of, all the recordings of new biomedical subjectivities. It is a vagueness that is popular in today's academic world run as it is by the changing fashions and fortunes of grant-giving bodies, for it permits the study of almost everything, but commitment to nothing—and hence is a valuable strategy for the retention of patronage. This is not to say that Rose is openly opportunistic, but he does seem to suggest that one can separate the empirical analysis of contemporary life from larger questions of collective human direction and purpose. He keeps his hands clean.

It is as a means to having his cake and eating it too, that he invokes the authority and would-be personal example of Foucault. Although *The Politics of Life Itself* is fairly quiet on methodology compared to other of Rose's publications, his Foucauldianism is consistent with that to be found in his and Rabinow's "Biopower Today." There, while lauding the analytical utility of

Foucault concepts of biopower and biopolitics, it is confessed that Foucault's actual comments on the subjects were "limited and sporadic." This admission serves to justify the omission of any detailed discussion on how Foucault actually conceptualized these terms and thought of them politically (as opposed merely to explicating them historically according to the poles of populations and individuals). The suggestion is always that Foucault was somehow himself apolitical and was only concerned that people should not be made to submit to any historically devised moral consensus.[28] While this gives strength to an open pluralistic future of possibilities and opportunities, it fudges the fact that during his lifetime Foucault was far from apolitical. He may not have written much on his own personal politics, but he was certainly not without them, especially in relation to initiating the voice of resistance among prisoners or psychiatric patients whom he sought to bring back into the discourse that othered them. Well known, too, is his deep interest in (and ambition to theorize) the Iranian revolution which, contrary to many of his contemporaries, he saw as a new and exemplary form of collective will that could not be thought of as emanating from categories such as class struggle or economic oppression.[29] Moreover, the whole of his approach to biopolitics and biopower as well as to the epistemic ruptures of the past were quintessentially political searches for new ways to comprehend power and politics outside of the then reigning Marxism of Sartre, Althusser, and others. Thus Foucault's position stemmed from his resistance to other positions, which were then much more influential than they are now. He never shied away from taking a stand.

The historicization of Foucault himself, which is necessary in order to understand the political immediacy behind his terms, is not on Rose's agenda. He casts Foucault simply as a Nietzschian—ostensibly apolitical—analyst who managed to transcend conventional sociological structuration and idealization. There is a sleight of hand in this move, since it involves at the same time the expression of dissent from certain crucial elements in Foucault's thinking. Explicitly, Rose departs from Foucault's "history of the present," partly on grounds of its professed familiarity and overuse, but more importantly, because it associates the user with Foucault's radical views of epistemological mutation around the idea of "life" and its normatization. He insists that Foucault's claim of a rupture at the beginning of the nineteenth century "is overstated," and tethers this view to implicit dismissal of Foucault's understanding of the notion of "life" as generally open, and defined only through specific epistemological constellations and complexes of power at specific moments in time.[30] For Foucault, "life" was neither concrete nor essentialist, but a discursive correlate whose emergence proceeded hand in hand with the establishment of biology as an autonomous science.[31] The meaning of "life" for Foucault is caught in an ever-changing nexus of power/knowledge. Thus, for Foucault, it is impossible to explore the meaning of "life" without at the same time investigating the historically correlating forms of knowledge and power. While he accepted Canguilhem's understanding of normativity as something created by the inner dynamic of an all- and ever-changing life, "life" itself he saw as something "to normatize" from the outside.[32] Rose, by denying, or at least downplaying the idea of a historical rupture, and adopting Canguilhem's thinking on normativity as emerging out of life itself, frees himself from undertaking the kind of historical analysis of the nexus of knowledge/power that Foucault regarded as central to all his projects, including that of biopower.[33] Rose tends to believe that such a genealogical analysis is impossible in today's "complicated and complex" world.[34] This spares him the messy historical and immanently political task that most interested Foucault: the analysis of power *dispositifs* and discourses, or the study of the heterogeneous complex of beliefs, practices, and technologies that he perceived as continuously affecting and shaping all contemporary life, and, significantly, that intrinsically linked past and present in life's definition and practice.[35] Such complexity, which is also compounded by the many new bio- and neuro-techniques for making life visible, thinkable, and able to be experienced today, and by the new means to economic exploitation and control related to these techniques for manipulating life, is doubtless what inclines Rose to give up on even trying to define "life." Hence his opting simply for empirical observation and description. He claims in defense of this approach that at the same time as life-processes can be accessed via conscious manipulation, the question of "what is life?" is increasingly answered in enshrouded notions of probability and risk. Thus life shows an increasing tendency to indeterminacy.[36] Following Canguilhem's suggestion that the answer to the question "what is life?" needs to be found in the life of living beings themselves, Rose embarks on a task of "vitalist" empiricism, which draws conclusions from the ways of thinking and acting of living beings in the politics of life itself. "My aim," he declares,

> is not so much to call for a new philosophy of life, but rather to explore the philosophy of life that is embodied in the ways of thinking and

acting espoused by the participants in this politics of life itself. What beliefs do they themselves hold or presuppose about the special qualities of living things? What forms of differentiation of life and nonlife do they enact . . . ? What differences exist in the obligations that they accept towards entities at one side or the other of that divide?[37]

In other words, forget about searching for the "meaning of 'life' "; focus instead on "the problem of 'life' "—on the spaces, practices, and persons that problematize it.

Thus Rose folds the authority and ostensible personal examples of Foucault and his teacher Canguilhem into his own neoliberal political script. In the name of both (and that of postmodern theory which developed out of Foucault's work) he justifies his anti-moral stand, as well as his empirical endeavor. It is a position founded on an enthusiasm for descriptive scientific empirical analysis, born out of what Karin Knorr-Cetina calls an "exaggerated emphasis on instrumental reason and information," which also happens, she says, to "empower subjectivity thinking and cast doubt on social thinking."[38] Empirical description can thus stand above politics as morally motivated action, instead of confessing that it *is* political. It is a problem that might have been mitigated, but hardly avoided, had Rose substituted "practice" for the more provocative sounding "politics" in his book's title. For while empirical analysis sustains the illusion of enhanced understanding of contemporary reality, such as the construction of new subjectivities that Rose brings out so well, it keeps us locked within a subjective (historically constructed) social science gaze that can never stand outside its own form of certainty, and whose mere scientizations of life purport to tell us what life "really is." Its effect is to suggest that intellectuals need only gather and analyze more and more information without questioning or challenging its political implications. But this scientific pretension is based on the assumption that science is not itself morally inclined, or that it does not have as its ultimate goal—its ultimate reason for existence—the purported *bettering* of human life. Rose, in backing away from questions about where we go from here, or what we do with all the recordings of the world as we now find it, thus negates his own scientificity, offering us testimony only to postmodernity's disencouragement to moral certainty and assertion. In this respect, *The Politics of Life Itself* stands in stark contrast to Esposito's *Bíos*, which, without any such political inhibitions, enters fully into the black box of biopower.

The Politics of Life Itself is thus left open to conventional social and political critique. While critics might mistakenly confuse Rose's description of somatic individualism with apology for iniquitous bio-consumerism, they are nevertheless right to worry over his ethical stand. After all, for Rose the politics of life itself focuses on "our growing capacities to control, manage, engineer, reshape, and modulate the very vital capacities of human beings as living creatures."[39] Critics can hardly be blamed, therefore, for concerning themselves with the nature of those interventions, or with expressing dismay over Rose's seeming lack of such concern. For example, that cash-strapped patient self-help groups can become prey to instrumentalized international bio-tech corporations is an issue that, although taken up by self-help groups themselves as morally problematic, can only be pursued by Rose, at best, as part of the description of our times.[40] Critics might worry, too, that so much of his attention is on individual subjectivities, and scarce nothing on the social or communal forms of life that these new biologically enhanced individuals might or should entertain. His liberal openness and intellectual casualness in this regard is unsatisfactory. That the individual-subjective category spawned through the growth of technologies of the self now rules over previous social-structural ones in explaining and sustaining the political order tends to forget that individuation is itself a social phenomenon and, further, that, as Esposito has it, the individual "is not definable outside of the political relationship with those that share the vital experience."[41] Indeed, sociology itself is abandoned here—at least the sociology that was founded on certain notions of what "the social" was about, and sought (always as an inherently moral enterprise) to explicate it.[42]

Yet, for all Rose's mindedness to skirt morals and politics through his partial adoption of Foucault (i.e., Foucault without history), he is left within the sociological lexicon. This is belied by his use of the normative categories of "human rights," "individualism," "liberal democracies" and so on—old categories that, as Esposito points out, inappropriately "continue to organize current political discourse."[43] Rose's deployment of "biocitizenship" perhaps best conveys this entrapment. Heuristically, the term is well intended (no less than its cognate "biosocialities" coined by Rabinow).[44] By it, he and Carlos Novas (the co-author of the chapter in *The Politics of Life Itself* that was first published in *Global Assemblages*) sought to mark a clear distinction from the notion of "social citizenship" as articulated by T. H. Marshall.[45] But the very use of "citizenship" with the prefix "bio" to highlight difference with today's somatic individualism and consumerism (via cosmetic surgery, Viagra, Prozac and the all the rest) is to remain *within* the sociological parameters of Marshall.[46] As far as the analysis of biopower goes, the problem with Rose's version of it is that it originates *in* these categories before it seeks to overturn them—a problem common to most social science analyses, as Esposito makes clear.

More profound, and the main criticism of *The Politics of Life Itself* to which Esposito's *Bíos* could be said to hearken, is the impossibility of asserting an affirmative biopolitics based on Foucault's writings without first penetrating the logic of Foucault's notion of biopower. Without opening that philosophical black box, as Esposito urges, there can be no basis for the kind of optimism to which Rose's would-be Foucauldian analysis aspires. Without laying out the logical inconsistencies and possibilities of Foucault's concept, no affirmative understanding of it can be established. Indeed, in the absence of that move an affirmative biopolitics is vacuous. Without it, furthermore, Rose cannot explain the world he endeavors to describe. We might therefore turn the tables on his and Rabinow's comment on Hardt and Negri, as describing everything but analyzing nothing.[47] From the vantage of Esposito's *Bíos*, Rose analyzes everything but explains nothing, or at least nothing that goes beyond individual subjectivities emerging and shaped by the new biotechnologies.

Esposito's description of the contemporary strategies of biopower is limited to the introduction of *Bíos*, where he presents four extreme international examples: France in November 2000, when the courts ruled the right not to be born of an individual with serious genetic lesions; Afghanistan, one year later, when the oxymoronic notion of "humanitarian bombardment"—the superimposition of a declared intention to defend life upon the production of actual death—was widely televised; Henan province, China, in February 2003, when it became public that upwards of allegedly a million and a half persons had become HIV and HCV sero-positive after their blood, from which the plasma had been extracted to sell to the rich, was re-injected into them; and finally, April 2004, when a UN report revealed that around 10,000 infants in Rwanda were the product of mass ethnic rapes committed during the genocide. All these events, Esposito claims, have escaped traditional political explanation, and through the notion of "biopolitics" have found a new complex of meaning beyond the merely descriptive. For him these events are representative of most major political occurrences today. Through them we experience ever more intensely the indistinction between power and life. Moreover, our bodies that do the experiencing are no longer those of individuals or of sovereign nations, but rather, of a world that is both "torn and unified."[48] Such political occurrences, he maintains, are locked within a strange double-bind logic. On one hand, we witness a growing confluence between the domains of power (or law) and that of biological life. On the other, an equally close implication seems to derive with regard to death. The question, posed by Foucault in the 1970s and which today is more pertinent than ever, is why does a "politics of life" always risk reverting to a "politics of death"? Foucault, allegedly, never fully answered this question; according to Esposito he refused to decide between these two hermeneutic options—refused to decide, that is, between a biopolitics that produces subjectivity or produces death, or between "a politics *of* life and a politics *over* life."[49] Thus in Foucault's writings "the category of biopolitics folds in upon itself without disclosing the solution to its own enigma."[50] Foucault sensed this, thinks Esposito; beyond his "manifest intentions" there is "an impression of insufficiency," an "underlying reservation . . . as if Foucault himself wasn't completely satisfied by his own historical-conceptual reconstruction, or that he believed it to be only partial and incapable of exhausting the problem."[51] Esposito may not be the first scholar to suggest this, but he is the first not to dodge the issue by running to a biopolitics that is either absolutely positive or indecisive (like Rose) or totally negative (like Agamben). His mission is to solve the riddle of the two opposing hermeneutics, or at least to find a mode of acceptance between them, with the ultimate aim of suggesting a new subjectivity for our (unquestionably) biopolitical times.

The beauty of *Bíos* is that what it takes of Foucault's thought it takes seriously in order to extend it philosophically. Not here Foucault simply for the purpose of critiquing him, nor for mere explication or interlocution. Instead, Esposito probes Foucault's semantic logic through a close listening to his conceptualization of biopower. He does not sneer at the fact that Foucault's times are not ours, despite proximities, or that Foucault died before most of the biopolitical events, circumstances, and technologies that preoccupy Rose became apparent. While sharing Rose and Rabinow's regard of Foucault's concept of biopower as neither trans-historical nor metaphoric, "but precisely grounded in historical, or genealogical analysis" (as Rabinow and Rose admit), Esposito goes further by entering into Foucault's sources to unpick biopower as "a productive *dispositive*."[52] He is good at revealing how, for Foucault, "life" does not belong either to the order of nature or to that of history. "It cannot be simply ontologized, nor completely historicized, but is inscribed in the moving margin of their intersection and their tension."[53] The meaning of biopolitics (he quotes from Foucault's *History of Sexuality*) is sought " 'in this dual position of life that placed it at the same time outside history, in its biological environment, and inside human historicity, penetrated by the latter's techniques of knowledge and power.' "

Esposito does not investigate the assemblages of contemporary biopower. Instead, he looks in detail at its birth in language, how it was configured over time, and how it continues today as a rhetorical puzzle. He does this in order to throw light on what he detects as Foucault's indecisiveness vis-à-vis the question of whether biopolitics follows from or temporally coincides with modernity. "Does [biopolitics] have a historical epochal or originary dimension?"[54] First, he reveals how Foucault's writings on biopower were but a segment of a discourse specifically linked to the word "biopower," and which stretched back to the start of the twentieth century. In conducting this historicization of the semantics of "biopower," Esposito's purpose is not to put Foucault's use of the term in its proper place as it were, nor trivially to put his ideas in their proper and appropriate context, as historians might imagine doing. Rather, it is to bring to light—for the first time—the lexical tradition of biopower and to reveal its "contingent and semantic intervals." Above all, it is to highlight Foucault's break with past lexical traditions. For Esposito this allows the enigma of biopower to be explored from new angles, and helps develop a critical perspective on what he identifies as Foucault's epistemological uncertainty with regard to the "relation between the politics of life and the ensemble of modern political categories."[55] His claim is that Foucault's uncertainty is attributable to the failure to use a more "ductile paradigm," one that is capable of articulating in a more intrinsic manner the compelling logical interconnection between modernity and biopower. To fill this "semantic void" he proposes the paradigm of immunization. Although metaphorically laden, it is a paradigm that he believes can restore the missing link in Foucault's argumentation and reveal biopower as specifically modern, for it is only in modernity, he insists, that individual self-preservation (inherent to the political and medical meaning of immunization) becomes the presupposition of all other political categories, from sovereignty to liberty. "Only when biopolitics is linked conceptually to the immunitary dynamic of the negative protection of life does [the] biopolitics of life reveal its specifically modern genesis."[56]

The paradigm of immunization (*immunitas*) as an interpretive key for political philosophy is taken "completely elaborated" from Nietzsche's works—that same source for the conceptual language of Foucault's categories, including "biopower."[57] From the moment that Nietzsche proclaimed "the soul as the immunitary form that protects and imprisons the body at the same time," Esposito claims, "the most innovative part of twentieth-century culture begins to make implicit use of the paradigm."[58] Nietzsche is thus regarded as an "extraordinary seismograph of the exhaustion of modern political categories when mediating between politics and life."[59] For Esposito at least, understanding the "immunitary semantics at the center of modern self-representation" is the means to arrive at a more comprehensive analysis of the radical transformation of the political into biopolitics.[60] More than that, these semantics are the stuff for conceptualizing a space for political thinking around the social, or what he refers to as "*communitas*."

It is hard here to do full justice to the *immunitas/communitas* concept and its application, not least because the main body of the theory behind it is contained in Esposito's previously published studies, most importantly *Immunitas* (2002). The basic idea is that, just as the body operates with immunities and requires such immunities to continue life, so communities to survive as healthy regenerating systems also require immune subjects (people who do not fit in, who are "other"). But, just as in the body, too much immunity in the community engenders its self-destruction. It is the equivalent of the autoimmune system in overdrive attacking the cells of the body and killing itself. Required for a healthy creative community, Esposito maintains, is the essential interplay between *communitas* and the *immunitas*. It was this essential interplay that was missing in the Third Reich, the sustaining of which was predicated on the excessive killing of non-Aryan and other members of the community who, according to Esposito's immunitary paradigm, were in fact necessary for its survival. The absolute newness in that regime, he argues (echoing Bauman and others), "lies in the fact that everyone, directly or indirectly, can legitimately kill everyone else."[61] Thus death became the motor of the entire social mechanism and "carried the biopolitical procedures of modernity to the extreme point of their coercive power, reversing them into thanatological terms." The result was "an absolute coincidence of homicide and suicide," a fact—crucial for Esposito—that places it "outside of all traditional hermeneutics."[62] As he labors to prove, no political philosopher from Plato to Heidegger and Arendt has had the vocabulary necessary to explain such mechanisms and strategies. Foucault is the near exception, though he limited himself by putting racial language and concepts at the heart of biopower. Esposito analyzes each philosopher in turn and finds them all wanting, not least, by implication, Agamben and the other modern philosophers of bio-political governmentality. As Esposito sees it, there has been no real response to the question

> literally of life and death that biopolitics open in the heart of the twentieth century and that continue to be posed differently (though no less intensely) today. Certainly, the most pervasive attitude has been to repress or even ignore the problem. The truth is that many simply believed that the collapse of Nazism would also drag the categories that had characterized it into the inferno from which it had emerged.[63]

Hence he sees the black box of biopolitics remaining closed. Necessary to open it and render biopower affirmative is the penetration of the semantic logic and strategies of the Nazis, and the overturning of that regime's biothanatological principles of life, body and birth. The latter, he submits, need deconstructing and then reconstructing with respect to their deadly results. Only then will they reveal their originary and intense sense of *communitas*. This is no simple matter. It involves, among much else, turning inside out the philosophical concept of "flesh" elaborated by Merleau-Ponty; taking up Freud's ideas on the biopolitical superimposition of birth and nation; Simondon's theory of never-ending individuation; Spinoza's concept of natural rights, and Deleuze's late thoughts on affirmative biopolitics in his "elliptical and incomplete" last work, *Pure Immanence* (2001).[64] The long and the short of it is that in order to understand political action in our world and to avoid another thanatological catastrophe we need to change the language of politics and, above all, its definition of the subject.

Manifestly Esposito's agenda is not a simple instrumental one for political action through biopower. Rather, it is a demand for deep sensitivity to the history and language of biopower. Those who accuse him of providing no "practical philosophical alternatives" to the direct contestations of life of the sort with which he begins *Bíos* entirely miss the point.[65] If, like Rose, Esposito declines any invitation to specific political agendas, it is not because he thinks it is academically suspect, but rather, unlike Rose, because he fears that without the re-investigation of our current political categories, including those of political action, we will continue to unleash the lethal power/potential of biopolitics. For the philosopher as for the sociologist the task is not that of "proposing models of political action that make biopolitics the flag of a revolutionary manifesto or merely something reformist."[66] However, for Esposito, unlike for Rose, this is not because that is too radical a concept—too subjective, ideational, or romantic. Rather, it is because it isn't radical enough. What is required today, he argues, is not to think of life as a function of politics (which has anyway become impossible because of the way the categories have folded into each other), but to begin to think politics within the same form of life *as* life.[67] This is a commitment to a reformulation of thinking about being, and of finding a new way of "talking" about it. Behind it lurks the question that Rose never asks and would advise us against asking, namely, what is the ultimate purpose of all this "making life"; what ultimately do we want collective humanity to be? More practically, Esposito's call is to the recognition that the categories of contemporary sociology and political philosophy are themselves often inside older biopolitical conceptualization. If Rose is against reductionism in the analysis of biopolitics, Esposito is against reduction to any form of political philosophy that tries to explicate biopower while remaining locked within its categories. He invents the term "impolitical" to indicate this irreducibility. This then connects to his politics of non-transcendence—his belief that there can be no transcendability; existence does not transcend, he insists, there is only the decision of an individual to be *just there*—to be constantly creating and recreating existence. Such, for him, is an affirmative biopolitics that is no longer *over* life but *of* life. The pity is that it takes so long for him to clear the ground to stake his claim for a new subjectivity founded on a new trans-semantic language that always has the collective within it. It leaves him with a mere twenty pages for the dénouement.

Bíos is not an easy book to come to terms with, especially for those with little background in philosophy. Esposito's address is first and foremost to his professional colleagues, and above all to those of them engaged in debate over bio-political governmentality. Too, Esposito's text is not made the easier (though it is made the more fascinating and compelling) by his meticulous devotion to language and his defense of it over forms of structural analysis. It is not easy, but then why should it be? Indeed, *how* could it be? To get outside the whale never is. What *is* easy is facile snipping at Esposito's "high brow academic philosophy," and his production, allegedly, of "a verbal stunt in a semantic universe without gravity."[68] Another reviewer indicts him for spending too much time going back to Nietzsche and to the Euro-centric thanatopolitics of Nazism, instead of attending to changes in contemporary biology and neo-Darwinianism, and to changes in political economy in its relation to neoliberal theory, policy, and practice.[69] These critics have a point, but since there is in fact no one way to open up the black box of biopower,

Esposito's aim and pursuit are not unimportant. And so long as the semantic logic of the thanatopolitical practices of the Nazis remain untheorized we will, like it or not, always run the risk of a return. That language does *create* realities we have known at least since the literary turn, although whether language is the only thing that makes identity in today's world of biological enhancement is another matter. For Esposito, unpicking the bio-juridical logic of politics, particularly Nazism, lies at the very heart of formulating an affirmative biopolitics.

Yet also lying at that heart is his longing for a new trans-individual semantics that in fact follows in its origin Canguilhem's notion of normativity out of life itself. Canguilhem's notion does not subject life to the transcendence of a norm, but rather makes the norm the immanent impulse of life.[70] To this extent, Esposito shares company with Rose and, like him—albeit from a totally different perspective and for wholly different reasons—in fact fudges Foucault whilst proclaiming scrupulous devotion to his thought. *Bíos*, as Timothy Campbell insists, is "nothing short of a modern genealogy of biopolitics that begins and ends in philosophy."[71] But that's the problem: the breadth of Foucault's thinking on biopower cannot be done justice to through a focus entirely on political philosophy and its language. Foucault's genealogical method was an attempt to move beyond exclusively discursive regimes, by relating their appearance and change to elements external to discourse and knowledge. In thinking on biopower over the course of his life, he moved away from his initial analysis of the immanent rules and possibilities of discourse to an investigation of the necessary interaction with elements outside it, which he believed ordered knowledge at any given moment. He spoke to this in an interview in 1976:

> I believe one's point of reference should not be to the great model of language (*langue*) and signs, but to that of war and battle. The history which bears and determines us has the form of a war rather than that of a language: relations of power, not relations of meaning. History has no "meaning," though this is not to say that it is absurd or incoherent. On the contrary, it is intelligible and should be susceptible of analysis down to the smallest detail—but this in accordance with the intelligibility of struggles, of strategies and tactics. Neither the dialectic, as logic of communication, nor semiotics, as the structure of communication, can account for the intrinsic intelligibility of conflicts. "Dialectic" is a way of

evading the always open and hazardous reality of conflict by reducing it to a Hegelian skeleton, and "semiology" is a way of avoiding its violent, bloody and lethal character by reducing it to the calm Platonic form of language and dialogue.[72]

Esposito's *Bíos* cannot therefore claim to be a coherent continuation of Foucault's thinking on biopower, any more than this can be said of Agamben's *Homo Sacer*. True, Foucault did not prioritize the specific semantics of Nazi biopower and its biothanato principles, which Esposito sees as the precondition for a possibility of a new affirmative regime of biopower. It is quite in order, therefore, for Esposito to aim to solve Foucault's apparent epistemological uncertainty with regard to biopower, particularly in its Nazi disguise. But the question remains whether a philosophical investigation of his kind is able to solve such "uncertainties." Was Foucault really unable to formulate an affirmative connotation of biopower, or was it that his understanding of biopower and his genealogical method to investigate it were never intended for that purpose?

Bíos, no less than Rose's *The Politics of Life Itself*, fails to capture the inherent historicity of Foucault's thinking on biopower. Driven by the compulsion to turn biopower away from its negative past and towards an affirmative future, it too forecloses on the "messiness" that was constitutive of Foucault's view of "life"—a view perceived as methodologically empty and open to meaningfulness only through historically specific epistemological constellations and knowledge/power. Semantic engagement with biopower from the perspective of political philosophy cloaks quite as much as the mere empirical description of it in contemporary life. Although *Bíos* lies outside the would-be apolitical objectivity of *The Politics of Life Itself*, it does not in the final analysis arrive at its measure through political philosophy. By what it does as much as by what it doesn't do, it, like Rose's work, nevertheless compels serious attention to biopower—the engagement now essential for comprehending the biopolitical present that we have in one way or another become inescapably a part of. Neither book opens the black box of biopower as Foucault might have done, but both, in their wake, leave it substantially cracked.

10

The New Poverty of Theory
Material Turns in a Latourian World

AGAINST THE BACKDROP OF the privatization of public universities in Britain and grave concerns over the fate of the humanities in higher education throughout the anglophone world, this essay was written (again with Claudia Stein) out of a heady mix of literary vision and historical exasperation. The vision was Margaret Atwood's in *Oryx and Crake* (2003) and its sequel, *The Year of the Flood* (2009). Both novels seemed to us frighteningly accurate depictions of the present embellished with credible predictions of the future to flow from it. They deal with the desperate lives of the few remaining survivors of a global disaster precipitated by entrepreneurs in the genomic redesign of humans and animals. The main characters flash back to a time before the catastrophe when the graduates of the most prestigious science university, the "Crick and Watson," lived vacuously fulfilled lives under constant surveillance in high-security compounds, while the graduates of the last few remaining totally degraded colleges for the humanities have little prospect than that of making up advertising copy for the all-powerful pharmaceutical and bio-corporations.

This hardly seemed to us like fiction, and not only in light of trends in biology, such as the announcement of the first annual Francis Crick Memorial Conference, to be dedicated to the posthuman topic of "Consciousness in Human and Non-Human Animals."[1] In the United Kingdom, which in many ways has been the testing ground for applications of neoliberal economics to higher education, it comes as no surprise that the latest governmental exponents of privatization pin their hopes for the country's salvation on investments in corporate biotechnology—meaning very largely more pharma industry.[ii] Involved is the old rhetorical ploy of Prime Minister Harold Wilson in the 1960s extolling the then "white heat of technology," and President François Mitterand in France in the early 1980s, using science and technology as rallying points to unite citizens around visions of economic progress.[iii] The difference today is that those who would govern us need pray all the harder for the economic payoffs, not just the party political ones.

But this is not the only or the most important difference. More fundamental is that hand in hand with today's devaluation of the humanities in favor of the sciences is a particular vision of being human—a vision that is now increasingly entertained and given legitimacy through various branches of the humanities and social sciences themselves. According to this, humans are not much different from animals, inasmuch as their rationality is compromised or determined by their neurobiology. It is a view, derived entirely from the natural sciences, that goes hand-in-glove with the conception of today's world pushed for by affect theorists and others in the humanities on the incapacity of humans to act (or even needing to act) rationally or responsibly in our increasingly "messy" times.

And so to the source of our exasperation: an article published in one of the leading history journals (in February 2010) encouraging exactly this view.[iv] Written by Patrick Joyce and meant as a historiographically cutting-edge overview of the "material turn" in history, the article called for historians to look to the social sciences for methodological direction. In particular, it encouraged them to look to the Actor-Network Theory (ANT) of the French ethnographer of science Bruno Latour, along with the theory of affect closely allied to it. According to Joyce, the adoption of these theories from the social sciences would permit social historians to reignite the political flames of the 1970s that allegedly had been extinguished by the cultural turn of the 1980s and 1990s (the turn with which Joyce himself had been involved). The article seemed to us a sign of its epistemic times. But at a moment when the continuance of academic history was seen by many to be hanging in the balance and wholly dependent on the whims of ideologues of economic utility, how could a historian call for a *theoretical* posturing of politics around the mediation of that very ideology—in effect a call to embrace the values and virtues behind the catastrophe depicted in Atwood's dystopias? The other-planetness of it all seemed to us staggering.

Nevertheless, there was good to be gained by it. The crucial point it raised—precisely because it didn't—was the ever more desperate need for historians to get an ethical grip on themselves, or become aware of their political position. They needed to see, contra Joyce, that theory is never disembodied thinking that floats ahistorically in time unaffected by the moment of its emergence but, rather, is always *in history*. Flowing from this benefit of the article was that of recognizing the importance of assessing critique in history, as well as re-valuing the potential of history *as* critique. Brought home to us by reading Joyce was the question of what had happened to critique in the course of its move from social to cultural history. And what would happen to it now in the course of the move proposed by Joyce out of his new-dawned perception of the past "darkness of deconstructivism"? Through the work of François Cusset and others we had become aware that postmodernity in its Anglo-American expression was very much a reflection of neoliberal ideals and virtues. What, then, could be guiding this new move by a former exponent of postmodern theory? What was his agenda, and how did this derive from and sustain the current episteme?

It was not until after we wrote the essay that we encountered Steve Fuller's critique of Actor-Network Theory, written over a decade before.[v] In it he made clear that ANT was "little more than a strategic adaptation to the democratization of expertise" and not at all the politically radical move that ANT enthusiasts assumed was mirrored in their taking it up.[vi] Fuller's article was a refreshing confirmation of our suspicion of ANT as a form of mediated neoliberalism. Had we seen it earlier we might not have been quite so cautious in our commentary on those proposing ANT's deployment in history-writing. As it was, some of the colleagues who read our manuscript chastised us for being too polemical in drawing an analogy between the ethical significance of Joyce's ANT-based enthusiasm for (for example) studying the making of a "paper empire" in nineteenth-century India, and our paralleling of that to Adolf Eichmann's paper empire in the Third Reich. Perhaps, our colleagues suggested, reference to the Stasi in the former German Democratic Republic might prove a more measured means to drive home the argument for historians to exercise greater ethical awareness in the adoption and application of theory. We agreed and were persuaded for a time, but after reading Fuller there seemed less need for temerity. Which is not to say that Fuller's paper eliminated the need for ours; his audience were the less-than-critically-aware enthusiasts of Latour in social studies of science, whereas ours were historians likely never to have heard of Latour and who were in danger of taking him up as yet one more happy turn in history-writing, blithe to the implications of the adoption.

In many ways this essay is the appropriate one on which to close this volume. It is the latest thought on the journey so far, and one that returns starkly to the concerns raised in Chapter 1. Writing it was for me a reminder of why I had turned to the study of history in the first place: the will to make apparent that which is not, including the unveiling of the methodological means by which we undertake the unveiling. My targets have changed, and my toolkit has been refreshed, along with a better understanding of the crucial importance of locating myself in the business of historical critique. Consistent, though, I think, has been the effort to dig beneath the surface of any set of historical, historicized, and historiographical claims in order to expose their politics. For reasons returned to in this essay, the neo-Marxist approach of the 1970s, with its focus on the mystified mediations of capitalist ideology in science and medicine, had to be tempered in the 1980s and 1990s in light of postmodernist insights. But, as Joan Scott has recently reminded us, this tempering does not undermine either the ethical or the political purpose of historical critique to uncover that which has become concealed, obscured, or enchanted.[vii] All that the move to postmodernist critical thinking meant was greater awareness of the constraints upon *all* thinking, or greater sensitivity to what Nietzsche had announced, that none of us are ever the authors of our own thoughts. Revealing these constraints may not serve the old ideals of effecting major social and political change or cultural transformation, but it can at least enlighten us to the nature of the hedges around it, and those in need of burning.

The one thing that *is* different, and which underscored the writing of this essay, is the limited and shrinking space for the sort of critique indulged in here that cuts across other peoples' texts to reveal the epistemic ground they occupy. We account for this undermining in terms of the dominance of neoliberal philosophy in the academy. If we are right, then what is explored in this essay may *literally* be the last chapter: the going under of an endangered species.

Yet, who can tell? Who can tell what epistemic shifts lay around the next corner? As in the Isaac Asimov story of the man who travels back in time, accidentally kills a butterfly, and then returns to his own time to find words, meanings, and political orders radically altered or reversed, who knows what tiny circumstances and thoughts uttered now or in the future will morph into something totally undreamed of? Although the pessimist in me says that the way things look just now the species "historical critique" is decidedly Atwood bound, historians have little to lose and much to gain in trying to keep it alive. The academic discipline of history as we have known it may not survive the flood that currently overwhelms the humanities in general, but critical historical practice should and must. Historians have only to become more aware of how the biological and neurobiological sciences now shape their present, and so appreciate the urgency of the need to engage critically with it as proud exponents of the view that no knowledge creation is ever outside of its historical fashioning. Passivity and amorality should have no place, while self-reflexivity and political passion should have a great deal more.

> *As the great governing political and historical narratives of the 1960s and 1970s have been displaced by a much more fluid and changing political and economic world, new sorts of politics as well as of academic theory and practice become necessary to engage with this new world.*
> —Patrick Joyce, 2010

> *If I saw a glass of wine repeatedly presented to a man, and he took no notice of it, I should be apt to think that he was blind or uncivil. A juster philosophy might teach me rather to think that the offer was not really what I conceived it to be.*
> —Thomas Malthus, *Principle of Population*, 1798

It has become a commonplace that we have left the "cultural turn" behind. On this, as Joan Scott observes, "left" and "right" now converge.[1] The one laments the turn's political shortcomings, while the other congratulates itself for its prescience in having opposed it all along. The sounds of lip-smacking mix with nostalgia and ennui; the cultural turn gets debunked for having failed to live up to, or deliver on, a practice of history constitutive of the politics of dissidence, struggle, and change of the 1960s and 1970s (a social-history measure and lament from the start).[2] Simultaneously, among neoconservatives especially, a "yearning for certainty, security and stability" engenders a parade of "new empiricism"—or, more accurately, anti-intellectualist "new descriptivism"—that trivializes and denounces the "linguistic turn" and, overall, contributes to what Scott deplores as "the premature obituary" on postmodern critical thinking.[3]

It is from this state of affairs that Patrick Joyce seeks to deliver the Anglophone historical community in his "What Is the Social in Social History?" *Past & Present*, February 2010. It is to be accomplished, he believes, by taking the material turn—"the most significant of all recent 'turns.' "[4] Oddly though, in view of the old social history that Joyce's culturalist interventions of the 1990s moved against, this new move also purports a return to social history. It will even, Joyce hopes, recapture some of the political energy that characterized social history in the 1960s and 1970s. Many readers may find this as hard to believe as we do—at least on the basis of the article itself. More to the point here, though, is that Joyce's material turn is very much a sign of our political times, and that these have *already* begun to bear upon history-writing. Thus the politics to which Joyce now reports require some attending. Historians may choose for themselves whether they wish to turn in the direction that he now urges, but they should at least be aware of what the turn entails.

Despite appearances, the move that Joyce advocates as a means to arrest the rot of cultural history does not represent a historiographical *volte face* on his part. In the first place, the social history to be returned to is not that for which some historians are nostalgic. (In the attention that Joyce pays to Indian history in his article, for instance, there is not a subaltern to be seen). And, second, the new turn is not a departure from cultural history. As Joyce admits near the end of his paper, the "much talked-about end of the cultural turn is a . . . misnomer," for "we are still in it."[5] The social history that he seeks to reconstitute around the pursuit of material objects is really more an extension of the cultural turn. In fact, the particular material object with which Joyce has become enamored—paper, as the basis for a historical project on the construction of a bureaucratic empire in nineteenth-century India—might be regarded as extending from his prior literary-cum-cultural turn inasmuch as it attends literally to the stuff upon which words are written. In effect, through the material turn (or, more precisely, "the inclusion of materiality in our thinking about the social") Joyce proposes a maturation of cultural history, albeit in the interest of a rather peculiar "social history."[6] However, as we will argue, this would-be extended ("materialized") cultural history not only abandons characteristic features of the "old social history," but also jettisons crucial aspects of the cultural turn. In the process, it contributes to the demise of the very thing that Joyce seeks to save, the academic discipline of history as a critical forum.

Importantly for Joyce, the dynamism of the old social history was not primarily dependent on its practitioners' engagement politically with the world around them. In his recollection, notions of inequality, exploitation, social justice, responsibility, community, and human agency scarcely feature. It is not conceded that social history (in ways politically different from, but parallel to, the cultural history that came after it) might have been *most* notable historically for its liberation of the common people from the condescension of a history written through the eyes of ruling elites, and, simultaneously, for its worry over the suffocation of individual human action in the face of a perceived sociologically restrictive and ideologically ossifying Marxist materialism. It was these considerations that prompted the socialist humanism of E. P. Thompson, with its perception of class and class consciousness as uneven inter-relational processes transpiring within changing material conditions.[7] In 1978, in defense of the politics of historical materialism in history-writing, Thompson famously assailed "the poverty of theory" or the "ahistorical theoreticism" of those historians who had come under the spell of the French Marxist philosopher Louis Althusser.[8] But for Joyce the political energy of the old Marxian social history stemmed not from this, but primarily from its take up of "the big theoretical questions of its time." Social history at "its best," he submits—and what has been increasingly lost to view in subsequent developments, he maintains—was *"a critical engagement with theory as well as with other disciplines, especially the social sciences."*[9]

Above all, it is engagement with theory in relation to the idea of "the social" and its intellectual progress that Joyce wants to promote through what he calls the material turn, "for it is in understanding the material world that . . . assumptions about the nature of the social have to a large extent blocked the development of cultural and social history."[10] Theory is central to this exercise in intellectual plumbing, and it is to its legitimating around the "productive" historical pursuit of material objects that Joyce aspires by flagging up the political "exhaustion" of the cultural turn. Admittedly, the justification is novel, at least among historians.[11] If Joyce's former colleague and graduate student James Vernon is at all typical of historians who have recently taken the material turn (allegedly back to "social history"), the move has hitherto been made almost in a spirit of *counter*-theory. Vernon, in the preface to his study of hunger ("that most material of conditions") proposes "no discussion of method, no rehearsal of the debates between social and cultural history," and no "metatheoretical frame" for a history that he casts as *"after (and I hope beyond) the cultural turn."*[12] Joyce, in effect, comes to the rescue; through theory the descriptive material turn will be intellectually anchored in social history. And it will be politically animated through the address to theory, for in Joyce's view engagement with theory *is* political—albeit, significantly, done alone in the study.[13]

Joyce's theorizing is aimed, in particular, at overcoming some of the conceptual dualisms "that have bedeviled the social sciences, and which still form a great part of their disciplinary common sense."[14] Culturally turned studies ultimately failed to collapse these dualisms, or they fell back into them after having made the attempt, he argues. Structure/agency, structure/culture, "society"/"culture," subject/object, and, above all, "meanings and representations on the one side [the stuff of cultural history], and materiality and social relations on the other" he regards as his targets (while the binary between social history and cultural history continues to hover in the background). The will to this endeavor, and its orientation, derive from what Joyce submits are recent "key developments in the social sciences." It is through these, he believes, that we can now begin new deliberations on social theory, and through that forge a new "direct connection between the politics of the present and the politics of academic, intellectual activity." What exactly these politics are is never explained, and Joyce keeps his own to himself in the course of criticizing the "eclectic and utilitarian way" that historians have hitherto drawn on other disciplines. By borrowing from the social sciences in a more rigorous way, he thinks, we might even be able to convey our own "particular kinds of understanding [back] to these disciplines."[15]

We fundamentally disagree with both Joyce's construction of "the problem" and his proposed solution. In our opinion, the idea that the practice of academic history needs to be (or ever can be) transformed through re-theorization and re-politicization misconceives the problem, while the rationale for its solution—the need to move beyond a cultural history impaled by the linguistic turn and/or by the dualism representations/materiality—is a red herring. What is required today is not a further "turning" in the academic discipline of history, but (as a first step at least) a better understanding of the mechanisms by which the discipline intersects with, and competes against, other contemporary cultural practices involved in the production of history or memory. As Ann Rigney has recently argued,

> The key issue is not so much whether academic historians should be doing something radically different from what they are already doing, but rather whether their knowledge and expertise can be brought into circulation in a multi-media world which is not at the academic historian's bidding and where blockbuster fiction has arguably as much power as academic history to shape widely held views of the past or seducing people into forgetting history altogether."[16]

In the face of this threat to the very survival of the discipline, along with others that are constitutive and now painfully immediate [see Chapter 1], it is not internal reform that is required, but rather, how informed judgments about the past can be "brought into circulation and made to intersect critically with other less well formed or downright erroneous mythical views."[17] Needed for this (as a second step) is critical engagement with the forces beyond academic history that now not only operate for making up the past (such as the Internet and other instruments for what Rigney identifies as both "quality information" and "the proliferation of errors"), but also contribute to undermining the foundations of academic history. It is by becoming critically aware of the nature and function of these powers that the enterprise might also be intellectually justified (a possible third step).

By "critical" we mean taking a stance vis-à-vis the epistemological forces that compose the world of the historian's own present, or that shape his or her own gaze. Guided here by Joan Scott, Judith Butler, and Hayden White, we understand critique as a form of self-awareness essential to history-writing.[18] Hence, to be critical is to more than merely look to the analytical categories that historians and others apply in seeking to make the functioning of the world intelligible—categories such as "the political," "the social," and "the economic," or the various binaries that Joyce points to. It is also to inquire into how one's own values and epistemic virtues contribute to the upholding or abandoning of these categories. "Theory," of course, is necessarily a part of this making up of the world, but on its own it is not a critique of it, as Joyce supposes. Indeed, the application of theory to history-writing without the accompaniment of critical self-awareness easily becomes merely support for the particular visions of the framers of any particular theory. A critical history must contribute to investigating the current set-up and running of the present, including the theories that are drawn to it. One needs to ask how such theories, produced in different academic disciplines, intersect with interpretations of the world outside academia.

This is a fundamentally different agenda from that of Joyce, who also has a different understanding of critique. His goal, as he puts it elsewhere, in the same volume of "manifestos for history" to which Rigney, Scott, and White contribute, is to re-forge a political identity for academic history—"an ethics of the discipline of history expressing a political identity."[19] Lamenting the loss of such an identity—in part because of the "vacuous" and "ecumenical" borrowing of theory in history-writing, rather than rigorous engagement with it—Joyce argues for prioritizing a "critical history" that rethinks "the many pasts of freedom in relation to its various presents and its future possibilities."[20] His particular (freedom) concern is "an understanding of the past as a contribution to what is [an] ongoing rethinking of the present of liberalism."[21] However, this "critical history" (not only because it is premised on a notion of social history as the history of liberalism) fails to take into account the extent to which it is itself a historical construct. It perceives itself as something standing outside its time, and which can simply embrace theory. The question Joyce never poses is why liberalism (be it in nineteenth-century India or twenty-first century Britain) is something particularly worthy of historical investigation, or for that matter, why liberalism means freedom for him? In short, he fails to interrogate the underlying assumptions of his critical-cum-neo-social-history. From our perspective on critique, what he offers is not critical history, so much as a contribution to the making-up of our particular times. It is precisely this that we wish to illuminate through the critique of his article, and in so doing offer an illustration, we hope, of history-writing *as* critique.

It is, then, with Joyce's assumptions and the logic of his argument, along with the nature of the identity that he seeks for academic history, that we are primarily interested. To approach these we need first to explore the theory upon which his argument for the material turn rests—that is, the social science he wants us to embrace in order to advance his version of critical history. While the vision of the world inherent to his theory is a cause for our concern, as important to us is how, as a means for making the turn-beyond-the-cultural-turn, it closes off alternatives that might in fact be more beneficial to the survival of academic history-writing as an intellectually worthwhile critical enterprise.

It is important to note that Joyce does not set his sights on *all* the social sciences. Those he wants us to engage with are mainly concerned with investigating the boundaries between the natural and the human sciences. Many scholars are now involved in this task, among them the spatial geographer and cultural theorist Nigel Thrift, in whose work, as we shall see, Joyce finds much to admire. But it is above all the French sociologist, philosopher and ethnographer of science Bruno Latour who is most prominent here. Joyce rightly identifies him as one of the major players in the construction of a new notion and analysis of "the social," and it is primarily for this reason, and in this connection, that we too focus on Latour. It is worth bearing in mind that Latour is not merely a theorist confined to science studies. A gifted but exceedingly slippery writer, he has since the 1980s been regarded as one of France's leading public intellectuals.[22] Consistently he has shaken conventional thought and the tools for it in politics, science, philosophy, education and, in at least one stunning study, the cultural history of biomedicine.[23] Not unlike Althusser in his time, Latour, in his, has done a great deal to reshape people's thinking across the board, although in his case the manner of thinking is largely derivative of management studies and information technology. It is from these sources that emerges the metaphor of "networks" embedded in Actor-Network Theory (ANT), Latour's elaboration of which in the 1980s first drew attention to him.[24] Although he now feels a degree of discomfort with the cumbersome term, the conception remains central to his thinking.[25]

In truth, ANT is less a "theory" than an observational practice intended to challenge the assumptions and expectations of conventional sociology. As elaborated within science studies with respect to laboratory "objects" or "things" (such as bacteria), ANT refuses to seek out meaningful patterns, to prioritize any particular agency, or to lend causal power to one thing over another in accounting for the stabilization of anything.[26] In the face of a sociology of human interactions which presupposes something inherent to the world and which believes it capable of being revealed or discovered, ANT posits a sociology of interaction between "things, or objects, or beasts" in which each "*actant*" (so called) is regarded as important as the other, and from the observation of which interactions no discernible outcome is to be preconceived.[27] Never still, actants are forever entering into new relations at the same time as changing existing ones. It is this galaxy of parts constantly in action that constitutes Latour's "sociology of associations" or "sociology of networks" or, more commonly now (although usually without explicit debt to Latour), the investigation of "assemblages."[28]

The immediate context for the elaboration of ANT was one in which science had already come to be seen as a cultural practice bounded by oral traditions and subjective systems of belief. Such an understanding can be traced back at least as far as the work of Ludwik Fleck in the 1930s, which, in turn, inspired the weighty "paradigm shifting" of Thomas Kuhn in the 1960s.[29] By the 1980s claims about the social production or "social construction" of supposedly "objective" natural knowledge were well entrenched, especially among those concerned with the history of epistemology. The extension of that thinking continues to the present, although much of it is radically at odds with the more ontologically orientated work of Latour and his associates.[30] This divergence largely stems from the fact that Latour's alliance with social constructivism was never entire. From the start ANT was in tension with the sociology of scientific knowledge (SSK) as prominently represented in the 1970s and 1980s by the "Edinburgh School" (itself founded, in part, in opposition to the neo-Marxist history of science forged by Robert M. Young in Cambridge, famous for challenging the dualisms between science/society, fact/value, and science/ideology as mediations of capitalist ideology). In the mind of Latour and his associates, the problem with SSK, and by extension any sociologically informed analytical practice such as social history, was its one-sidedness: at the same time as it provided a devastating critique of philosophical realism, it relied on the categories and explanatory goals of sociological realism—explaining nature by reference *to* society.[31] ANT was/is a means to avoid this methodological asymmetry in order to talk about science and society (in essentially realist rather than constructivist terms) in a manner not prejudicing or privileging the location of agency. The ideal ANT study is one that allows all the actants to speak simultaneously. None are silenced, for through that a prior evaluative relationship would be confirmed, as in the binaries science/context, nature/politics, and society/culture. It is out of reaction to this tendency to dualisms (as the defining characteristic of modernity, Latour argues) that emerges the need to abandon all distinction between "things" and "people," between nature, culture, and society, and between the "knowing subject" (as once called) and the object of inquiry.[32] The practitioner of ANT must simply observe the heterogeneous network, or the infinite and changing associations and disassociations between things, and let the things speak for themselves—almost like "facts" in conventional history-writing.[33] Further, and above all, s/he must abstain from all discussions of power and ideology, and avoid any ethical commitment, since this would pollute the clinical nicety of the investigation.

The "social," so far as it can even be spoken of within the Latourian scheme of "things," is neither a fixed entity nor something that has innate substance, power, or force whereby to shape the lives of individuals, as assumed in conventional studies of gender, race, class, nations, and states. Rather, "the social" is but the "affect" *of* "the social" as it emerges hand-in-hand with its "scientific" description or its "unfolding" by analysts.[34] Inherently fragile, it is ever fugitive, and is therefore inexplicable. Humans can explain it, but only one-sidedly through recourse to such reductionist categories as economics and ideology. What they ought to do instead, says Latour, is investigate the specific constellations and conglomerates consisting of humans and things (the actants in their networks of conversation)—and this not only in the human sciences, but in the arts, ecology, and everywhere. Society, therefore, does not exist for Latour because it is ever in-the-making by the associations of actants; the social is not something standing at the actants' back, as it were, ready and waiting to have its would-be universal laws and nature revealed, as conventional sociology believed. Moreover, social groupings—actants creating the life of groups—need continually to be reconstituted in order to be kept together; their inherent fragility being threatened at every moment with destruction and change—processes that Latour fondly regards as creative.

This is a fundamentally anti-humanist view that is now fashionably celebrated as posthumanist. It not only renders material things as significant as humans, but disdains all sociological attempts to explain inter-subjective interactions, individual calculations, and intentionality.[35] What is more, it is a critique of critique as an Enlightenment humanist project in the Kantian tradition. Critique is dismissed by Latour as "the sacred task [of] modernity" (a stand that compels him in *We Have Never Been Modern* (1993) to pedal furiously backwards in order to extricate himself from any association of his own critical thinking with critique).[36] Critique "has run out of steam," he submits elsewhere.[37] Indeed, he condemns critique today for bordering on "irrelevancy" due to its preoccupation with binary divisions. Critique is also misguided in its efforts to totalize: Marxism, feminism, eco-ism and so on, all lose sight of the concrete configurations of any situation. The conventional critical sociologist, Latour argues, is forever seeking to configure patterns, whereas a good ANT narrative or "report" should emphasize and be committed to the infinite details of the associations between "things, objects, and beasts."[38] "Critique" is "inadequate," and—conveniently for the ANT scholar—any sociological critique of ANT can immediately be dismissed as antiquarian, be it "modern" or "postmodern." ("Postmodernism," in Latour's view, is but "a symptom [of modernity], not a fresh solution").[39] In the place of the "obstacle" of sociological critique, and in the name of "empirical studies" and "purification," Latour forwards the "sociology of criticism" as a part of the panoramic observational enterprise—an enterprise now generally bracketed as a part of the "descriptive turn" and widely adopted in the social sciences for connoting non-obeisance to essentialisms and causality.

These ideas are reiterated and refined in Latour's *Reassembling the Social: An Introduction to Actor-Network Theory* (2005), the text to which Joyce mainly refers. Behind them, and intended to motivate the embrace of them, is the historicized conviction that the world today is surprisingly unpredictable, ever changing and highly contingent—"an enormous ocean of insecurity broken up by some little islands of stabilized forms."[40] With its rapid information flows and material exchanges (as on eBay), its celebrated "breakthroughs" in biology and neurology, its ethnic pluralism, multi-tasking, financial flexibility, and so on, it has the *appearance* (through the way Latourians write and implicitly historicize it) of being infinitely more complicated and problematic than it ever was in the past. Given, too, the end of communism, it is a world, according to Latour, that conventional sociology in its relentless search for patterns cannot account for—unlike the "sociology of associations," which doesn't try.

But in its ontological version of the contemporary technologized world, ANT in fact performs *for* it: it substantiates that we now inhabit a place where people and things must be kept in constant circulation, always "flexible,"[41] exchanged but never hoarded; a place, not of "systems" or even "webs," but of bits and bytes, and where society cannot be imagined as a whole or a solidity, but only in terms of fragmented groupings that must be kept fragmenting and multiplying; a place where human activity, perceived as a succession of involvements in new projects and networks, becomes not only the ideal of life, but the measure of human equality;[42] a place in which the obsession with material things not only leads people to define themselves *through* material things and to think that their betterment only comes through their ever-greater consumption of them, but also, to global credit crisis; a place in which governments turn against their own people whenever and wherever they refuse to conform to the principles of market growth (seldom skimping on security forces to help with this); a place where the idea and practice of human agency is controlled, dampened, and/or denied; and a place, alas, where critique (as opposed to endless description and mere commentary) is increasingly circumscribed through the imposition of managerial stratagems that effect not just the organization and "impact" of research "products" (as measured among academics through assessment exercises), but thinking about thinking and knowledge itself. In short, as others have pointed out, it is the world of neoliberalism. According to David Harvey, neoliberalism is a theory of economic practice that entails the

> "creative destruction," not only of prior institutional frameworks and powers ... but also of divisions of labour, social relations, welfare provisions, technological mixes, ways of life and thought, reproductive activities, attachments to the land and habits of the heart. In so far as neoliberalism values market exchanges as "an ethic in itself, capable of acting as a guide to all human action, and substituting for all previously held ethical beliefs," it emphasizes the significance of contractual relations in the marketplace. It holds that the social good will be maximized by maximizing the reach and frequency of market transactions, and it seeks to bring all human action into the domain of the market.[43]

For Harvey, writing from a Marxian perspective, neoliberalism, through its creation of a tiny but extremely influential and obscenely rich elite, requires analysis in terms of a new class struggle. That in the UK, for instance, the top one per cent of income earners have doubled their share of the national income from 6.5 per cent to 13 per cent since 1985 is for him a cause for urgent analytical intervention. Moreover, such intervention is all the more necessary because of the remarkably thorough and profound way in which consent to neoliberalism has been constructed.

Whatever Patrick Joyce's political intentions might be, his effort to redirect and reinvigorate history-writing through the promulgation of the now widely diffused and variously re-blended ideas of Latour and his associates is, we suspect, doomed to the spread of this consensus, or, worse, to the further demise of what Jacques Rancière argues for as healthy intellectual "dissensus."[44] This, we fear, will be the fate of Joyce's material turn, the virtues of which he seeks to illustrate through sketching a bureaucratic "paper empire" in Colonial India. This is not to suggest that he might be deliberately smoke-screening neoliberalism, any more than we would want to attribute this to Latour (although it is worth noting that Latour has been described as "a leading beneficiary of the emerging neoliberal order").[45] At best, "neoliberalism" is a shorthand that will always be found to be too narrow, too wide, or too vague to capture the totality of what we as academics and citizens have engaged in, lived, and are living. It is to say, rather, that, like Latour in his thinking, Joyce's adoption of core Latourian ideas for historical purposes is wholly of its times and, hence, if unwittingly, a validation of them. As such, some may wish to critique his adoption of these ideas in terms of a mystified mediation of the ideology of neoliberalism. But from where we stand today that type of reductive analysis—essentially of Frankfurt School Marxism—would be intellectually retrograde, as we will amplify below. Nevertheless, at risk of what some might see as conceptual naïveté, it is tempting to think of Joyce's paper empire project as a manifestation in history-writing of a kind of false consciousness, and to liken it in that respect to another such empire in history that was indeed horrifically "materialized"—that accomplished by the German SS officer Adolf Eichmann, whose logistical talents helped organize the mass deportation of Jews. Eichmann's administrative "paper empire," too, made a fetish of rigorously observing and following (paper) things around. As Hannah Arendt concluded in her thesis on the "banality of evil," it thereby wholly obscured from its enthusiastic participants the gross consequences of their behavior.[46] Joyce's projected study of an administrative paper empire in nineteenth-century India as a means to reveal connections or networks likewise leaves little space for reflection on the moral or ethical standpoint of the historian involved in any such endeavor. Indeed, it is difficult to see where a concept such as "injustice"—so central to the old social history—could be located in this chase of material things. Does not the effort to bring materiality into our thinking simply obscure those rendered subaltern within paper empires? At the very least, it withdraws attention from the standpoint of concern with the uneven distributions of power consequent upon the nature and movement of such material objects.[47]

In these respects Joyce's thinking and his historical project share ground with a now fairly substantial body of literature devoted to the pursuit of materiality and "things"—some of which Joyce draws on in support of his argument and for his particular Indian historical illustration. Most of this literature—not all of it historical—has come into being since the early 1990s. In contrast to most social and especially cultural history, it is commendable for the importance it attaches to economics, and for not regarding economics in a reductive and dominant manner. But, typically, in its concern to avoid the use of unproblematized categorizations and binaries such as society/economy, nature/culture, and base/superstructure, it also de-centres humans and their actions, including the historian as an ethical actor. Favored, instead, is discussion of entities such as the state (or, as in Joyce's nineteenth-century India project, the "technostate"), regarded in terms of the ontological concepts of emergence, constant change, flexibility, and instabilities. In general, out of a disinclination overtly to pass judgment, it celebrates the description of "complicated things," of which humans, reduced to biological entities, merely constitute a part.

It is this orientation that attracts Joyce to Nigel Thrift's *Non-Representational Theory: Space, Politics, Affect* (2008), a work he praises for its "post-humanist, anti-individualist, relational" qualities.[48] In particular, Joyce admires it for moving the agenda of inquiry onto the constitution of "the social," or the understanding of "sociality" "*beyond questions of meaning and representation*" (italics added). The latter, of course, was that which motivated Joyce's previous work on "class" and "the social" as essentially discursive constructs—the stuff of meaning language. In Thrift's work, however, Joyce sees attention being drawn to "affect" as "an approach that attempts to capture the 'onflow' of everyday life by attending to the biological and the precognitive rather than to consciousness."[49] The prioritizing here of biology and neurology, and the displacement of "consciousness" as the predominant concept for thinking about humans, is typical of the literature on affect, which runs essentially on Latourian steam.[50] It is also now typical of much of the cutting-edge scholarship making the "neuro-turn" in literary, anthropological, *and* historical writing (where it contributes significantly to the "new empiricism).[51] In the affect industry it is only in a more pronounced manner that reliance is placed on selective research from "non-linear biology, quantum physics, cognitive science, and cognitive and affective neuroscience, as well as the related social science discipline of developmental psychology."[52] As Constantina Papoulias and Felicity Callard have pointed out in a valuable critical analysis of the logic of affect theory, its theorists draw not on a model of biology-as-destiny (something that the social history of science and medicine fought hard against from the 1970s, and cultural theory did in the 1990s), but rather, distance themselves from it, at the same time as summonsing a (neuro)biology "envisaged as a kind of 'ground' for culture."[53] The biology that is glimpsed reductively and uncritically through bodies' affective experiences "is essentially presented as a creative space, a field of potentiality that, crucially, *precedes* the overwriting of the body through subjectivity and personal history."[54] This move, moreover, operates as a helpmeet for a distinctly *political* project. . . . affect theory provides the language for an imagining of a biology that, since shot through with "the dynamics of birth and creativity" [Thrift], can act as a prototype for a certain progressive politics, a spatiality that precedes and trumps all manner of calls to order. In its previous incarnation, theory offered a hope that modes of interpretation, or indeed praxis, could contribute to a vitalized social reality. But in affect theory, an emancipatory and optimistic dynamic exists already in the present, in the various contours of biology to which attention has now turned.[55]

This "emancipatory" move, to which Joyce now urges historians, is intended by affect theorists to overcome methodological and theoretical differences between the natural and human sciences—"'the silly splits between science and humanities,'" as one of their number puts it. It is as if (add Papoulias and Callard) "all that childish silliness can now be set to rest in this new dispensation of interdisciplinarity and mutual borrowing."[56]

Joyce admits that he does not know where such an approach in history-writing might lead. He only desires that we be open to such borrowings at a time when "the writing of history seems itself to be in the process of change, moving beyond the employment of history in order to find out who we are and who we might be, and towards something else."[57] This "something else" is unclear, he confesses, while linking it to the supposition that we are "moving towards a new political dispensation" in which the politics of identity that were central to the cultural turn (and Joyce's own contributions to it) may have less force than capitalism and liberalism. While pointing to "a collapsed world economy and an endangered planetary ecology," as well as to the "*re-engineering of the interface of the natural and the social apparent in modern science*," Joyce almost suggests that it is in the name of the "invaluable legacy" of the old social history that we now have a moral duty to "continually pose the question 'What is the social?'"[58] This may not be what it sounds like—a rationale *for* the "new political dispensation," now to be consolidated through a history-writing that adopts theory from the social sciences. It is, however, a good illustration of the kind of language and stream of thought common to Latourians and affect theorists: the logic of "a" to "b" to "c" surrendering to the non-causal flow of "conversations" between, and for, clichéd descriptions of the world.

But does any of this add up to an alternative to the "enervated" cultural history (without politics) that Joyce wants us to leave behind? We think not, believing rather that he has failed entirely to understand not only the nature of the cultural turn, but also its politics. On the basis of the latter misunderstanding he justifies a material turn that is still well within those politics. Fundamentally, he seems not to understand that cultural history did not so much run out of steam in the 1990s as undergo a political change that rendered its theorization increasingly supportive of neoliberal values, morality, and epistemology. As clear from François Cusset's study of how "French theory" came to transform intellectual life in the United States, the cultural turn was part and parcel of a host of wider socioeconomic and cultural shifts, locally experienced and politically contingent.[59] Historians and others who made the cultural turn were indeed being radical in doing so, but, as we can see in hindsight, they were also focusing ever more narrowly on the mere description of the world around them; their conduct, subjectivities, and sensibilities being, as Foucault would have said, more and more governed by the prevailing socio-economic discourse. Put crudely, as others have, Thatcher's famous declaration that there is "no such thing as society" was not incidental to the historical moment for the acceptance of poststructuralist literary theory. Neoliberalism in the Anglophone world, with its cultural privileging of utilitarian individualism, was constitutive of the intellectual conditions of possibility for it.[60]

This is not to say that the theorizing or intellectualizing involved in the cultural turn in history, or in the humanities and social sciences more generally, was "neoliberal" in intent, any more than it is to suggest that it was unproductive, or that it was one-directional, or that it was "bad." Rather, it is to say that it was political through and through (as the "culture wars" made plain).[61] The theory involved, contrary to Joyce's inclinations and memory, was not merely esoterically "theoretic," as if existing in some ahistorical and apolitical space transcendent of its sociocultural moment. With regard to representations of the body and representations of "the other," prominent intellectuals such as Paula Treichler and Douglas Crimp in feminist, queer, and visual studies marched on the streets, often consciously implementing Foucault as a part of an activist politics [see Chapter 5]. Scholars such as these were also sometimes aware of the worrying connections between postmodern "postures" and neoliberal cultural politics. To uncritically embrace "postmodernist fragmentation and dispersion," Treichler sharply remarked in the early 1990s, "runs the risk of duplicating the move to a market-driven consumerist model of human populations in which the fragmentation of conventional identities [has been honed to] . . . a fine art."[62] It is true, as Joyce submits, that in the 1990s these activist politics and awarenesses began to flatten out in cultural studies, and that the distance between a political understanding of postmodern times and the assertion of postmodern theory became ever greater. Postmodernism increasingly became an industry of thinking without any intentional political engagement. Concern over forms of community and humanity, as in discussions on AIDS or subaltern studies, were increasingly marginalized.

It is precisely from this fate that Joyce wishes to deliver us. But, as we have hinted, his inputs from the social sciences do not present a new way forward; they are merely a continuation of the cultural history that he now shares with others in disparaging for its lack of political and critical appearance. The challenging of binaries and so on, that are held to afford a "new opening" in history-writing, are but a form of subscription to the social world that was already being written into through the cultural turn, and is furthered through the theorizations of Latourians and other cultural theorists influenced by the hype around "affect." For all Joyce's professed worry over the new political present, he does not critically question it; rather, through the theoretics of materiality he writes uncritically *into* the new political constitution of Anglophone knowledge, and keeps the politics of history-writing separate from it. But even at a superficial level there seems little reason to buy into this proposed "alternative" to cultural history. Given the current global financial crisis, mass unemployment, the collapse of pension funds as well as that of whole monetary systems, and given (germane to the present discussion) the ghettoizing of the humanities in favor of the natural sciences, engineering, and business and management studies as the only knowledges worth having, a skeptic might be forgiven for thinking that a return to the values and politics of the old social history would be far more appropriate (if not to concerns over class as a material force, to follow Harvey).[63] We share something of this view, but we do not propose it as an "alternative" to Joyce, for the simple reason that we have no desire to throw out the baby of postmodern critical thinking with the bathwater (Joyce). The "poverty of theory" we see in Joyce's new material turn is wholly different from that which Thompson saw in the Althusserians; epistemically, and hence epistemologically, our times are radically different and require other means of critical comprehension and engagement.[64]

Thus, to Joyce's would-be "alternative," we propose something else: re-engagement with the notion of "critique," especially as elaborated by Foucault, who, after all, was interested in history as critique. As both Butler and Scott have made clear, the idea of "critique," which was also elaborated by non-postmodern thinkers, is different from "criticism" or mere fault-finding (as in Joyce's approach to contemporary cultural history).[65] In the rush to judgment, as Theodor Adorno pointed out, the fault-finder invariably claims a higher sovereignty over the criticized and, crucially, assumes an independence from the analytical categories deployed.[66] The critic fails to see how the categories involved in criticizing are themselves historically composed, and how they, in fact, constitute a festishization of isolated categories. "Critique," in contrast to "criticism," Adorno argued, asks after the occlusive constitution of the fields of categories themselves. Foucault's notion of critique has much in common with this. "Criticism," he suggested in his famous 1984 essay on "What Is Enlightenment?" (responding to Kant's even more famous answer to the question of 200 years earlier), must be "a historical investigation into the events that have led us to constitute ourselves and to recognize ourselves as subjects of what we are doing, thinking, saying."[67] In contrast to Kant, whose answer proposed an exploration into the *limits* of reason, Foucault aimed at laying bare the very coming into being of such concepts. Those central to modernity, such as reason, he argued, were not transcendental but historical formations that required their historicization.

Critique, then, does more than merely call foundations into question, denaturalize political and epistemological hierarchies, and establish "perspectives by which a certain distance on the naturalized world can be had."[68] Above all, to follow Foucault's reach beyond Marxist critical theory, it seeks to bring into relief the very framework of the evaluation itself. It constitutes not compliance with given categories, but an interrogatory relation with the field of categorization—referring, at least implicitly, to the limits of the epistemological horizon in which practices are formed.[69] This kind of postmodern critique, as Hayden White points out in his "Afterword" in *Manifestos for History*, resolutely insists on "the present belonging to history," "the irreducible *historicity* of all things," and the "de-transcendentalisation of every regime of truth and knowledge."[70] From this understanding of critique the accusation of "ahistorical theoreticism" means, unlike for Thompson in his indictment of the Althusserians, the failure to realize and attend to the history of the present, including that embedded in theory. In practice it means constant self-questioning on the historian's part of the constructed constraints on, and the possibilities for, his or her thinking. It demands engagement with the norms in which the historian's self is formed, rather than merely seeking to resist them or supply some alternative to them (as in conventional Marxism, *or* as in the revisionist historian's ideal of professional salvation through ever more turns in history writing).[71] In such ways postmodern critical theory moves beyond Enlightenment thinking with its dependency on rational solutions external to the analyst him- or herself, and beyond the analyst's ostensible directions for progress and change ("turns"). Instead, it poses a "practice that has self-transformation at its core," central to which is the constant risking of the self as a judging subject, or the unstabilizing of the self as an analyzing subject.[72] Crucial, too, as White remarks, it must entail "not only seeking to discern the limits of our current ideas about history, but also the turning of a historical consciousness thus critiqued to the criticism of history's relations with the other disciplines of the human sciences and arts."[73]

Joyce, for all his professed debts to Foucault, remains wedded to an Enlightenment notion of critique based on the power of reason.[74] In contrast to White, he believes that for "the problem" of contemporary history-writing there are rational solutions which can be gained from the social sciences and readily be applied to the task of forging a new political identity for the academic discipline of history.[75] Talking against binaries he sees as a means to this end, but he fails to realize that such talking merely buys into another form of power (in the Foucauldian sense)—another set of "unspeakable" evaluative categories and presuppositions. These, as we have indicated, are in fact the would-be ontological truths of our times. Stuck within these, Joyce is unable to enter into an ongoing interrogatory relationship with the powers that create these ontological truths. To expose the limits of the current epistemological horizon he would have to de-stabilize himself as an analyst, self-reflexively examine his own moral categories (not least those underpinning his definition of liberalism), and abandon the idea of producing shiny new "alternatives." By failing to engage with postmodern critical theory—that is, with the knowledge effects within which critique operates, and which structure and restrain subjectivity—Joyce only contributes to the neoconservative "premature obituary" on postmodern thinking lamented by Scott.[76] In the final analysis, out of a species of what has come to be referred to as contemporary "discipline envy" in the humanities, he does little more than hold a mirror to the social sciences.[77]

But what *should* a critical history that uses materiality as a theoretical tool look like? As we see it, it would have to ask, first of all, what is involved in making this move in terms of its own blind spots? In Joyce's case this would mean asking "to what am I blinded by arguing for the inclusion of materiality in our thinking about the social"? What are the norms and epistemic virtues propagated in the visionary connectionist world? And, second, it would have to ask what it means to borrow from some of today's social sciences. We have suggested that Joyce's blindness consists in not questioning the concepts or methodological tools he draws from the social sciences—above all, that of "materiality." Noticeably, he never tells us for what this particular tool is useful (or, for that matter, never defines it), let alone compares it to other of its uses in history-writing, past and present. Further, we have suggested that any borrowing from the social sciences today (however well meant) is not a borrowing from the social sciences at all, but rather, an indirect lifting from the natural sciences that now heavily inform them, especially biology and the neurosciences. Ultimately, therefore, any such borrowings only consolidate another binary, that between the sciences and the humanities, with the sciences firmly on top. In other words, the borrowing exercise does political work, and it is of a kind that is not a great deal different from that in postcolonial studies, which, famously through "critical" practice, also left subalterns voiceless.[78]

We agree with Joyce that any critical history must engage explicitly with theory, and we couldn't agree more with his sentiment, quoted at the outset, that to engage with the world today we need "new sorts of politics as well as of academic theory and practice."[79] But we couldn't agree less with his conception of "the problem" facing academic history today, and his proposed solution. To us his ahistorical and apolitical assumption that theory is separable from the social world, rather than constitutive of it, is as problematic and worrying as his idea that we can simply pluck approaches from the social sciences and apply them to history-writing. Joyce's turn to materiality in the name of re-theorizing "the social" in history-writing comes down to, at best, an idle diversion, and, at worst, a dangerous distraction while the academy burns, wiki-history thrives, and "history" flourishes in the service of political causes right, left, and centre. His agenda and the historical project he proposes only obfuscate (as it they reproduce) central features of the political and economic dispensation inherent to the current malaise in history. Moreover, they blunt the tools of critique that might enable us to dampen these fires and at the same time justify the value of academic history as a heuristic for critical engagement with the present. Joyce's material turn—his "theoretical route to critical history"—as a means to move beyond the perceived political sterilities of cultural history—is but a pale red herring delivered up at a time when the season for turnings is past.

NOTES

Preface

1. Pickstone, "Brief History," 7. His comment was mainly prompted, he tells me, by the ignorance economy operating among today's makers of health policy.
2. Foucault, "Truth and Power," 114.

Chapter 1. The End?

1. For examples—themselves referring to an abundance of other recent literature—see N. Rose, *Politics of Life*; A. Clarke et al., *Biomedicalization*; Ong and Collier, *Global Assemblages*; Lock and Nguyen, *Anthropology of Biomedicine*; Downey and Dumit, *Cyborgs and Citadels*. For more particular studies, see Petryna, *When Experiments Travel*; Biehl, *Will to Live*; Morgan, *Icons of Life*; and Gregg, *Virtually Virgins*.
2. N. Rose and Novas, "Biological Citizenship"; N. Rose and Noval, "Neurochemical Selves"; Cooter, "Biocitizenship."
3. N. Rose and Abi-Rached, "Birth of the Neuromolecular Gaze"; N. Rose, "Screen and Intervene."
4. Adelson, "Visible/Human/Project," 359.
5. Waldby, *Visible Human Project*.
6. Consistently, Nikolas Rose has been critical of those scholars who speak anachronistically of the social evils of biology, such as racism. For him, moreover, the engagements that people now have with their biology are conceived in largely positive terms: *Politics of Life*. See also Rabinow, "Artificiality and Enlightenment."
7. Fuller, *New Humanist*.
8. Gray, *Cyborg Citizen*; Stadler, "Neuromance."
9. N. Rose, *Politics of Life*, 26; N. Rose, *Inventing Our Selves*.
10. Fishman, "Making of Viagra." Another instance of de-psychologization is to be found in the shift in the regard of homosexual men from "psychiatric deviants" to "a pathological risk group," a shift that began before the AIDS epidemic in the 1980s: McKay, "Imagining 'Patient Zero,' " 89ff.

11. Agar, *Liberal Eugenics*; Agar, *Humanity's End*.

12. The economics of biomedicine and biotechnology have yet to receive the attention they deserve. The *novelty* of "large scale capitalization of bioscience and mobilization of its elements into new exchange relations" has been highlighted by Rabinow and Rose ("Biopower Today," 203), and the *transformation* of the valuation of science among scientists themselves as a result of the inroads of venture capitalism since the late twentieth century has been observed by Shapin (*Scientific Life*). But, overall, in Anglo-American critical discourse, preoccupations with "French theory," on the one hand, and the interconnected decline of sophisticated Marxist analysis, on the other, have left the economic elephant in the room largely unobserved and, until now, almost unobservable. Exactly how "biopolitics became bioeconomics" (N. Rose, *Politics of Life*, 32–39) remains sketchy. A major exception is Cooper, *Life as Surplus*. See also Styhre and Sundger, *Venturing into the Bioeconomy*. On the genomic "revolution" as a process of biocapitalist commoditization of life, see Rajan, *Biocapital*; and Fortun, *Promising Genomics*. For a listing of some of the most recent sociological literature on bioeconomy, biocapital, and biovalue (including bioprospecting and biopiracy and with attempts at definitions), see A. Clarke et al., "Biomedicalization," 7–10. For another way of thinking about biocapital (in terms of class), see Harvey, *Neoliberalism*. On contemporary scientists as entrepreneurs, see below.

13. Scheper-Hughes, "Last Commodity"; N. Rose, "Value of Life." Paralleling and reinforcing the commoditization of the body has been the commoditization of spirituality, on which see Carrette and King, *Selling Spirituality*.

14. I am grateful to Adolorata Iorizzo for reminding me that it was Francis Bacon, in *The Advancement of Learning* (1605) and elsewhere, who first strategically forwarded the notion of medicine as a "benefit to mankind."

15. Powers, *Galatea 2.2*, 308.

16. For example, see *The Fair Society* by the biologist Peter Corning; *The Master and His Emissary*, by the psychiatrist Iain McGilchrist; and *Incognito* and *The Tell-Tale Brain* by, respectively, the neuroscientists David Eagleman and Vilayanur Ramachandran. Like Hatemi and McDermott's *Man Is by Nature a Political Animal*, such works purport to build on a deep neurobiological understanding of human nature in order to provide biological reductionist views of history and political behavior in what Corning calls our new "biosocial context."

17. Jasanoff, *Designs on Nature*, 7, 14–16. Cf. Hopkins et al., "Myth of the Biotech Revolution"; and Marks, "What Is Biotechnology?"

18. A. Clarke et al., "Biomedicalization," 6.

19. Quotation from Safranski, *Nietzsche*, 287.

20. Nietzsche, *Vom Natzen und Nachteil der Historie für das Leben*, 8. See also Safranski, *Nietzsche*, 320.

21. Hammer and Champy, *Reengineering the Corporation*, 31; Liu, *The Laws of Cool*, 17; the phrase "sloughing off of yesterday" is Drucker's in *Managing in Turbulent Times*, 5, 43–45.

22. Among historians and philosophers of science, see, for example, Daston and Galison, *Objectivity*; Hacking, *Historical Ontology*; Ogilvie, *Science of Describing*; and Shapin and Schaffer, *Leviathan and the Air Pump*.

23. It is also helpful to see postmodernism as a "worldview based upon a distinctive conception of history": White, "Postmodernism and Textual Anxieties," 27.

24. Useful on the contributions of postmodernity to thinking about the body and history is Shildrick, *Leaky Bodies*. On key differences with the epistemology embedded in the social history of medicine, see Cooter, "Traffic in Victorian Bodies," and, with specific reference to the categories of "the social," "history" and "medicine," Chapter 3, below.

25. Forman, "(Re)cognizing Postmodernity," 1, 6.

26. N. Brown, "Shifting Tenses."

27. On Foucault's methods, see Dean, *Critical and Effective Histories*; Smart, *Foucault*; Foucault, *Power/Knowledge*; Foucault, "Nietzsche, Genealogy, History"; and Foucault, *Archaeology of Knowledge*. See also N. Rose, *Inventing Our Selves*, chapter 1: "How Should One Do the History of the Self"; and Koopman, *Genealogy as Critique*.

28. By the "history of the present" Foucault did not mean the history of "now," but rather the pursuit of the temporal specificities for different concepts and theories at any moment in historical time. See Smart, *Foucault*, chapter 2: "Questions of Methods and Analysis," 39–63.

29. For Marx on the historicity of all concepts, see Marx and Engels, *German Ideology*, and Young, "Marxism and the History of Science"; on the modernist project, see White, "Postmodernism and Textual Anxieties," 29.

30. N. Rose, *Politics of Life*, and see the discussion in Chapter 9, below. For a critique of the claims to novelty and the transformative power of biosociality and biocitizenship, see Raman and Tutton, "Life, Science, and Biopower."

31. Lorimer, "More-Than-Representational"; Thrift, *Non-Representational Theory*.

32. For critiques of affect theory from the perspective of its dependence on neuroscience, see Leys, "Turn to Affect," 440; Papoulias and Callard, "Biology's Gift"; and Korf, "A Neural Turn."

33. Quotations from Thrift, "Pass It On."

34. Rose, "Human Sciences in a Biological Age," 2.

35. Thrift, "Pass It On."

36. Veeser, *New Historicism*, "Introduction," xi.

37. This view, to which Marx and Engels gave air in *The German Ideology*, gains strength from Butler, "What Is Critique?"; Butler, "Critique, Dissent, Disciplinarity"; and Scott, "History-Writing as Critique," esp. 25.

38. As Sander Gilman has recently warned, this new functionalism in history-writing has not only serious historiographical implications, but dangerous political ones, such as permitting new representations of "obesity" (not understood *as* a representation) to perform acts of social stigmatization: "Representing Health and Illness," 295–96.

39. For examples of historians and others in the humanities taking the neurobio turn, see Stafford, *Echo Objects*, discussed below in Chapter 4; Smail, *On Deep History*; Smail, "Essay on Neurobiology"; Konner, *Evolution of Childhood*; Freedberg, "Memory in Art"; and Bryson, "Neural Interface." An announcement for a workshop at the Ludwig-Maximilians-Universität in Munich in 2011 reveals how neurobiology, within the ideology of "multidisciplinarity," is readily incorporated into historical study. The workshop proposed that "disciplines can make major advances when they synthesize their ideas and methods with those of other disciplines. This workshop focuses on the ways in which neuroscience might help us understand history (and, ideally, vice versa). Following the lead of Daniel Smail...." (http://h-net.msu.edu/cgi-bin/logbrowse.pl?trx=vx&list=H-Sci-Med-Tech&month=1102&week=b&msg=w8geKkxPspFEPF8W7h WyTA), cache 9 February 2011. Other examples of this shifting of discussion onto the neural plane include the Royal Society of Arts' report by Rowson, "Transforming Behaviour Change," and a call for papers for a conference at St. Anne's College, Oxford, on "Persons and Their Brains," 11–14 July 2012 (http://www.iranamseycentre.info/persons-and-their-brains-conference), cache 10 December 2011.

40. Koopman notes how Foucault's anti-totalizing agenda now confronts "a (perhaps desperate) resurgence" of grand master narratives of the past: "Foucault Across the Disciplines," 8. Like other defenders of Foucauldian methodology, however, Koopman does not connect this to neurobiological turns.

41. R. Smith, *Being Human*, 6. Smith argues that "there is no position outside the historical forms which human life takes for an absolutely objective or eternally valid view." He therefore seeks to put back into historical context the ways of thought "in which the particular human dimension they describe emerged as an object of knowledge" (3). "If we do not come to understand these contexts," he adds, "we will never come to understand ourselves" (7).

42. Sweeney and Hodder, *Body*, 3.

43. Adam Bencard, e-mail to author, 30 April 2008. On how post-genomic biology promotes itself, see the journal *Proteomics*.

44. Adam Bencard, e-mail to author, 30 April 2008.

45. In the anglophone world, unlike in Continental Europe, the human sciences or "moral sciences" (*Geisteswissenschaften*) were separated from the natural sciences

(*Naturwissenschaften*). In the English language "science" typically means only the "natural sciences," the unfortunate consequence of which is that biology easily becomes the dominant means of answering the question "What is human nature?" The separation also fosters the positivistically inclined belief that the natural sciences are *explanatory*, while the social sciences and the humanities are *interpretive*—and "merely" so.

46. This also creeps in from less post-postmodern positions. See, for example, Kirk and Worboys, "Medicine and Species," which, in the interest of leveling the importance of animals with humans in the history of medicine, promotes an obligation for scholars to re-engage on a larger-than-usual and less anthropocentric and essentialized canvas. However, they provide no argument for why we should, beyond the novelty of it, or the fact that historians hitherto haven't. Animal- or plant-centricity tends always to the view that a discourse on animals or plants can be parallel to that on humans, forgetting that both are human discourses. Like talking on rocks, animal history (or non-human exceptionalist history) can only be a project entertained by humans, as Nietzsche made clear at the outset of *Vom Natzen und Nachteil der Historie für das Leben*. An example is the Society for Plant Neurobiology, founded in Florence in 2005. According to its website, plants are not only "as sophisticated in behavior as animals," but are also "capable of refined recognition of self and non-self" (http://www.plantbehavior.org/neuro.html).

47. Leys, "Turn to Affect," 440.

48. R. Smith, *Being Human*, 23; Stadler, "Neuromance," 143. The new neuro study of "human nature," Stadler suggests, is inherently reactionary in being deployed by those with a vested interest in extolling the brain as the material and cultural centre of being, and hence effecting romantically to *re*-cohere understanding of the world and man's place within it. Nikolas Rose, referring to essentialist notions of "human nature" and "humanity" embedded in the persona of the medical agent, opines that "the historical relativists among us might as well save their breath, for [their] . . . empirical and historical observations are radically insufficient to interpret [what he refers to as 'somatic ethics']," *Politics of Life*, 97. On how the concept of "humanity" was deliberately narrowed over time, and on contemporary possible reversions to "pre-scientific" attitudes to it, see Southgate, " 'Humani nil alienum,' " 73.

49. Mitchell, "Art, Fate, and the Disciplines," 1026–27.

50. Daston, "Historical Epistemology," 282.

51. Collingwood, *Idea of History*; Carr, *What Is History?* In late nineteenth-century Germany, by contrast, such deliberations on history-writing were commonplace. See Iggers, *German Conception of History*; Oakes, *Weber and Rickert*; and Schnädelbach, *Philosophy in Germany*.

52. I expand on this theme in "La médecine dans la pensée historique contemporaine." On the familiar and superficial idea that history-writing is shaped by its times and its authors own experiences, see Burrow, *A History of Histories*.

53. I am hardly the first to argue for the need for historians to get present to the present (see, for example, Scott, "Evidence of Experience"), but materially and epistemically the present is no longer that of 1991 or earlier. Historians of literary theory, or informed by it, have been rather more alert to this need; see, for example, Jenkins, *Re-Thinking History*, and Ankersmit, "History and Post-Modernism."

54. As Halttunen has pointed out, "the function of historical narrative, like that of any other ideologizing discourse, is not transparent representational but moral and political": "Cultural History and the Challenge of Narrativity," 166.

55. Eagleton and Beaumont, *Task of the Critic*, 152. Presentism has recently taken on a new and more politically sinister form in the work of various historiographers concerned with "presence"; see the discussion in Chapter 4, below.

56. Foucault defined an episteme as the historical a priori that grounds knowledge and its discourses and thus represents the conditions of possibility for it within a particular epoch; "retrospectively as the strategic apparatus which permits of separating out from among all the statements which are possible those that will be acceptable within, I won't say a scientific theory, but a field of scientificity, and which it is possible to say are true or false. The episteme is the 'apparatus' which makes possible the separation, not of the true from the false, but of what may from what may not be characterised as scientific": Foucault, "Confession of the Flesh," 197.

57. White, "Public Relevance of Historical Studies." See also Cooter and Stein, "Can Historians Change Philosophy."

58. Jordanova, *History in Practice*, 203.

59. On facts as artifacts produced in human-material practices, see Haraway, "Situated Knowledges"; and Daston, "Historical Epistemology," 284–85. On "facticity," see also Veyne, *Writing History*, 32ff., and the sources cited in notes 76 and 112, below. In the 1930s, Ludwik Fleck made clear that "objective information" is as impossible as "objective truth," since information is always orientated toward an aim; it is partial and biased by the very essence of its existence and exchange. See Bonah, " 'Experimental Rage.' " On the assumptions and taboos in conventional history, see Ermarth, "Closed Space of Choice," 52–58.

60. Butterfield, *Whig Interpretation* ([1931], Butterfield, *History and Human Relations*, 103; Marwick, *Nature of History*; Marwick, "Two Approaches"; Fulbrook, *Historical Theory*; Mommsen, "Moral Commitment." See also Evans, *Defence of History*. Similarly, John Tosh in his *Why History Matters* can only *assert* the moral view that our world would be "better" governed and administered if a "better" understanding of the past were available to both decision makers and the public. I don't necessarily disagree, although we might recall that Hitler and his cronies certainly felt they had a better understanding of the past than those in other political regimes, past and present. At issue for historians making such moral claims for professional history-writing is what actually *is* their desired present and future.

61. White, "Response to Arthur Marwick," 242; see also Cracraft, "Implicit Morality"; Vann, "Historians and Moral Evaluations"; Rüsen, "Historical Objectivity"; and Jenkins et al., *Manifestos for History*.

62. For example, Macmillan, *Uses and Abuses of History*, who calls for historians to avoid "theory." As Scott points out, "theory" in such texts tends to be associated with leftist politics, and "objectivity" perceived as its antidote ("History-Writing as Critique," 23). The criticism of "theory" as something conducted at the expense of "practice" goes back to the philosophy of Francis Bacon; see Southgate, " 'Humani nil alienum,' " 67. On the insistence of "context" above all else in history-writing, see Rosenberg, "Mechanism and Morality," 182.

63. Scott, "History-Writing as Critique," 22.

64. Ermarth points out that the problem with the rationalist agendas of historians is that "they have allowed them and their cultural functions to remain unquestioned for too long. Meanwhile, unnoticed, the world has moved on, and in it our mental tools, derived so substantially from conventional history, are increasingly inadequate.... The options are not worse, just different." "Closed Space of Choice," 63. Or, as Southgate says, the "philosophy of history has become detached from its practice" and vice versa: " 'Humani nil alienum,' " 67.

65. "Objectivity," as the American historian Peter Novick observed a quarter of a century ago, has been one of the central sacred terms of professional historians, like "health" for physicians, or "valor" for the profession of arms. Thus, for many historians, "what has been at issue [in the debate over objectivity] is nothing less than the meaning of the ventures to which they have devoted their lives, and thus, to a very considerable extent, the meaning of their own lives": *Noble Dream*, 11, see also 564ff.

66. Ogilvie, *Science of Describing*. On the achievement of the Renaissance to raise "experiment" to a principle of research, see Weber, "Science as a Vocation," 10. On the scientificity of historians, see White, "Public Relevance of Historical Studies"; and White, "Response to Marwick," 236–37.

67. Directly or indirectly, since at least the late eighteenth century, this rationale has served as the basis for pressing social reform and legitimating social activism. See R. Porter, *Gibbon*. Social historians of medicine routinely invoke this rationale, as, for example, Rosenberg, in his claim that "medicine need not be what it is": *Our Present Complaint*, 130. No historian, so far as I know, has sought critically to deconstruct this ideal; to do so would threaten most of what the academic discipline stands for.

68. Daston and Galison, *Objectivity*, 363.

69. Ibid., 41.

70. Ibid., 349, 355.

71. Ibid., 355. Latour, drawing on the work of Isabelle Stengers and Vinciane Despret, points out that the "popular ideal of science [as] made of a mute disinterested scientist letting totally mute and uninterfered with entities run automatically through

sequences of behaviour" is contrary to the reality of the *"passionately interested scientist who provides his or her object of study with as many occasions to show interest and to counter his or her questioning through the use of its own categories . . . neither distance nor empathy* defines well-articulated science. . . . To the contrary, one must have as many prejudices, biases as possible, to put them at risk in the setting and provide occasions of manipulation for the entities to show their mettle": "How to Talk About the Body," 218, 219.

72. This problematic is akin to that which Bourdieu identified as "the logic of practice": *Logic of Practice*. I am grateful to Stephen Casper for drawing this to my attention.

73. Nor does Shapin when observing the same phenomenon in science in his *Scientific Life*.

74. A claim also made by Bourdieu in *Homo Academicus*. Historically, historians and natural scientists have shared something of the same logic as producers of knowledge, as Roger Smith argues in *Being Human*. It is also the case that many historians and social scientists still commit to a nineteenth-century-like scientific faith in ("mechanical") objectivity. Today, on what actually counts for knowledge in the humanities as opposed to in the natural sciences, see Post, "Debating Disciplinarity," 756.

75. Daston and Galison, *Objectivity*, 363 (emphasis added). A similar point, although without reference to epistemic virtues, is harbored in Haraway's notion of "situated histories and situated naturecultures": "Situated Knowledges." It is also one of the main themes in Pickstone, *Ways of Knowing*. On the non-disappearance of "older forms of cultural classification of bio-identity," see also Rabinow, "Artificiality and Enlightenment." As Nietzsche put it with regard to religious belief: "God is dead: but given the way of men, there may still be caves for thousands of years in which his shadow will be shown": *The Gay Society* (1882), quoted in Carrette and King, *Selling Spirituality*, 13.

76. See, for example, Poovey, *History of the Modern Fact*; Daston, *Biographies of Scientific Objects*; Davidson, *Emergence of Sexuality*; and Hacking, *Historical Ontology*.

77. Deleuze, *Foucault*, 25–26. Cf. De Certeau's analysis of Bourdieu and Foucault in *Practice of Everyday Life*, chapter 4: "Foucault and Bourdieu," 45–60.

78. For Foucault, one of these constellations was "biopower" (on which see below, Chapter 9).

79. Eagleton, *Illusions of Postmodernism*; Cusset, *French Theory*. On the admitted complexities of defining "postmodernism," see Featherstone, *Consumer Culture and Postmodernism*.

80. Foucault, "Questions of Methods," 79. See also Smart, *Foucault*, 39–63.

81. Hitherto, for the most part, only historians of historiography have systematically addressed postmodernity, favorably or unfavorably: see, for example, Evans,

Defence of History; Eley, "Is All the World a Text"; and Iggers, *Historiography in the Twentieth Century*, 118–33.

82. Forman, "(Re)cognizing Postmodernity," 3, who draws a crucial distinction between "postmodernism" and "postmodernity," and the fallacy of thinking the former the cause of the latter.

83. Daston and Galison, *Objectivity*, 376.

84. On invisibility-making through scientific "revolutions" as formulated by Kuhn, see Krüger, "Does a Science Need Knowledge of Its History," 222.

85. Indeed, more recently, Daston has suggested that, vis-à-vis science studies at any rate, the future lies in philosophy: "Science Studies and the History of Science."

86. On "modernity" *in* and *of* medicine, see Cooter, "Medicine and Modernity." In the sense of being the fearful state of experiencing the new, modernity might be said to be the condition of possibility for the modern discipline of history, at least in Germany where it took shape in the early nineteenth century. See Iggers, *German Conception of History*. The same fearful state, I suggest below, may also prove to be the condition of possibility for its demise.

87. Cited in Labisch, "Von Sprengels 'Pragmatisher Medizingeschichte,'" 243.

88. Among exceptions is that by the historian of ideas, Frank Manuel, who, observing that no festschrift was forthcoming upon his retirement, produced his own: *Freedom from History*. Another exception is Rousseau, *Nervous Acts*. Most stunning, however, is Young's *Darwin's Metaphor*.

89. Preeminent among exceptions are Steedman's *Landscape for a Good Woman*; Steedman, *Past Tenses*; and Steedman, *Dust*. A more objectivist reflection on historical identity is Eley's *Crooked Line*. On historical imagination as always inside its political times, see N. Z. Davis, "A Life of Learning."

90. Cited in Ferber, *Bioethics in Historical Perspective*.

91. Mintz, "What's in a Word," 227.

92. See Burke, *Death and Return of the Author*; Tallis, *In Defence of Realism*; R. McDonald, *Death of the Critic*.

93. Laslett, *The World We Have Lost*.

94. Bonfield et al., *The World We Have Gained*.

95. See Whitfield, "A Genealogy of the Gift."

96. Weber, "Science as a Vocation." 12.

97. Cooter, "Crisis."

98. The patient first came into historical focus in 1976 in the essay by the sociologist Nicholas Jewson, "Disappearance of the Sick-Man." In the social history of medicine the manifesto for attending to patients was Roy Porter's "The Patient's View." I discuss the latter's anachronism in "NeuroPatients in Historyland" (see also Chapter 8, below). On Tuskegee, see Reverby, *Infamous Syphilis Study*.

99. Work on the medical marketplace, initiated by Roy Porter, was largely indebted to *The Birth of a Consumer Society*, by McKendrick, Brewer, and Plumb. On the origins and historiographical location of this interest, see Brewer, "The Error of Our Ways." Among Porter's contributions, see his *Health for Sale*. For a recent study that folds the past into contemporary individualist consumptions of health and health products, see Ueyama, *Health in the Marketplace*. The same move to commoditization and the marketplace explanation occurred in the history of popular science; see Nieto-Galan, "Antonio Gramsci Revisited."

100. In an interview in the *Lancet* in 1997 Porter confessed that among the books that inspired him most as a historian was E. P. Thompson's *Making of the English Working Class* (1963): Interview, 1410.

101. On the latter, see Pickstone, "Rule of Ignorance."

102. I was reviews editor of *Social History of Medicine* from 1986 to 1992.

103. See, for example, Sudhoff, "Aus meiner Arbeit," cited in Rütten, "Karl Sudhoff," 99. See also Stein, "Divining and Knowing."

104. Linkner, "Resuscitating the 'Great Doctor.'" However, Sigerist clearly lived up to supporting socialized medicine in Canada and the Soviet Union, and constantly argued for universal healthcare in America. See Fee and Brown, *Making Medical History*.

105. E. R. Brown, *Rockefeller Medicine Men*. An important book for our thinking at the time was Braverman's *Labour and Monopoly Capital*.

106. For a review of the history of medicine in Britain in the 1970s and 1980s and its connection to the history of science, see Pickstone, "Development and Present State of History of Medicine in Britain."

107. On Kuhn and Kuhnianism, see Fuller, *Kuhn*.

108. On the various incarnations of science studies, see Shapin, "Here and Everywhere," and Pestre, "Thirty Years of Science Studies." On its "demise," see Daston, "Science Studies and the History of Science." On the science wars, see K. Parsons, *Science Wars*; Sardar, *Kuhn and the Science Wars*; Jardine and Frasca-Spada, "Splendours and Miseries"; Ross, *Science Wars*; and Labinger and Collins, *One Culture*.

109. Teich and Young, *Changing Perspectives*.

110. It was subsequently perceived as a "pioneering contribution to Diaspora studies" and published as *When Paddy Met Geordie*. The history of this undertaking is recounted in the book's preface, which also reflects on what it means to do history as a West Coast Canadian.

111. Reprinted in Young, *Darwin's Metaphor*. Young's classic statement "Science *is* social relations" appeared in *Radical Science Journal*, 1977. On the Radical Science Collective (established in 1971) and on Young as Radical Science's Bernal, see Werskey, "Marxist Critique."

112. Young, "Can We Really Distinguish Fact from Value." Fleck's famous essay of 1935 was not widely available before its 1979 publication, *Genesis and Development of a Scientific Fact*. The concern with "facts" and "objectivity" as filtered through Foucault and Georges Canguilhem came a decade or so later: see, for example, Daston, "Baconian Facts," and the sources cited above (note 59). A new edition of Canguilhem's *Normal and the Pathological* (an analysis of the philosophy of biology under conditions of the collapse of fact/value) by the postmodernist publisher Zone Books appeared in 1991.

113. "How do I form my thoughts, and how do my thoughts form me?" Nietzsche asked in *Beyond Good and Evil* (1886), quoted in Safranski, *Nietzsche*, 55–56.

114. The Victorian novel as the exception is explored in Sparks, *Doctor in the Victorian Novel*.

115. In this tradition, persuasive at the time, was Werskey's *Visible College*; Arditti et al., *Science and Liberation*; Easlea, *Liberation and the Aims of Science*; H. Rose and S. Rose, *Radicalisation of Science*.

116. The most notable literary studies of science, by Levine, *One Culture*, and Beer, *Darwin's Plots*, come in for criticism for their complicity with the consciousness of science in Sparks's *Doctor in the Victorian Novel*. See also Cordle, *Postmodern Postures*. On the implicit complicity of social studies of science, see Fuller, "Why Science Studies Has Never Been Critical of Science."

117. Arendt, "What Is Authority," 93.

118. Golinski, *Making Natural Knowledge*; Bury, "Social Constructivism"; Armstrong, *New History of Identity*.

119. Golinski, *Making Natural Knowledge*, 128–30. This methodology, he added, located me, as a historian, at the heart of a hermeneutic circle that depended for its legitimacy on accepting its own Marxian theoretical presuppositions (rather than proving them). He was right.

120. Eagleton and Beaumont, *Task of the Critic*, 149. Similarly, "the rejection of all philosophies of history is itself a philosophy of history insofar as it sets a limit on the kinds of meaning that history might be conceived to have": White, "Postmodernism and Textual Anxieties," 27, n. 2.

121. Cf. Nieto-Galan, "Gramsci Revisited." Nieto-Galan points out that "Gramsci seems to have been diluted by the more powerful force of the so-called postmodern approaches" (455), although he also notes that in 1976, Perry Anderson, one of the founders of the *New Left Review*, criticized Gramsci for "never properly theoriz[ing] the site of specific mechanisms of bourgeois hegemony, and fail[ing] to ground a proper revolutionary strategy as a result" (471).

122. Jay, *Dialectical Imagination*.

123. Cooter, "Power of the Body." On the impact of Douglas's work in science studies, see Shapin, "Citation for Mary Douglas." While "cultural meaning" found a

further outlet in Jacobs's *Cultural Meaning of the Scientific Revolution*, "the body" became *the* word for a generation of historians and literary scholars (see Chapter 4, below).

124. Bored by my study of the Irish, I joined a *Madness and Civilization* reading group in Durham in 1971 and soon found myself applying for fellowships in Cambridge in order to pursue the history of rationality in relation to madness. The reading group was led by Peter Barham, a Tavistock psychologist who subsequently produced, among other scholarly works, *Forgotten Lunatics of the Great War*. The interest in madness (i.e., in mediations of our own psychology and thought processes) was then fashionable and led to my "Phrenology and British Alienists." Like John Forrester, whom I got to know in Cambridge in this connection, *Madness and Civilization* left me puzzled, but at the same time moved and intrigued. Forrester, *Seductions of Psychoanalysis*, 286.

125. Other contemporary theorists were also fairly remote to me then, perhaps the most important of whom was Haraway, who gained early support from Bob Young. Her deconstruction of the category "human" dates from the 1980s. For a revealing historicization of *Worlds in Collision*, see Gordin, "Abgrenzung und Demokratie." On the Foucault/Chomsky "debate," see Rabinow, "Introduction," *Foucault Reader*, 3–7.

126. Some of the references to this work and its historiographical location are contained in Cooter and Pumphrey, "Separate Spheres and Public Places."

127. The move to orthopedics came by way of an invitation from John Pickstone. He was then gathering resources for what was to become one of the United Kingdom's most vibrant centers for the history of science, medicine, and technology. One of those resources was the Mancunian pioneer of modern orthopedics Sir Harry Platt, who was then nearing his one hundredth birthday and resting on a rich store of local sources and political knowledge of the sort with which to launch a historical project. Taking it up was the only quasi-calculated careerist move I ever made, for the reason that writing on "fringe" medical subjects, I was informed, was no way to curry the favor of the United Kingdom's main funding body for the history of medicine, the Wellcome Trust. To them I needed to demonstrate that I could do "the orthodox stuff" too (and without lefty orientation, it was hinted). Only later did I discover that there was a historical connection between orthopedics and phrenology: the "Orthophrenic Institute" set up in Paris in the 1820s. It is referred to in my *Surgery and Society*, 12.

128. See, in particular, Cooter and Sturdy, "Science, Scientific Management."

129. On neoliberalism, see Harvey, *Neoliberalism*. His definition (3) is quoted below in Chapter 10.

130. Quotation from Theroux, *Ghost Train*, 4.

131. With regard to science, this is the duty of the historian posited by Nelkin, "Promotional Metaphors," 30.

132. Thomas Holt puts this well in his response to Joan Scott: "It is a necessary reminder that neither my identity, my history, nor my experience is autonomous from systems of coercion within a larger history. It is a necessary factor in what contribution I might make to my craft and to the academy because they will likely arise precisely from my critical consciousness about my location—a location history has determined—at the institutional boundary between the powerful and their victims": "Experience and the Politics," 395–96.

133. Nussbaum, *Not for Sale*; Nussbaum, "Skills for Life."

134. Megill, "Fragmentation and the Future of Historiography."

135. See, for example, Giroux, "Schooling and the Culture of Positivism," 264.

136. Especially after the publication of Novick's *Noble Dream*.

137. Žižek, "Plea for Leninist Intolerance," 544.

138. See the special issue of *Critical Inquiry*, Summer 2009. I follow Robert Post and Steve Fuller in defining a discipline as "not merely a body of knowledge but also a set of practices by which that knowledge is acquired, confirmed, implemented, preserved, and reproduced." A discipline is thus itself a kind of epistemic virtue—a regulative ideal—that is sustained, among other things, by peers acting as gatekeepers through refereeing publications and vetting applicants for academic jobs. Post, "Debating Disciplinarity," 751; Fuller, "Disciplinary Boundaries," 302. Cf. Osborne, "History, Theory, Disciplinarity"; and J. H. Hall, "Cultures of Inquiry and the Re-Thinking of Disciplines."

139. Iggers, *Historiography in the Twentieth Century*, 121.

140. Most notably, Jones, *Languages of Class*, 1983.

141. Iggers, *Historiography in the Twentieth Century*, 133, 140, 144. On "pluralism" as ideology, see Rooney, *Seductive Reasoning*.

142. Grafton, "Britain: The Disgrace of the Universities." In the United Kingdom the effects were manifest in the government's adoption in 2010 of the report chaired by Lord Browne on changes in the funding of higher education ("Securing a Sustainable Future for Higher Education"), for critical commentary on which see Collini, "Browne's Gamble"; Head, "Grim Threat"; Luckin, "Crisis"; and Vernon, "The End." See also Evans, "Wonderfulness of Us."

143. See "Campus Cuts" and "How the University Works," *Chronicle of Higher Education*, 27 June 2010, at http://chronicle.com/blog/campuscuts/21. Grafton's article, "Britain: The Disgrace of the Universities," annoyed some American colleagues for suggesting that the situation was rosier in the United States than in the United Kingdom. I am grateful to Bruce Moran for alerting me to this, and to Pamela Gilbert for broadening my knowledge of the situation in the United States. See also Grafton,

"History Under Attack"; Grafton, "Can the Colleges Be Saved"; and Cusset, *French Theory*, 41ff., on how the leaders of American universities—once captains of erudition—became the gurus of "learn to earn."

144. Power, *Audit Society*.

145. Rosenstone, "Space for the Bird to Fly," 17.

146. Beevor, "Eyes on the Prize." 79.

147. This is not to slight the task of communicating intellectual problems to untutored, but receptive, minds; as Max Weber appreciated, that is perhaps the most difficult pedagogical task of all ("Science as a Vocation," 4). But simply canvassing votes is another matter. At the end of the day, the salvation of professional historiography "does not lie in the vulgarization of its practices," as Hayden White has argued, but rather in "amending our notions of history's importance as a field of study": "Public Relevance of Historical Studies," 335.

148. For a trenchant critique, see Collini, "Impact on Humanities Researchers." On the need to contest the rhetoric and values of the administrators of higher education, see Vernon "The End."

149. In many ways the urtext for this now commonplace popular conception of humanness is Friedrich von Hayek, *The Sensory Order*, first published in 1952. My thanks to Steve Fuller for reminding me of this.

150. Barnes, "Elusive Memories of Technoscience," quoted in Forman, "(Re)cognizing Postmodernity," 5 (emphasis added).

151. Forman, "(Re)cognizing Postmodernity," 7. On how professional scientists have increasingly become experts in the service of political and economic power, see also Pestre, *Science, Argent et Politique*; Nowotny et al., *Public Nature of Science Under Assault*; and Jasanoff, *States of Knowledge*. On the transformation of medical research in the United States in the 1960s and 1970s from a public service ideology to a research-for-profit mentality, see Dutton, *Worse Than the Disease*.

152. I am grateful to Roger Smith for forcing this thought.

153. Overy, "Historical Present," 34. Or, as another author puts it, referring to the humanities in general, the kind of "learning that disturbs and disrupts ... that cannot be relied on for ulterior purposes and yet is wholly necessary for keeping open the options of being human, that cannot be defended on the grounds of what is it good for because no one can know what it is good for until it has been explored, examined, and weighed in each generation": Jeffrey Sammons, "Squaring the Circle" (1986), quoted in Post, "Debating Disciplinarity," 760. Such statements tend, however, to privilege the autonomy of the historian; cf. Scott, "Evidence of Experience," 783.

154. Forman, "(Re)cognizing Postmodernity," 9.

155. Ibid., 13, n. 16.

156. See Fuller's review of Smail's *On Deep History*; Fuller, "Putting the Brain at the Heart of General Education."

157. Fuller, "Warwick 'Human Futures.'" For similar, see Rose, "Human Sciences in a Biological Age," 5–6; Kinner, "Groundhog Day;" and Newton, "Truly Embodied Sociology."

158. Guha, *History at the Limit of World History*, quoted in Nieto-Galan, "Gramsci Revisited," 468.

159. For example, Warwick University's vice-chancellor, Nigel Thrift (a new-wave human geographer and affect theorist with an interest in the history of time and space), who uses a Fuller-like historicized essentialist biology to promote contemporary neoliberal political economy. See, for example, Thrift, "Pass It On."

160. For stunning examples, see McGilchrist, *Master and His Emissary*; and Smail, *On Deep History*. For various critical perspectives on contemporary "brain-centricity," see Vrecko, "Neuroscience, Power, Culture"; Stadler, "Neuromance"; Tallis, *Aping Mankind*; Crawford, "Limits of Neuro-Talk"; Korf, "Neural Turn"; and Cromby et al., "Neuroscience and Subjectivity."

161. Fukuyama, *Origins of Political Order*, which builds on the sociobiology of E. O. Wilson. See Wade, "From 'End of History' Author." Wilson's *Sociobiology* was first published in 1975. Foucault also posited an "end of history" but for an entirely different reason: the end of the "era of man" as conceived by Saint Simon and others in the nineteenth century. See Kurzweill, "Michel Foucault."

162. See Head, "Grim Threat."

163. Cf. Edward Said in 1981 on "today's intellectuals . . . not interested in dealing with the world outside the classroom": *Representations of the Intellectual*, 458. Presumably, Said was targeting the most ethereal of postmodernists, the historical legitimacy of which is testified to in Cusset's analysis (*French Theory*).

164. "Integrative medicine" is the new label for "holistic" or "alternative" or "complementary" medicine, but with "evidence-based medicine" pitched as the rhetorical enemy. "Translational medicine," driven by the search for pathways to domains of "utility," is likely a consequence more by stagnation in pharmaceutical innovation and by patterns of funding than by new patterns of collaboration between laboratory scientists and clinicians—something announced every decade or so as the "decisive breakthrough" in cancer research. My thanks to Illana Lowy for this critical sharpening. On the history of "teamwork," see Cooter, "Teamwork."

165. Weingart, "Interdisciplinarity," cited in Post, "Debating Disciplinarity," 755. For a very different justification of interdisciplinarity (as a means to promote non-hierarchical social relations), see Piaget, *Main Trends*.

166. On the merits of "dissensus" over intellectual consensus, see Rancière, *Aesthetics and Its Discontents*, 115ff.

167. Latour, "How to Talk About the Body," 217.

168. Weber, "Science as a Vocation," 3. On such "cross-pollutions" being relegated to the dustbin of history, see also Stadler, "Neuromance," 152.

169. Žižek, "Plea for Leninist Intolerance," 544.

170. Latour, "How to Talk About the Body," 217.

171. Cooter, "Traffic in Victorian Bodies," 526. On the history of "medical humanities" and its re-fashioning, see Dolan, "Second Opinions"; and H. J. Cook, "Borderlands." Cook's wisdom emerges, he says, from the perspective of one who, as former director of the Wellcome Trust Centre for the History of Medicine at University College London, has had "personal responsibility for other people."

172. Quotation in Overy, "Historical Present," 34.

173. In the United Kingdom this ever-greater scientization is apparent, not least, in the encouragement of the major funding bodies to studies of "Science and Culture" and "Science and Politics"—in other words, for academics to be the functionaries of politicians justifying techno-science entrepreneurism. See the websites of the Humanities and Education Research Council, the Economics and Social Science Research Council, and the Leverhulm Trust.

174. Kuhn, quoted in Iggers, *Historiography in the Twentieth Century*, 120.

175. Although Rosenberg is one of the few historians of medicine to express in print the current need to pause over the shared values and assumptions of fellow practitioners, he can only assert that "to be effective historians must maintain their disciplinary identity, their own criteria of achievement and canons of excellence": "Anticipated Consequences," 203. The basis of this faith is neither questioned nor qualified.

Chapter 2. Anticontagionism and History's Medical Record

i. Wright and Treacher, *Problem of Medical Knowledge*.

ii. For a recent appreciation of the article, see Stern and Markel, "Commentary." For some of the subsequent literature on, and continuing debate over, the history of anticontagionism, see Hamlin, "Predisposing Causes"; Baldwin, *Contagion*; and Pelling, "Meaning of Contagion." Engagement with my own intervention appears in Coleman, *Yellow Fever*, 188–94.

iii. Rosenberg, "Erwin H. Ackerknecht." Ackerknecht's political views are not entirely clear. According to Illana Lowy, he was "totally shaken" by the alleged killing by Stalin's agents in a Paris hospital in 1938 of Trotsky's son, Leon Sedov (1906–38), "to the point of being disgusted by politics. He ended up rather conservative and extremely pro-Israeli (though he was not Jewish)." E-mail to the author, 24 June 2010.

iv. Delaporte, *Disease and Civilization*.

v. See Mildenberger, "Die Geburt der Rezeption."

vi. For a lucid account of the commonalities and differences between the critical thinking of the Frankfurt School of Critical Theory and Foucault, see Scott, "History-Writing as Critique."

vii. See, for example, the response of Adrian Desmond and Jim Moore to the reviewers of their *Darwin* (1991) in the *Journal of Victorian Culture*, 152. For a more considered contemporary response to Foucault in relation to the crisis in Marxist thinking in the 1980s, see Smart, *Foucault, Marxism and Critique*.

viii. Foucault himself disdained such labeling. In *The Archaeology of Knowledge* he wrote, "Do not ask who I am and do not ask me to remain the same: leave it to our bureaucrats and our police to see that our papers are in order" (17).

ix. When Palmer moved to Toronto, he invited Thompson to lecture there. Thompson by then was ill with prostate cancer, and it was the social historian of medicine and health activist Sam Shortt, MD (then the Hannah Professor of the History of Medicine at Queen's University), who ministered to him. Thompson was in the hands of a good social constructivist: see, for example, Shortt, "Clinical Practice and the Social History of Medicine."

x. Scott, "The Evidence of Experience"; Latour, "How to Talk About the Body," 213.

xi. On Stacey and the commitment to the ideology of egalitarian rights to health, see Mold, "Patient Groups," 507.

1. Ackerknecht, "Anticontagionism."

2. Easton and Guddat, *Writings of the Young Marx*, 400.

3. Ackerknecht, "Anticontagionism," 567.

4. Ibid., 589.

5. On the now extensive secondary literature in the history of medicine and in social history, see Pelling, *Cholera*, 4. For social histories of cholera in different national and regional contexts, see Rosenberg, *Cholera Years*; Bilson, *A Darkened House*; Evans, *Death in Hamburg*; and Durey, *Return of the Plague*. Dissenting from Ackerknecht's interpretation in an empirically justified tone similar to Pelling is G. P. Parsons, "British Medical Profession and Contagion Theory."

6. Pelling's other point, that prior to the 1832 cholera outbreak most practitioners in Britain were neither contagionists nor anticontagionists, but either were uncommitted, were contingent-contagionists, or were disbelievers in the reality of cholera, has been substantiated by Durey, *Return of the Plague*, esp. chapter 5: "Cholera, Medicine and the Medical Community," 101–34.

7. Pelling, *Cholera*, 302.

8. For contemporary comment on the passions raised in the debate, see Winterbottom, "Thoughts on Quarantine," 62.

9. S. Smith, "Plague," 530.

10. On the seventeenth-century demarcation of "science" from "pseudoscience," see Mendelsohn, "Social Construction of Scientific Knowledge," 3–26, and on the demarcation problem generally, see Wallis, *Margins of Science*; and Hanen et al.,

Science, Pseudo-Science and Society. On the changes in seventeenth-century economy, see Appleby, *Economic Thought and Ideology*, esp. 245; Dickson, "Science and Political Hegemony in the Seventeenth Century," and Young, "Why Are Figures So Significant." For a lucid critique of positivism, see McCarthy, *Critical Theory of Jürgen Habermas*, esp. 5–8, 40–52.

11. See Cooter, "Deploying "Pseudoscience"; and Young, "Getting Started on Lysenkoism." At least one contagionist likened the methodology of the anticontagionists to that of phrenology's "inventor," F. J. Gall: Gooch, "Plague," 241–42.

12. Cited in Diamond, "Anthropology in Question," 425. Or, as Horkheimer put it: "The very concept of 'fact' is a product—a product of social alienation; in it, the abstract object of exchange is conceived as a model for all objects of experience in the given category": *Eclipse of Reason*, 81–82.

13. Quotation from S. Smith, "Contagion," 137.

14. Pelling, *Cholera*, 299; cf. Cantor, "Critique of Shapin's Social Interpretation," esp. 256.

15. Pelling, *Cholera*, 60, 63, 9, 20, and passim. See also Luckin's review of the book, 566.

16. Without such consideration one could not hope to explain the following except by calling it an untruth or a coincidence: "In 1817 I published a small treatise on Atmospherical Disease; in 1819 I was examined on the subject of pestilential disease by a Commission of the House of Commons, who were at that time investigating the subject of contagion and the quarantine laws. This circumstance led me to the knowledge of the opinions and writings of Dr Maclean [on whom see below] whose extensive book is perhaps the first of any consequence that appeared since the period alluded to on the true nature of pestilence: and it is somewhat remarkable that that author and myself should have been writing on the same subject, and with similar views at the same time, while as yet totally ignorant of the occupation of each other": T. I. M. Forster, *Facts and Enquiries*, 31.

17. These two points are made explicit in Rosenberg, *Cholera Years*, 7–8.

18. On the former point, in 1819, for instance, the anticontagionists lobbying for quarantine relaxation were dubbed "senseless fools" by George Canning and the practice of quarantine upheld partly on the grounds that its abolition would damage English commerce by causing foreign ports to impose stricter quarantine on British vessels in the event of an epidemic. See Granville, *Autobiography*, vol. 1, 172, 363–64. (In the absence of detailed knowledge on whose commercial interests were thought to be protected by quarantine and whose by abolition, it is difficult to speak with certainty; more important to realize is that we do not have to, at least not in direct terms.) On the latter point, see J. C. McDonald, "History of Quarantine"; and Pelling, *Cholera*, 26–27.

19. Historians of cholera similarly trivialize the knowledge by rehearsing its social and political rationalizations, believing that in the absence of "true" epidemiological knowledge in the first half of the nineteenth century, these rationalizations constitute a "social explanation" for the knowledge and the belief in it. Under the guise of "contextualizing" medical mis-belief, the constituents of the knowledge get left out. It is an indication of how Pelling's views have changed since the writing of her book that in her review of Morris, *Cholera 1832*, she deplores that "following Kuhn and Ackerknecht [Morris] finds that social factors have a role only when the scientific solution is not clearly given or when the significance of a 'discovery' fails to be realized": *Annals of Science*, 203.

20. See, for example, Pickstone on Bichat's physiology: "Bureaucracy, Liberalism and the Body"; Lawrence, "Nervous System and Society"; Cooter, "Power of the Body"; Cooter, *Cultural Meaning of Popular Science*; and, above all, Figlio, "Chlorosis and Chronic Disease," which is written from the explicitly political perspective of medicine as a part of capitalist social relations. See also Figlio, "Sinister Medicine."

21. This bears in mind the point of Merleau-Ponty's "phenomenological positivism" that "man does not live only in the 'real' world of perception. He also lives in the realms of the imaginary, of ideality, of language, culture, and history. In short there are various *levels of experience*": Edie's introduction in Merleau-Ponty, *Primacy of Perception*, xvi.

22. Like other Marxist terms, "mediation" takes its meaning from the context of Marxist usage, where it usually denotes that which is ideologically interposed between reality and social consciousness in a single thing, as, for example, an object of art. It is an indirect reassembly (never a simple reflection) in an alternative form of the ideological or conceptual apparatus of certain social relations, which functions actively to promote a certain perception of social reality. "Nature," for example, is never encountered unmediated because it is always fashioned by people's consciousness. It is somewhat tautological to refer to a "mystified" mediation, since in most cases what is mediated has by that very process had its political nature disguised. However, there are degrees of mystification rendering some mediations more opaque than others. See R. Williams, *Marxism and Literature*, 95–100; and Jay, *Dialectical Imagination*, 54 and passim.

23. Ackerknecht, "Anticontagionism," 565.

24. Douglas, *Purity and Danger*, esp. 2, 94, 35. See also Temkin, "An Historical Analysis of the Concept of Infection."

25. [Gooch], "Contagion," 525. In fairness, it should be added that this author (a contingent-contagionist) thought that contagionists were just as misguided in having a "submissive dependence upon prescriptive rule." Another author referred to anticontagionists as those "breathing the leveling principles of the day": Winterbottom,

"Thoughts on Quarantine," 62. My use of the term "marginal" derives from both the sociological interpretation, as recounted in Inkster, "Marginal Men," and from a closely connected anthropological usage (but referring more to placelessness in relation to accepted classification schemes) recounted in Douglas, *Purity and Danger*, 96–97. Morris (*Cholera 1832*, 182) locates the anticontagionists in the medical profession as "lower-status groups ... men from the middle classes but who shared little of the authority of government, the India men, the surgeons, and the Scots all tended to go in to print as anticontagionists."

26. Most of the social historians of cholera deal with this; see, in particular, MacLaren, "Bourgeois Ideology." For a more nuanced critical investigation of the social control argument, see Durey, *Return of the Plague*, chapters 7 and 8. For a typical contemporary illustration of the rhetorical value of cholera, see *Directions to Plain People*, and for many similar examples, see Longmate, *King Cholera*. On the mediate use of physiological knowledge to organize people's consent to urban industrial capitalism, see Cooter, "Power of the Body."

27. S. Smith, "Education,"43ff., an attack on aristocratic education.

28. Quotation from Nesbitt, *Benthamite Reviewing*, 93. Cf. Pelling, *Cholera*, 8: Smith's "most radical public utterances were the incitements he offered to a (literate) working-class audience in 1847, not to violence or a disregard of Parliament, but to orderly agitation against vested interests."

29. Gooch, "Plague," 236; Douglas, *Purity and Danger*, esp. 21–22.

30. Examples include T. I. M. Forster, *Facts and Enquiries*, 76; and S. Smith, "Plague," 520–21. Later, when the dominance of a bourgeois reform order seemed more certain, Smith sought to minimize the importance of the language difference between contagionists and anticontagionists by referring to it as merely semantic: "Vast and immeasurable as the differences appear to be between contagionists and anti-contagionists, if regard be had merely to their language, yet if attention be paid only to their ideas, to this end and to this only, narrow as the compass is, the whole controversy is reduced.... It must be manifest that since both sects are perfectly agreed about the facts, the dispute can be only verbal. If the one would consent to restrict their use of the term contagious, for which there is the best authority and ancient custom, to those diseases which arise from a specific contagion, and would call those which arise from every other poison infections, there would be an end to this apparently interminable, and in many respects mischievous, controversy": *Treatise on Fevers*, 365–66.

31. T. I. M. Forster, "On the Atmospherical and Terrestrial Commotions," 113. One of the most frequently referred to "proofs" of the non-contagion of epidemics was the infrequency of disease transmission from patients to staff in hospitals. This was contrasted with the frequency of transmission from household to household in

low-lying slum areas. On this and other planks in the anticontagionists' platform, see Morris, *Cholera 1832*, 176–84. As Pelling points out (*Cholera*, 23), S. Smith was "seeking to establish the contagious character as the exception rather than the rule; to put the onus of proof on the 'contagionists'; and to eradicate a supposed habit of mind which saw a contagionist explanation of two consecutive events in the field as 'intuitively obvious,' regardless of the absence of evidence for the relations being one of cause and effect."

32. S. Smith, "Contagion," 152. By calling Maclean "extraordinary," Smith divorces his theory of epidemics from the social context of Maclean's and his own deployment of the knowledge. On Smith, see Poynter, "Thomas Southwood Smith," and Bowring, ed., *Works of Jeremy Bentham*, vol. 11, 35; on Maclean (1788–1824), see *Oxford Dictionary of National Biography*. Although it is likely that Maclean also had Benthamite connections, it would be wrong to identify him within the Benthamite camp of the mid-1820s, for he neither shared the same evangelical moral fervor, nor was as unequivocally committed to the virtues of bourgeois political economy.

33. See Morris, *Cholera 1832*, 183, and Rosenberg, "Florence Nightingale on Contagion," esp. 117–20.

34. S. Smith, "Contagion," 143, 142; T. F. Forster, *Brief Inquiry*, 4; see also T. I. M. Forster, *Facts and Enquiries*. On the failure of anticontagionists to give any account of the origins of epidemic diseases, see Pelling, *Cholera*, 24.

35. S. Smith, "Plague," 509, 505–6, 515.

36. S. Smith, "Contagion," 142 and passim. A distinction needs making between what anticontagionists articulated and what, in being mediate, was mystified to them. What they were frequently led to argue was that the atmosphere was the source or cause of epidemics, and they were thus subsequently challenged as having provided not an explanation for the cause of epidemics like cholera, but merely providing a woolly cloak for medical ignorance. In what follows, it is at a level beneath this recoverable history (and with no recourse to contemporary statements taken at face value) that the mediate meaning of the atmosphere is sought. With the atmosphere understood as the critical concept, there can be (though for entirely different reasons) agreement with Pelling on the historical distortion involved in perceiving medical reality in terms of contagionism/anticontagionism. Moreover, in bringing forward the importance of atmosphere, a number of contingent-contagionists, who otherwise are only to be accounted for with reference to some extraneous scientific rationality, can be brought within a framework of social understanding. The radical and editor of the *Lancet* Thomas Wakley, for example (though a supporter in the 1820s of other kinds of countercultural knowledge) gave no more support to anticontagionism than he did to contagionism. However, in an unsigned article in his journal in 1831 the importance of atmosphere was emphasized: "For my part, I admit that I can more easily

comprehend the propagation of certain epidemics by contagion, than I can by any other means, *when unaccompanied by sensible atmospheric changes*; and if I reject contagion in cholera, it is because whatever we have in the shape of fair evidence, is quite conclusive as to the non-existence of any such principle": Alpha, "Remarks on the Cholera Morbus," 108. Wakley, keenly aware in 1832 (in the midst of substantial rioting against medical men and medical schools), submitted that "the scientific character of English medical practitioners [is] at stake": Editorial, *Lancet*, 1 February 1832, 706.

37. Maclean, *Summary of Facts and Inferences*, 156.

38. Although atmosphere supplies more of an *anti*-conception to reality's social construction than an alternative to it, insofar as it posits that the reigning view of social relations is mistaken, it might be seen as having for bourgeois revolutionaries much the same critical potential that dialectical materialism held for socialists.

39. S. Smith, "Contagion," 142.

40. Ibid., 143, 141.

41. See Jameson, *Prison-House of Language*, esp. v.

42. For example, S. Smith, "Contagion," 134. The need for the certainty of rigid classification schemes became all the greater with the outbreak of cholera.

43. Maclean, *Summary of Facts and Inferences*, 155.

44. See Durey, *Return of the Plague*, 176ff.; Durey, "Bodysnatchers and Benthamites"; S. Smith, "Use of the Dead to the Living"; and Richardson, *Death, Dissection and the Destitute*.

45. Although historical work has yet to be conducted on the kinds, and extent and nature of the alternatives to "orthodox" medicine in nineteenth-century Britain, there is little doubt that a considerable section of the population rejected the authority of the orthodox professionals, just as many in need of spiritual aid rejected orthodox religion. See Pickstone, "Medical Botany"; F. B. Smith, *People's Health*; and Barrow, "Socialism in Eternity."

46. S. Smith, "Contagion," 135–37, 148. As a liberal humanist, Smith did not hold the "gullible masses" responsible for maintaining this "delusion," but blamed the reactionaries in the medical profession who, as a result of poor training and the acceptance of authoritarian dogma, were "unfitted" to their task.

47. As Andrew Scull has written in the context of another aspect of medical history, "the Evangelicals were content merely to try and moralize the individual within the existing social framework. The Utilitarians sought to moralize the social framework itself": *Museums of Madness*, 58.

48. Rosenberg, *Cholera Years*, 228.

49. There is, for example, a striking difference in the tone and intent of Southwood Smith's writings on anticontagionism in 1825 and those of 1830–31, as there is

between the social function of these writings and those of his deployed by the Board of Health in the 1840s. Like the passage quoted above (note 30), which is taken from the only place in Smith's *Treatise on Fevers* (1830) dealing specifically with the contagionist-anticontagionist controversy, his article on "Spasmodic Cholera" in the *Westminster Review* of 1831 is remarkably conciliatory toward contagionists, admitting not only that the atmospheric theory of disease causation "has been questioned by many, and on grounds of no trifling weight," but also, that the evidence for contagionism is "entitled to much attention" (which he duly gives). It is also worth noting the functionalist conception of fevers that prevails in the book—so much so that in places it is hard to tell where the social metaphors leave off. It would not be difficult to account for these differences with reference to (i) changes in the social structure and changes in Smith's social ambitions and interests, and (ii) to some extent constitutively, with the generation of new interests and experiences within the institutional "fine structure" of medicine. It could be argued, perhaps, that once the class interests represented by Smith had begun to be socially ascendant, continued support for deviant anticontagionist knowledge came to be seen as counterproductive to new class-consolidating interests both inside and outside the medical profession.

Chapter 3. "Framing" the End of the Social History of Medicine

i. Patrick Joyce, for example, in "The End of Social History" (1995) and "The Return of History" (1998). I recollect Joyce lamenting at the time that few historians responded to his invitations to debate.

ii. The prime example is Aronowitz, *Making Sense of Illness*.

iii. Latour, *Never Been Modern*, 27.

iv. P. Joyce, "What Is the Social."

v. According to the historian of science Jo Gladstone, who had the opportunity to ask E. M. Forrester about what was supposed to have actually gone on in the famous scene in the Marabar Caves in *A Passage to India*, the author had not the foggiest idea. Personal conversation, c. 1995.

vi. Both Jordanova and Salter subsequently moved to King's College, London. John Arnold moved to Birkbeck College, London, and Mark Knights to Warwick University.

vii. *Journal of the History of Medicine and Allied Sciences* 63, no. 4 (2008).

1. Rosenberg, "Framing Disease," introduction to *Framing Disease*. The introduction was a revised version of Rosenberg's "Disease in History: Frames and Framers."

2. A sure sign was the commonplace appearance in the 1990s of "frames" and "framing" in applications for history of medicine funding. For further reflection on the term's vogue by the millennium's turn, see Worboys, *Spreading Germs*, 12–13, who

submits that to describe his approach he has "chosen to use 'construction' rather than the currently more popular terms of *social construction and framing*."

3. Wear, "Introduction," *Medicine in Society*, 1; Jordanova, "Has the Social History of Medicine Come of Age." Of Jordanova's four criteria for determining a field's "maturity," her most substantial one—the conducting of sophisticated debates to encourage, refine, and if necessary, radically alter interpretations—has remained largely unfulfilled within the social history of medicine.

4. See Steedman, *Dust*, esp. chapter 7: "About Ends: On How the End Is Different from an Ending."

5. Pickstone, *Ways of Knowing*, 6.

6. Jordanova has repeatedly drawn attention to the definitional problem with the word, most recently in *Nature Displayed*, esp. 101. In much of the historiography of medicine from Sprengel to Sigerist, "medicine" was debated in terms of whether it was a "science" or an "art." See Webster, "Historiography of Medicine."

7. N. Rose, "Medicine, History and the Present," 49.

8. Jordanova, "Social Construction of Medical Knowledge," 362. See also Warner, "History of Science and the Sciences of Medicine."

9. Pickstone, "Medicine, Society, and the State," 304.

10. Klein, "Crises of the Welfare States." See also Pickstone, "Production, Community and Consumption."

11. Kuo, "Swimming with the Sharks," 828.

12. Kuklick, "Professional Status and the Moral Order."

13. Lilford et al., "Medical Practice: Where Next."

14. A. H. Jones, "Narrative Based Medicine," 255. The use of the phrase in relation to ethics is symptomatic of the dominance in contemporary medical discourse of evidence-based medicine. See also, for example, Baum, "Evidence-Based Art."

15. R. Porter, review of Petersen et al., 178.

16. S. Williams and Calnan, " 'Limits' of Medicalization," 1616.

17. L. Turner, "Medical Ethics in a Multicultural Society." The political correctness of multiculturalism is another form of would-be universalistic morality. Rawls's notion of "moral consensus" compares with Richard Rorty's "solidarity" or "community" among Western philosophers, and has been denounced as "an exemplar of the nationalist philosophy of a new world": Billig, *Banal Nationalism*, 162ff.

18. Beck, *Risk Society*; Beck, "Reinvention of Politics," 1–55.

19. Wiesemann, "Defining Brain Death," 151.

20. S. Williams and Calnan, " 'Limits' of Medicalization."

21. See Jordan, *Activism*.

22. N. Rose, "Medicine, History and the Present," 50.

23. Pickstone, "Development and Present State of History of Medicine in Britain," 484.

24. Hunt, "Does History Need Defending," 241.

25. R. Putman, *Making Democracy Work*, 8; Marx, "Eighteenth Brumaire," 146.

26. Novick, *Noble Dream*.

27. Marwick, *The Nature of History*; Marwick, *New Nature of History*; Palmer, *Descent into Discourse*; Appleby et al., *Telling the Truth About History*; Evans, *Defence of History*; Eley, "Is All the World a Text"; Hobsbawm, "On History"; Samuel, "Reading the Signs"; Jones, "The Determinist Fix"; and Mayfield and Thorne, "Social History and Its Discontents."

28. The phrase is Cordle's: *Postmodern Postures*.

29. Iggers, *Historiography in the Twentieth Century*. See also Attridge et al., *Poststructuralism and the Question of History*.

30. Wallace, "Hijacking History." See also the regular feature in *Radical History Review* in the 1980s on the "abusable past."

31. Hewison, *Heritage Industry*; Lowenthal, *Heritage Crusade*; Bennett, *Birth of the Museum*; Jordanova, "Sense of a Past"; Jordanova, *History in Practice*; and Fox-Genovese and Lasch-Quinn, *Reconstructing History*, xvii.

32. Baudrillard, *In the Shadow of the Silent Majorities*. The social has been absorbed into the cultural, he argued, because "there is no longer any social signifier to give force to a political signifier" (19). Quoted in Casey, *Work, Self and Society*, 10.

33. P. Joyce, "End of Social History." For a lucid account of "the ending" in terms of its epistemological bankrupting, see Kent, "Victorian Social History." With reference more to the epistemological wasteland of socialist historians in the wake of the end of the Marxist epic, see Samuel, "On the Methods of *History Workshop*," and the discussion in Steedman, *Dust*, 79ff. For a more mundane account, see Evans, *Defence of History*, 168ff.

34. Quotation from Beck, "How Modern Is Modern Society," 163. This is not to suggest that the Francophone philosophers and linguists were anti-Marxists. Samuel ("Reading the Signs") hinted that many were so, and the exposure in 1987 of one of the leading postmodern literary theorists, Paul de Man, as having had fascist connections during the occupation of Belgium during World War II (made much of by Evans in *Defence of History*, 233ff.) was seized upon by many as part of the hidden reactionary agenda of literary deconstructionists in their insistence on the irrelevance of authorial intentions in textual interpretations. But the attempt to lump postmodernists together on the right is no more convincing than the effort to lump them on the left (as in Marwick, "All Quiet on the Postmodern Front"). Many of the Francophone founders of discourse analysis *were* Marxists (for example, Régine Robin, Michel Pêcheux, Denise Maldidier, Jean-Baptiste Marcellesi, Jacques Guilhaumou, and Jean Pierre Faye), though after May 1968 they became uncomfortable with the workers' movement and the preconceived historical explanations offered by dogmatic Marxists.

There may be more truth to Marwick's assertion that "much of postmodernism appealed profoundly to those who were by no means politically radical." Marxist literary theorist Terry Eagleton said as much in 1983, accusing deconstructionists of being intellectual elitists who savored signifiers over "what ever might be going on in the Elysée Palace or the Renault factories." Cited in Novick, *Noble Dream*, 567, who also concedes that postmodern thought "was on the whole quite apolitical" (566). On the anti-Marxist tendency of poststructuralists, see also Bennett, "Texts in History," 66–67. Nevertheless, the assault on modernity under the banner of *différance* was a part of a critique of the reifying rationality culture of late capitalism. Moreover, as the reaction to the "linguistic turn" suggests, there was nothing non-political about adopting avant-garde literary theory in the wake of the perceived exhaustion of the "political" in the 1970s.

35. Daniel Bell's classic *The End of Ideology: On the Exhaustion of Political Ideas in the Fifties* (1960) appeared laughably anachronistic after 1968, but in 1988 it could be credibly reissued. For the context of Bell's work, see Dittberner, *End of Ideology*. Albert Camus apparently first used the phrase "end of ideologies" in 1946 referring to absolute utopias such as Marxist ones which destroy themselves. On the history of the utopian politics of an "end to politics," see Rancière, *Shores of Politics*.

36. "The *young conservatives* embrace the fundamental experience of aesthetic modernity—the disclosure of a decentered subjectivity, freed from all constraints of rational cognition and purposeness, from all imperatives of labor and utility—and in this way break out of the modern world. They thereby ground in intransigent antimodernism through a modernist attitude. They transpose the spontaneous power of the imagination, the experience of self and affectivity, into the remote and the archaic; and in Manichean fashion, they counterpose to instrumental reason a principle only accessible via 'evocation': be it the will to power or sovereignty, Being or the Dionysian power of the poetic. In France this trend leads from Georges Bataille to Foucault and Derrida. The spirit [*Geist*] of Nietzsche that was reawakened in the 1970s of course hovers over them all": "Modernity Versus Postmodernity," quoted in Wolin, "Introduction," Habermas, *The New Conservatism*, xxi–xxii.

In France, the so-called new philosophers were André Glucksman, Alain de Benoist, Bernard-Henri Lévy, and Pascal Bruckner. All were opposed to totalitarianism in Europe, the seeds of which they saw in Marx and Hegel's philosophy, and some were deeply involved in human rights in Bosnia and elsewhere. My thanks to Eve Seguin for this information.

37. See Foucault, *Power/Knowledge*, 51, 57–58, 76, 89. Foucault eschewed the label "postmodernist," and legitimately so, for as Nikolas Rose points out, his "rejection of unities was not done in the name of a post-modern metaphysics that celebrates diversity, [but rather] in the light of a more sober and, dare one say, more historical

conviction than that which is much less determined, much more contingent, that we think": "Medicine, History and the Present," 70; see also Best, *Politics of Historical Vision*.

38. See Joseph, "Derrida's Spectres of Ideology."

39. Foucault, *Discipline and Punish*, 25–26; During, *Foucault and Literature*, 3. On the place of Foucault within the intellectual transformation of the 1980s, see Kent, "Victorian Social History," and Dean, *Critical and Effective Histories*.

40. Foucault, *History of Sexuality*, vol. 1, 7–12.

41. Palmer, *Descent into Discourse*.

42. Laclau, *New Reflections*, see esp. chapter 2: "The Impossibility of Society"; Law, *Power, Action and Belief*, 3. See also Sim, *Derrida and the End of History*.

43. For two excellent examples of the application of postmodern modalities to history-writing, see Walkowitz, *City of Dreadful Delight*; and Daniel Pick, *War Machine*.

44. Haraway, *Simians, Cyborgs, and Women*; Said, *Orientalism*; Said, *Culture and Imperialism*. Darnton's remark was made at a conference on "Dissolving the Boundaries: Historical Writing Towards the Third Millennium," University of Warwick, July 1997.

45. Jenner, "Body, Image, Text."

46. Some explication of this "mess" is provided by P. B. Clarke, "Deconstruction."

47. Burke, *Death and Return of the Author*.

48. Eley and Nield, "Why Does Social History Ignore Politics," 267.

49. D. Porter, "The Mission of Social History of Medicine," 359. Cf. Charles Webster's call in 1983 for a social history of medicine "that would place its primary emphasis on the changing pattern of health of the population as a whole": "The Historiography of Medicine," 40.

50. Jordanova, "Social Construction of Medical Knowledge," 363.

51. Among them, by Gilman: *Difference and Pathology; Disease and Representation; The Jew's Body; Inscribing the Other; Health and Illness*; and *Making the Body Beautiful*.

52. Responding to one of the American critics of Adrian Desmond and Jim Moore's *Darwin*, Moore remarked, "The 'F' word today is flung about by critics like navvies flourish theirs. . . . It would be a solecism to suppose that either of us owes anything to Foucauldian 'archaeology' or 'epistemic shifts.' Desmond himself has never cracked a book by Foucault; Moore has repeatedly faulted Foucault-like accounts. . . . We didn't need his advice on sucking epistemological eggs, nor indeed did the scholars whose work made *Darwin* possible." *Journal of Victorian Culture* 3 (1998), 152. The latter point is also made by Jordanova in "Social Construction of Medical Knowledge," 368–69, and see Jones and Porter, *Reassessing Foucault*.

53. Cooter and Sturdy, "Of War, Medicine and Modernity," esp. 5.

54. Jenner, "Body, Image, Text," 143–54; see also Jenner and Taithe, "Historiographical Body." For evidence of the "mopping up," see the books sympathetically reviewed by Wiener in terms of their contribution to modern British history: "Treating 'Historical' Sources."

55. P. Joyce, "Return of History," 212, n. 18, cited in Green and Troup, "The Challenge of Poststructuralism/Postmodernism," 297.

56. The case for the study of the laity is made in Usborne and de Blecourt, "Pains of the Past."

57. See Jenner, "Body, Image, Text," 143. A more recent example is Roy Porter's *Bodies Politic*, which, while nodding in the direction of "the body as text," is preoccupied with "contextualizing [visual material] within the wider cultural pool" (12). In this respect, as in others, *Body Politics* is, as Porter submits (35), a sequel to his previous social histories of the sick and the sick trade.

The most explicitly Foucauldian studies of the body were conducted not by social historians of medicine but by historically minded sociologists of medicine such as David Armstrong, Nikolas Rose, Deborah Lupton, and Bryan Turner: Armstrong, *Political Anatomy of the Body*; Lupton, *Medicine as Culture*; Lupton, *The Imperative of Health*; N. Rose, *Governing the Soul*; B. Turner, *Medical Power and Social Knowledge*. See also Featherstone et al., *The Body*. A notable exception among social historians writing on the body (and drawing heavily on sociology, anthropology, and political theory) was Duden, *The Woman Beneath the Skin*; see esp. chapter 1: "Toward a History of the Body" [and the discussion in Chapter 4, below].

58. Rosenberg, "Framing Disease," xiv.

59. For example, Aronowitz, *Making Sense of Illness*.

60. Rosenberg, "Disease and Social Order in America."

61. Novick, *Noble Dream*. Reference to many of the reviews of the book are noted in its discussion by Megill: "Fragmentation and the Future of Historiography."

62. See Norris, *Against Relativism*, 102. On Lyotard, see also the special issue of *Parallax*.

63. Cited in Douglas, "Social Preconditions of Radical Skepticism," 81.

64. McMillen, "Science Wars." The spark for this particular conflagration was Geison's de-mythologizing, contextualizing, and social constructivist-tending biography of *Louis Pasteur*.

65. Rosenberg, "Framing Disease," xv. As it emerged, substituting "frame" for "social" was all the more politically expedient for those applying for research funding in this charged environment.

66. See, in particular, Goffman, *Presentation of Self*. "Framing," of course, had and has other lives in medicine and elsewhere; see, for example, Collins and Pinch, *Frames of Meaning*; and A. Edwards et al., "Presenting Risk Information."

67. For illustration of this use of framing as a descriptive category within an alleged historical case of "framing" for the avoidance of attributing blame or attaching responsibility to any particular interest group, see Feudtner, "Minds the Dead Have Ravished." Feudtner draws explicitly on Rosenberg's concept of framing to avoid what "seems irresponsible, to dismiss shell shock as 'myth' or 'social construct' " (380).

68. Schaffer, "A Social History of Plausibility," 133.

69. Rosenberg, "Disease and Social Order," 29 (emphasis added).

70. Cited in Jameson, *Postmodernism*, 397.

71. Ibid., 398.

72. N. J. Fox, "Derrida, Meaning and the Frame," 134–35. See also Frow, who has deployed "the frame" to mark off a literary space that establishes "the particular historical distribution of the 'real' and the 'symbolic' within which the text operates." For Frow the frame organizes the "inside" and the "outside" of a text and the relations between them; the function of the frame is culturally dependent. Frow, "Literary Frame," 25.

73. Cf. Figlio, "Second Thoughts," 165.

74. A partial and peculiar exception is Harley, "Rhetoric and the Social Construction of Sickness," to which Palladino has replied in his "And the Answer Is . . . 42."

75. Douglas, "Social Preconditions of Radical Skepticism," 86. "The intellectual position of the relativists," she adds, "can be shown to be contingent on their sense of futility or immorality of exercising power and authority, and this contingency rests in turn on their place in a social structure." Sloterdijk, *Critique of Cynical Reason*. On the signs of Francophone return to the political, see *Democracy and Nature* 7 (2001), especially the contributions by Boggs, "How Can We End the End of Politics"; Tormey, "Post-Marxism"; and Fotopoulos, "Myth of Postmodernity."

76. Palmer, "Is There Now, or Has There Ever Been, a Working Class," 100. See also Appleby et al., *Telling the Truth About History*, 230 [and Chapter 4, below].

77. See, for example, Jones and Wahrman, *A Cultural Revolution*.

78. Bonnell and Hunt, *Beyond the Cultural Turn*; Fox-Genovese and Lasch-Quinn, *Reconstructing History*. See also Valentine, "Whatever Happened to the Social."

79. Eley and Nield, "Farewell to the Working Class"; and Eley and Nield, "Reply."

80. Burton, "Thinking Beyond the Boundaries." See also C. Porter, "History and Literature."

81. Casey, *Work, Self and Society*, 11. Casey's study is also one of the most lucid on the relations between postmodern theory and the politics of late capital.

82. Among examples of this work, not cited below, are Aisenberg, *Contagion*; and Burney, *Bodies of Evidence*.

83. Poovey, *Making a Social Body*. Ian Burney's debts to Poovey are made explicit in his contribution to the roundtable discussion "Making of a Social Body."

84. Foucault, "Ethics of Pleasure," 261, cited in Kendall and Wicham, "Health and the Social Body," 9–10. Donzelot, *L'invention du social*.

85. O'Connor, *Raw Material*. Cholera, breast cancer, amputations, and monsters are among the tropes she critically explores.

86. J. Epstein, "Signs of the Social," 483.

87. O'Connor, *Raw Material*, 215.

88. Ibid., 214. These comments echo Stuart Hall's observation that "anybody who is into cultural studies seriously as an intellectual practice must feel, on their pulse, its ephemerality, its insubstantiality, how little it registers, how little we've been able to change anything." Quoted in Treichler, *How to Have Theory in an Epidemic*, 3.

89. Driscoll, *Reconsidering Drugs*.

90. The best discussion of these matters is still N. Rose, "Medicine, History and the Present."

91. By Warner, who perceives their study as having been delegitimized, not by the force of postmodernism, but by too great an emphasis on the history of science. "History of Science," 173.

Chapter 4. The Turn of the Body

i. Palladino, "Medicine Yesterday, Today, and Tomorrow," 542; Cooter, "Disabled Body"; Cooter, "Dead Body"; Cooter, "Ethical Body."

ii. On the former, see Pickstone, *Ways of Knowing*; and Pickstone, "Working Knowledges Before and After Circa 1800," 497; on the latter, see Daston and Galison, *Objectivity*, discussed in Chapter 1.

1. R. Porter, "History of the Body Reconsidered," 236; R. Porter, "History of the Body," 212, 226.

2. Foucault, *Discipline and Punish*, 25.

3. Among the latter I would number Laqueur, *Making Sex*; Poovey, *Making a Social Body*; and (later) Gallagher, *Body Economic*.

4. B. Turner, *The Body and Society*, 1; Hancock et al., *Body, Culture, and Society*; Shilling, *Body and Social Theory*.

5. Featherstone, *Consumer Culture and Postmodernism*.

6. N. Rose, *Politics of Life*, 26.

7. Latour, "How to Talk About the Body," 206.

8. P. Joyce, "End of Social History," 83.

9. Long, *Rehabilitating Bodies*.

10. Such at least has been the claim of many authors since the 1970s; quite why they sought objectively to describe the origins of university disciplines in this way is beyond the bounds of this essay.

11. Foucault rendered "biopower" and "biopolitics" most explicit in his lectures at the Collège de France 1975–1976 (English trans. 2003) [see below, Chapter 9].

12. Foucault, "Politics of Health in the Eighteenth Century." For a lucid exposition of the transformation Foucault describes, see During, *Foucault and Literature*, chapter 2: "Medicine, Death, Realism."

13. On the origins, and Foucault's use, of the concept of normativity, see Sinding, "Power of Norms"; and Ernst, "Normal and the Abnormal." Exactly when "normal" was first used to mean "typical" (and thereby "naturalized" and rendered immutable) is unclear. Hacking, *Taming of Chance*, 166, cites Balzac's use of it in 1833. On the place of statistics in the creation of normativity and objectivity, see T. Porter, *Trust in Numbers*; and Desrosières, *Politics of Large Numbers*. For an example of the complexities that could be involved in this process, and how individuals might come to regulate themselves through normatizing technologies in medicine (specifically the clinical thermometer), see Hess, "Standardizing Body Temperature."

14. N. Rose, *Inventing Our Selves*; N. Rose, *Politics of Life*; Conrad, *Medicalization of Society*; cf. A. Clarke et al., *Biomedicalization*.

15. Cooter, "After Death/After-'Life.' "

16. The social history of medicine was not alone in this understanding in the 1970s and 1980s; sociologists shared the same biologically essentialist view, as indeed did most feminists.

17. R. Porter, "Patient's View."

18. M. Davis, *City of Quartz*, 302.

19. Rosenberg, "Holism," 347, 355, n. 41. By the late 1980s, the new genetics was also thought to be contributing to the collapse of the boundaries between "the natural" and "the social" and, hence, the collapse of the Enlightenment narrative about liberating humans from the constraints of "nature." Rheinberger, "Beyond Nature and Culture."

20. M. Bloch, *Royal Touch*.

21. Geertz, "Thick Description"; Douglas, *Purity and Danger*; Douglas, *Natural Symbols*; Douglas, *Implicit Meanings*.

22. Cooter, "Power of the Body"; Jenner and Taithe, "Historiographical Body."

23. I am grateful to Chris Millard for reminding me of this. See also Bencard on the "phenomenology and the flesh" in "History in the Flesh," 168ff.

24. Gallagher and Laqueur, *Making of the Modern Body*, vi.

25. The "New Historicism" was devoted to contextual readings of cultural and intellectual history through literary texts. It was not especially body orientated, and its debts to Foucault were inclined more to his discussion of subjectivities and technologies of power (mechanism of repression and subjugation) than to biopower. See Veeser, *New Historicism*.

26. Laqueur and Gallagher's *Making of the Modern Body* first appeared in 1986 as a special issue of *Representations*.

27. This also became an awareness in social studies of science where constructivism led to worry over trust in "the social." It came to seem that all that was left to discuss was "discourse," with everything as text; see Zammito, *Nice Derangement of Episteme*.

28. Scott, "Evidence of Experience," 781.

29. Butler, who was very much a part of the linguistic turn to the body in the first place, well illustrates through this move the Foucauldian point [made in Chapter 1] that constellations of thought constantly change over time but also continue to serve older constellations. A cynic might say that, as such, she is as much a part of the problem as its "realist" solution.

30. Scott, "Evidence of Experience," 779.

31. Ibid., 779–80. See also, however, the critique of this paper by Thomas Holt, which accuses Scott of essentializing discourse and paying insufficient attention to the material and political world in which discourse operates: "Experience and the Politics."

32. "Performativity" in relation to "representationalism" also became an issue in the history of science around the same time. There, however, the "representational idiom" was understood as the orthodox ("objective") representing of nature in science transferred to the historical and contemporary study of scientists as disembodied knowledge. This was countered by a more reflexive mode of studying scientists, which focused more on human agency, including that of the academic observer. Drawing on this work, Andrew Pickering sought to sketch out "a basis for a *performative* image of science, in which science is regarded a field of powers, capacities, and performances, situated in machinic captures of material agency ... [and] to understand scientific practice within such a performative idiom," i.e., "material performativity." As in feminist cultural study of the body, this technique did not exclude the representational idiom, but was rather "a *rebalancing* of our understanding of science away from pure obsession with knowledge and toward a recognition of science's material powers": Pickering, *Mangle of Practice*, 5. See also Hezig, "On Performance"; and Zammito, *Nice Derangement of Epistemes*, 151–82.

33. Flynn, *Existentialism*, 100.

34. Kemp, *Seen/Unseen*, 2.

35. According to Frank Ankersmit, one of the contributors to the discussion on presence in *History and Theory*, the "epistemological reorientation [of postmodernity] was not a distraction from returning to ontology *but rather its precondition*": " 'Presence' and Myth," 350.

36. Bentley, "Past and 'Presence,' " 349.

37. Runia, "Presence," 195.

38. Gumbrecht, "Presence Achieved in Language," 318; see also Gumbrecht, *Production of Presence*.

39. Gumbrecht, "Presence Achieved in Language," 317.
40. Hacking, *Historical Ontology*.
41. Peters, "Actes de presence," 372.
42. Ibid., 372–73.
43. Bentley, "Past and Presence," 349.
44. The attempt to incorporate presentionalism into the historicization of the historicized body, and thereby resolve the tension between the textual and the experiential, was the purpose of Adam Bencard's doctoral dissertation, "History in the Flesh." It succeeded in problematizing the gap between the discursive construction of the body and the common experience of it. But, ultimately, it could not reconcile these two irreconcilable modes of intellectual discourse: the historical, on the one hand, and the metaphysical or philosophical (and psychological), on the other. Because the latter are *a*historical concepts (see Hacking, *Historical Ontology*; R. Smith, *Inhibition*; and R. Smith, *Being Human*), attempts to bridge the gap between language and reality in history through their use only generates disconnects with historical time and place. The only way out of this conundrum is to posit a redesign of history itself, a reconfiguration that would involve "nothing short of the reconceptualization of the past—indeed of time itself," as one of the contributors to *History and Theory* had proposed as the goal for the future (Bentley, "Past and Presence," 349). This is fine and well for the philosophy of history, or for inquiry into the nature of the past itself such as the presentationalists were seeking. But it is less well suited for historical practice. At best, within history-writing as a form of critical inquiry, it can only leave open the question of the relationships that might possibly exist between knowledge, experience, and epistemology. At worst, for the history of the body in historical writing, it evaporates contextual exigencies. Bencard's dissertation also overlooked that a large body of literary theory on the "metaphysics of presence" had been unfolded by Derrida and others since the 1970s (e.g., Derrida, *Grammatology*)—a fact rendering all subsequent discussions of "presence" in one way or another already embedded in literary theory and not, therefore, an escape from it, as Bencard hoped.
45. Stafford, *Visual Analogy*, 10.
46. Stafford, *Echo Objects*, 1 (emphasis added).
47. Ibid., 1–2.
48. Ibid., 175–76.
49. Most prominently, perhaps, the Renaissance art historian David Freedberg, who sits on the editorial boards of *Arts et Neurosciences, Paris*, and the *Journal of Neuroesthetics*. See his "Memory in Art." Other examples include Bryson, "Neural Interface"; Onians, *Neuroarthistory*; and Konner (a leading anthropologist), *Evolution of Childhood*.
50. Fuller, review of Smail, *On Deep History*, 389.

51. Judovitz, *Culture of the Body*, 1.
52. Forman, "(Re)cognizing Postmodernity"; Cusset, *French Theory*.

Chapter 5. Coming into Focus

i. Cooter and Stein, "Visual Imagery and Epidemics."

ii. Gilman, "Representing Health and Illness."

1. Gilman, "How and Why Do Historians of Medicine Use or Ignore Images," 9.

2. Quotation from Mirzoeff, *Introduction to Visual Culture*, 6. On the status of the visual in contemporary medicine, see K. Joyce, *From Numbers to Pictures*; K. Joyce, "Body as Image"; Pauwels, *Visual Cultures of Science*; Kevles, *Naked to the Bone*; Sturken and Cartwright, *Practices of Looking*, esp. chapter 8: "Scientific Looking at Science"; Dolan and Tillack, "Pixels, Patterns and Problems"; Söderqvist, "To Give Global Genomes a Local Habitation"; and "Microarrays." On the status of "the visual" in the cultural history of medicine, see Hofer and Sauerteig, "Perspektiven einer Kulturgeschichte der Medizin."

3. On the "age of AIDS," see Brandt, "Emerging Themes," 210; Garoian, "Art Education"; and Garfield, *End of Innocence*. In relation to health posters, see Studinka, *Poster Collection*, 5.

4. Mitchell, " 'Critical Inquiry,' " 612.

5. On this view for language, see Fissell, "Making Meaning from the Margins," 378ff.

6. Quotation from Hutchinson, *Poster*, 1. AIDS posters are the notable exception; in addition to the sources discussed below, see, for example, McGrath, "Health, Education and Authority."

7. Robert-Sterkendries, *Posters of Health*; Renaud and Bouchard, *La santé s'affiche au Québec*; cf. Cooter and Stein, "Visual Imagery and Epidemics."

8. Most are devoted more to illustration than text, and to the history of graphic design; among the most informative on the history of poster production are Hutchinson, *Poster*; Barnicoat, *Concise History of Posters*; Barnicoat, "Poster"; Rickards, *Rise and Fall of the Poster*; Hillier, *Plakate*; and Gallo, *Geschichte der Plakate*.

9. On political posters, see Rickards, *Posters of Protest*; J. Thompson, "Pictorial Lies"; and Sauer, "Hinweg damit"; for war posters: Rawls, *Wake Up America*; Hardie and Sabin, *War Posters*; and Rickards, *Posters of the First World War*; and for commercial goods: Gallo, *Geschichte der Plakate*; and Helfand, *Quack, Quack, Quack*.

10. Abdy, *French Poster*, 3.

11. Timmers, *Power of the Poster*, 12. Three genres are explored: "Pleasure and Leisure." "Protest and Propaganda" and "Commerce and Communication."

12. Most notably, in the English-speaking world, by Gregory (1724–73) in his *Observations on the Duties*.

13. The exceptions are Grant, *Propaganda*, esp. chapter 5, "Health Publicity, 1919–1939"; and Toon, "Managing the Conduct."

14. Brandt, *No Magic Bullet*, 168.

15. Ibid.

16. Ibid.

17. Jordanova, "Medicine and Visual Culture," 91. For a different criticism of this type of historical use of images (serving only a "discourse of appearances"), see Delaporte, "History of Medicine."

18. Gilman, *Picturing Health and Illness*, 16.

19. Brandt, *No Magic Bullet*, 184.

20. Ibid., 193, n. 7; see also Brandt, "Emerging Themes," 204.

21. Sontag, "Posters: Advertisement, Art," viii.

22. Ibid., vii. For the different postmodern conception of the "public," see Warner, *Publics and Counterpublics*; on the literature on the postmodern reconceptualization of "public health" as appealing to the individual body as a collective one," see A. Clarke et al., "Biomedicalization," 16ff.

23. The major text, critical of these methods, was Packard, *Hidden Persuaders*; see Gibbons, *Art and Advertising*, 55. Quotation from Sontag, "Posters: Advertisement, Art," viii. For Sontag's intellectual and political context, see Sturken and Cartwright, *Practices of Looking*, 151–78.

24. See Sturken and Cartwright, *Practices of Looking*, 51ff.; and Sherman, "Quatremère/Benjamin/Marx."

25. Sturken and Cartwright, *Practices of Looking*, 252.

26. Sontag, "Posters: Advertisement, Art," ix.

27. Sontag, *AIDS and Its Metaphors*, 161. For the history of this campaign, see Berridge, *AIDS in the UK*; Fields and Wellings, *Stopping AIDS*; and Fields et al., *Promoting Safer Sex*.

28. Sontag, *AIDS and Its Metaphors*, 163.

29. Brandt, "Emerging Themes," 210.

30. Sontag, *AIDS and Its Metaphors*, 165.

31. Adorno and Horkheimer, "Culture Industry," 33, quoted in Sturken and Cartwright, *Practices of Looking*, 167.

32. Sevecke, *Wettbewerbsrecht und Kommunikationsgrundrechte*, 24.

33. Toscani, *Werbung ist ein lächelndes Aas*, 58. The photograph in *Life*, entitled "Final Moment," was by the American photographer Therese Frare.

34. Gibbons, *Art and Advertising*, 79.

35. Toscani, *Werbung ist ein lächelndes Aas*, 58.

36. See also Döring, *Gefühlsecht*, 128–29. The British newspaper the *Guardian* was forced to defend itself in an editorial of 24 January 1992: see Salvemini, *United Colors*, 92–93.

37. Klein, *No Logo*, 279–309; see also Lasn, *Culture Jam*; and Schiller, *Culture Inc.* This practice was coined as "cultural jamming" in 1984 by the San Francisco audio-collage band Negativland. ACT UP was established in New York in 1987.

38. The ACT UP poster, designed by Andrew Dibb, is reproduced in Beckett, "Protest Politics," 5.

39. Klein, *No Logo*, 281–82.

40. Watney, *Policing Desire*; Gatter, *Identity and Sexuality*, 82ff.; Berridge, *AIDS in the UK*, 56.

41. Watney, quoted in Gilman, *Picturing Health and Illness*, 115. See also Watney and Gupta, "Rhetoric of AIDS."

42. Treichler, "AIDS, Homophobia, and Biomedical Discourse," 263–64.

43. Ibid., 277.

44. Davidson, "Epistemology and Archaeology."

45. Foucault, *Discipline and Punish*, 25.

46. Sturken and Cartwright, *Practices of Looking*, 97; see also Burchell et al., *Foucault Effect*; and Dryfus and Rabinow, *Foucault*.

47. N. Rose, *Politics of Life*.

48. See, for example, Duden, *Woman Beneath the Skin*; Davidson, "Epistemology and Archaeology"; Cartwright, *Screening the Body*; and Treichler et al., *Visible Women*.

49. N. Rose and Miller, "Political Power," 174. See also Hacking, "Archaeology of Michel Foucault." For a comprehensive overview of the changes in the historiography of public health in the English-speaking world emerging from Foucault's concept of power, see Lupton, *Imperative of Health*, esp. chapter 4, "Communicating Health," 106–30. The notion of "making up" derives from Hacking, "Making Up People."

50. Sturken and Cartwright, *Practices of Looking*, 350; N. Rose, *Politics of Life*, 50–54.

51. Crimp and Rolston, *AIDS Demo Graphics*; Crimp, *Melancholia and Moralism*. See also Miller, *Fluid Exchanges*; and Klusacek and Morrison, *Leap in the Dark*.

52. Gilman, "Beautiful Body," 159.

53. Ibid., 122.

54. A virtually impossible task in any case, since the National Library of Medicine, the source of the posters, provides no such information.

55. Gilman, "Beautiful Body," 163 (emphasis added).

56. Ibid., 163.

57. Ibid., 117.

58. Dikovitskaya, *Visual Culture*; Malcolm, *Approaches to Understanding*; Sturken and Cartwright, *Practices of Looking*; Mirzoeff, *Introduction to Visual Culture*.

59. Raymond Williams, cited in Dikovitskaya, *Visual Culture*, 1; see also S. Hall, *Representation*, and the interviews with Mirzoeff and Mitchell in Dikovitskaya, *Visual Culture*.

Notes to Pages 131–135

60. For the assertion of this view, see S. Hall, *Representation*.

61. Rogoff, "Studying Visual Culture," 16. See also Jay, *Downcast Eyes*, 7, who comments on the practice of seeing in terms of the "saccadic movements" of the human eye.

62. In the social history of medicine this narrowness was rarely the case; consideration of newspaper ads and comic books was not unknown. See, for example, Hansen, "Medical History for the Masses."

63. See, in particular, Treichler, "AIDS, Africa, and Cultural Theory"; and Treichler, "Beyond *Cosmo*."

64. Treichler, "AIDS, Homophobia, and Biomedical Discourse," 292, n. 8, provides a list of sources for her thinking on AIDS and visuality.

65. Mitchell, "Pictorial Turn."

66. Dikovitskaya, *Visual Culture*, 1.

67. Treichler, "How to Have Theory in an Epidemic," 2, points out that the arrival of AIDS coincided with a period of attention to language.

68. Ibid., 11.

69. Ibid., 139.

70. Ibid., 39.

71. Ibid., 223.

72. Ibid., 272.

73. Of course much else bore on this shift, not least psychoanalysis, especially as reworked by Jacques Lacan, on whom see Sturken and Cartwright, *Practices of Looking*, 74.

74. Mirzoeff, *Introduction to Visual Culture*, 1.

75. As Annales historian Marc Bloch reflected in the 1940s, "[Sterility] is only the price that all intellectual movements must pay sooner or later, for their moment of intellectual fertility": Bloch, *Historian's Craft*, 13.

76. At the XVI International AIDS conference on 14 August 2006, Bill Roedy, chairman of Global Media Aids Initiative (also president of MTV Networks International), gloated over the public relations benefits to industries involved in such work. "Media have such a huge role to play in this fight," he said, "and as a member of the media industry I can fully admit we're not doing enough [applause]. Media can actually be a force of the good. When is the last time you have heard media can be a force of the good? Well, here media can be a force of the good." http://www.kaisernetwork.org/health_cast/uploaded/files/081406_ias_media_transcript.pdf, 11.

77. The campaign can be viewed on http://infections.chapsonline.org.uk/Home.

78. Interview with Sir Nick Partridge, Chief Executive of the Terrence Higgins Trust, 22 June 2006.

79. This is usually posited as a challenge to the older view maintained in art history of only reading the visual *from* the visual. See Biernoff, *Sight and Embodiment*, 3.

Chapter 6. Visual Objects and Universal Meanings

i. My thanks to Sarah Hodges and Stephen Casper for reminding me of this. See also Anderson, "Subjugated Knowledge." Within the domain of biomedicine, a brilliant alternative reading to conventional global history is Anderson, *Collectors of Lost Souls*, for my appreciation of which, see Cooter, "Review of Anderson."

ii. "A theory that has ceased to have any connection with practice is art": Horkheimer in Adorno and Horkheimer, *Towards a New Manifesto*, 100.

1. See Iggers and Wang, *Global History*, esp. chapter 8: "Historiography After the Cold War, 1990–2007." "Transnational," "international," and "world" history are frequently used synonyms for "the global."

2. See Zeller, "Spatial Turn in History"; Finnegan, "Spatial Turn"; Cosgrove, "Landscape and Landschaft"; and Massey, *For Space*. Although Foucault in the 1970s predicted that intellectuals would come to privilege time over space, the contrary turned out to be more the case, especially in the burgeoning field of theory in spatial geography: Foucault, "Questions on Geography."

3. Dikovitskaya, *Visual Culture*; Sturken and Cartwright, *Practices of Looking*; Mirzoeff, *Visual Culture Reader*; Mirzoeff, *Introduction to Visual Culture*; Malcolm, *Approaches to Understanding Visual Culture*; Walker and Chaplin, *Visual Culture*; Elkins, *Visual Studies*; Pink, *Future of Visual Anthropology*. On the application of visual culture studies to historical study, see Paul, *Visual History*; and Dommann, "Vom Bild zum Wissen." On the "pictorial turn," coined by Mitchell in 1992, see Nikolow and Bluma, "Science Images," 36ff.

4. For example, Huber, "Unification of the Globe by Disease." Cf. Zinser, *Rats, Lice and History*; and McNeill, *Plagues and Peoples*.

5. For example, C. A. Jones and Galison, *Picturing Science*; Pang, "Visual Representation"; Latour and Weibel, *Iconoclash*; Daston and Galison, *Objectivity*; Cartwright, *Screening the Body*; Hopwood, "Pictures of Evolution"; Hansen, *Picturing Medical Progress*; Lightman, "Visual Theology"; Shteir and Lightman, *Figuring It Out*; and Hüppauf and Weingart, *Science Images*.

6. For example, on the condom as a material object, see Vitellone, *Object Matters*. For more historically focused studies on scientific and medical objects in global contexts, see Schaffer, "Instruments as Cargo"; H. J. Cook, *Matters of Exchange*.

7. Throughout this essay we use "AIDS posters" as shorthand for "HIV/AIDS" posters. This is the generic term used by vendors, collectors, and exhibitors for posters relating not just to AIDS specifically and the need for precautionary measures such as condoms, but also to issues such as homophobia. Our use of "poster" follows Hutchinson, *Poster*, 1: "essentially a large announcement, usually with a pictorial element, usually printed on paper and usually displayed on a wall or billboard to the

general public." However, the meaning of "the public" in this connection was to some extent challenged by AIDS posters (see below, note 56).

8. Studinka, "Foreword," 5.

9. For illustrations of AIDS/HIV posters, see Cooter and Stein, "Protect Yourself"; Cooter and Stein, "Positioning the Image of Aids"; Rigby and Leibtag, *HardWare*; Field et al., *Promoting Safer Sex*; Field and Wellings, *Stopping AIDS*; Döring, *Gefühlsech*; "AIDS Plakate International"; and the websites of the institutions mentioned below (note 15) holding the largest collections of AIDS posters.

10. Classen and Howes, "Museum as Sensescape," 200.

11. Ong and Collier, *Global Assemblages*.

12. We do not therefore engage here with the claim made by various theorists, that it is now impossible to talk of AIDS/HIV without referring to mutually metamorphosized models and theories of globalization. According to some, it is now impossible even to conceptualize "globalization" without also thinking in terms of the AIDS pandemic. See Altman, "Globalization and the AIDS Industry"; and R. Brock, "An Onerous Citizenship."

13. Cf. Rancière, *Politics of Aesthetics*. Throughout this essay we adhere to the crucial distinction established by Forman between "postmodernism" *as a body of thought critical of modernity* from "postmodernity" *as an era* in which we still live. Further, we follow him on the fallacy of thinking the former the cause of the latter: "(Re)cognizing Postmodernity."

14. We reflect on this problem in Cooter and Stein, "Visual Imagery and Epidemics," 173.

15. Wellcome Library, London, Library of the National Institute of Health, Bethesda, Maryland, and the Deutsches Hygiene Museum, Dresden. A collection of 625 AIDS posters from forty-four countries is held at UCLA, and can be accessed through http://digital.library.ucla.edu/AIDSposters/. There is perhaps another paper to be written on the interior politics of such purchases within economic climates of retrenchment, and on the demands this then places on the kind of advertising deployed for the exhibitions (in the Hamburg case, a website image of a punchy young woman conveying gender and alternative life styles), and on the actual display of the objects themselves in the interest of maximizing passage through turnstiles. These are not our concern here, though it might be noted that financial stringencies connect to the contemporaneous articulation of a wider problematic on the purpose and function of museums, internationally; see Cuno, *Whose Muse*.

16. The former, designed by Yossi Lemmel of Israel and photographed by G. Korisky, 1993, is reproduced in Döring, *Gefühlsech*, 145.

17. On the past and present *ambiguous* status of the condom as both a legal and morally approved hygienic product, and as an illegal and morally disapproved means to birth control, see Treichler and Gates, "When Pirates Feast." See also Vitellone, *Object Matters*.

18. Döring, *Gefühlsech*, 13. The 1996 exhibition was partly organized around AIDS; its other three themes were "Heads," "Bodies," and "Human Rights."

19. The 1896 exhibition took place three years before Roger Marx formulated the idea for such exhibitions in the journal *Les Maîtres de l'Affiche*, and proposed a *Musée moderne de l'Affiche illustrée*. See Timmers, *Power of the Poster*, 12–13. On the late nineteenth-century poster movement in Germany, see Aynsley, *Graphic Design*, 30ff. Aynsley mistakenly dates the Hamburg exhibition 1893, and misattributes *Das Moderne Plakate* (1897) by the curator of the Dresden Museum für Kunst und Gewerbe as the first German book on posters (31, 35); the first such book was that by Brinckmann, *Katalog der Plakat-Ausstellung*.

20. For cultural politics in Hamburg and Brinckmann's role as a patron of the arts, see Kay, *Art and the German Bourgeoisie*. On Brinckmann, see Spielmann, *Brinckmann*.

21. Brinckmann, *Katalog der Plakat-Ausstellung*, 92. Similar motives lay behind "Mr. Robert Newman's Promenade Concerts" (later known as the "London Proms") to bring "quality" music to the masses at low cost (1 shilling per concert), the first of which was held in August 1895. For contemporary expression of similar views in Germany and Britain, see Hoffmann, "German Art Museum"; and Koven, "Whitechapel Picture Exhibition." A further part of the purpose of poster exhibitions was to educate people to the technology of graphic design. For example, in 1931, the Victoria and Albert Museum's Exhibition of British and Foreign Posters asserted that "this Museum is concerned less with the economic aspect, the publicity value, of the poster than with its technical method and the artistic impulse which finds expression in the special means employed. From a Museum point of view, therefore, this Exhibition of Posters might almost equally well be described as an exhibition of lithographs and of lithographic technique." Quoted in Timmers, *Power of the Poster*, 19.

22. On the systematic collection of posters by national institutions as evidence of democratized and populist culture, and as challenge to the traditional arts, see Aynsley, *Graphic Design*, 30ff.; and Aulich and Hewitt, *Seduction or Instruction*.

23. Döring, *Gefühlsech*, esp. 187.

24. Ibid., 13–14.

25. Ibid., 13.

26. Interview with Döring, 22 June 2010.

27. Interview (Claudia Stein) with the assistant curator of the exhibition, Hendrik Lunganini, 21 June 2006.

28. On the technology for poster-making in the 1980s and 1990s, see Hollis, *Graphic Design*; Myerson and Vickers, *Rewind*.

29. Döring, *Gefühlsech*, 15. Such comments (including the idea of *Zeitgeist*) are strikingly resonant of those on "degenerate art" by the Nazis; see Barron, "Degenerate Art."

30. Döring, *Gefühlsech*, 15. Another way to interpret these views is in terms of overlapping "epistemic virtues" [as discussed in Chapter 1].

31. Ibid., 16.

32. Crary, *Suspensions of Perception*.

33. For Sontag's intellectual and political context, see [Chapter 6, above, and] Sturken and Cartwright, *Practices of Looking*, 151–78. On the Frankfurt School, see Jay, *Dialectical Imagination*.

34. For the images and historical commentary, see http://www.avert.org/his87_92.htm.

35. Watney, *Policing Desire*, 15–16. A good account of the turning is provided by M. Cook, "From Gay Reform to Gaydar."

36. The backlash is detailed in M. Cook, "From Gay Reform to Gaydar," esp. 204–14. Commenting on it, Watney, *Policing Desire*, 18, quotes Altman, " 'the risk to gay identity seems greater in countries such as Great Britain and the Irish Republic, where the gay movement has less legitimacy and seems less able to withstand a new ideological onslaught, backed by real fears and dangers.' We are now facing that onslaught . . . which threatens not only our health but our very social identity, as the term 'gay,' wretched away from the older pejorative discourse of 'homosexuality,' is reloaded before our very eyes with all the familiar connotations of effeminacy, contagion and degeneracy." On representations of gay men in the UK media at this time, see Howell, *Broadcasting*.

37. See, for example, the AIDS posters reproduced in Gilman, "Beautiful Body," 124–28.

38. For a statement on how Western medicine achieved and sustained this "Biblical"-like position in the twentieth century, see Armstrong, *New History of Identity*. For how it lost it through the debate over HIV as the cause of AIDS, see Fujimura and Chou, "Dissent in Science"; and S. Epstein, *Impure Science*. Subsequently, it was biomedicine in general, rather than the medical profession in particular, that came to define "life"—see N. Rose, *Politics of Life*.

39. See Cooter and Stein, "Visual Imagery and Epidemics," 170.

40. Gibbons, *Art and Advertising*, chapter 4: "Reality Bites." The campaign cost 70 million US dollars. See Sevecke, *Wettbewerbsrecht und Kommunikationsgrundrechte*, 24.

41. Quotations from, respectively, Watney, quoted in Gilman, "Beautiful Body," 115; and Treichler, "AIDS, Homophobia, and Biomedical Discourse," 263–64. On this subject, see also Watney and Gupta, "The Rhetoric of AIDS"; and Crimp, "Portraits of People with AIDS."

42. Nor was this for the first time in Germany; see Borneman, "AIDS in the Two Berlins." Quotation from "It's Prejudice That's Queer"—questions and answers. "For Internal Use by CHAPS/THT Staff Only," Terrence Higgins Trust, Internal Memo. Cited by permission of the Terrence Higgins Trust.

43. On the disconnection of the global marketplace from national politics, see Klein, *Shock Doctrine*; Beck, *Was ist Globalisierung?* (who regards "the global" as forwarded by liberal democracies in the course of their decline as politically autonomous nation-states); Saul, *Collapse of Globalism*; Appadurai, "Grassroots Globalization," 4.

44. See Bayly, *Birth of the Modern World*, 1–12.

45. Watney, *Policing Desire*; for biomedicine's role in this, see Lynch, "Living with Kaposi's Sarcoma."

46. Watney, *Policing Desire*; Gatter, *Identity and Sexuality*, 82ff.; Berridge, *AIDS in the UK*, 56.

47. Berridge, *AIDS in the UK*, shows that over the twenty-odd years since the syndrome first surfaced in the UK, there were at least four distinct phases to those power relations and their representations.

48. Baldwin, *Disease and Democracy*. See also Bastos, *Global Reponses to AIDS*; Garrett, *Betrayal of Trust*; O'Manique, *Neoliberalism and AIDS*.

49. Pisani, *Wisdom of Whores*.

50. Ong and Collier, *Global Assemblages*, 3.

51. Treichler, "AIDS, Homophobia, and Biomedical Discourse," 263.

52. "World Health Organization Launches Public Information Effort." The message has been reiterated annually since December 1988 when the WHO initiated "World AIDS Day."

53. By 1988, when *AIDS and Its Metaphors* was first published, Sontag could observe that at international congresses "the global character of the AIDS crisis was a leading theme," and add wryly that in these forums "the rhetoric of global responsibility" was naturally "a specialty": 177.

54. While media multinationals came largely to constitute the Joint United Nations Program on HIV/AIDS (UNAIDS, established in 1996), other giant multinationals-turned-philanthropic organizations, such as Gates, Viacom, and MTV Networks International, in their independent AIDS programs, came to spend far more money than the UNAIDS: see Tannen, "Media Giant"; and UNAIDS, "Joint United Nations Programme."

55. On the "semantic hegemony," see Beck, *Was ist Globalisierung*.

56. The former are not necessarily places more "public" than the latter. Many AIDS posters, contrary to the impression lent them through exhibitions such as that in Hamburg, were never seen outside of gay pubs, clubs, and toilets (and some were one-offs produced only for art shows). We begin to think they were "public" only because the specific groups to whom they were often targeted are dissipated in the archive or in the museum. Implicitly in these places—designed, of course, for preserving "public memory"—a new composite public is assembled for them. On the vicissitudes and

contradictions of "the public," see Warner, *Publics and Counterpublics*. On memory and history, see Cubitt, *History and Memory*.

57. For these politics in the practice of contemporary science, see Daston and Galison, *Objectivity*, chapter 7: "Representation to Presentation."

58. McNeill, "Rise of the West."

59. See, for example, Blaut et al., *1492*; and Gills and Thompson, *Globalization and Global History*.

60. Exemplifying this trend is the global historian Fernández-Armesto, "Global History."

61. Gray, *Cyborg Citizen*, 17.

62. See, for example, Kemp, *Seen/Unseen*, 2 [cited in Chapter 4]. For Kemp, seeking to reconstruct "some continuities and discontinuities between past and present," history stands for itself, rather than as a product of its times.

63. Dirlik, "Confounding Metaphors," 92.

64. Indeed, "the global" might be said to commit epistemic violence in its corralling some subjects as "global." Our thanks to Guy Attewell for this observation.

65. Dirlik, "Confounding Metaphors," 91.

Chapter 7. The Biography of Disease

i. These were: Tattersall, *Diabetes*; Hamlin, *Cholera*; Jackson, *Asthma*; and Scull, *Hysteria*—all published in 2009. The editors of the series were W. F. Bynum and his wife, Helen.

ii. For the latest, see Mukherjee, *Emperor of All Maladies*, winner of the *Guardian* First Book Award, 2011.

iii. Rosenberg, "What Is Disease?" For his earlier thinking on disease, see his "Introduction: Framing Disease."

iv. M. Edwards, "Put Out Your Tongue," 301.

1. Stolberg, " 'Abhorreas pinguedinem' "; cf. Oddy, *Rise of Obesity*.

2. See, for example, Kiple, *Cambridge World History of Human Diseases*.

3. A good example is Smail's attempt to explain past cultural practices in terms of sub-cortical negotiations: "Essay on Neurobiology."

4. Pickering, *Mangle of Practice*, 3.

5. Tattersall, *Diabetes*, 11.

6. Longmate, *King Cholera*.

7. Latour, "On the Partial Existence of Existing." See also, on bubonic and pneumonic plagues, Hacking, *Emergence of Probability*, 21.

8. Iwan Bloch, *Der Ursprung der Syphilis*; Sudhoff, "Origin of Syphilis."

9. Sudhoff, "Aims and Value of Medical History." I am indebted to forthcoming work on Sudhoff by Claudia Stein. See also Iggers, *German Conception of History*.

10. Fleck, *Genesis*; on Kuhn, see Fuller, *Kuhn*.

11. Cunningham, "Transforming Plague"; A. Wilson, "History of Disease-Concepts"; Stein, *Negotiating the French Pox*.

12. For example, Monica Green, "Letting the Genome Out." She is also the recipient of grants for projects on "Excavating Medicine in a Digital Age: Paleography and the Medical Book in the Twelfth-Century Renaissance," and "The Origins of Leprosy as a Physical Disease and Social Condition in Medieval Europe"—the latter intended to "create ways to open up dialogue between the history of medicine and the historicist scientific disciplines of paleopathology, genomics, and paleomicrobiology" (*History of Science Society Newsletter* 39, July 2010).

13. Robson, "Disorderly Genius," 36.

14. See Kay, *Book of Life*; and Bencard, "Life Beyond Information."

Chapter 8. Inside the Whale

1. Ashcroft, "Bioethics and Conflicts"; C. Elliott, "Six Problems with Pharma-Funded Bioethics"; C. Elliott, "When Pharma Goes to the Laundry"; Lewens, "Commercial Exploitation of Ethics"; McMillan, "Is Corporate Money Bad for Bioethics."

2. Toulmin, "How Medicine Saved the Life of Ethics."

3. MacIntyre, "Does Applied Ethics Rest on a Mistake."

4. Kleinman et al., "Introduction [to special issue of *Daedalus:* 'Bioethics and Beyond']."

5. DeVries and Subedi, *Bioethics and Society*; Hoffmaster, *Bioethics in Social Context*; W. J. Smith, *Culture of Death*.

6. Belkin, Review of *Bioethics*, 373.

7. Quoted in Stevens, *Bioethics in America*, 44; C. Elliott, *A Philosophical Disease*; see also John, *Bioethics*.

8. Bosk, *What Would You Do*; Pence, *Brave New Bioethics*; Hoffmaster, "Can Ethnography Save the Life of Medical Ethics."

9. Rosenberg, "Meanings, Policies, and Medicine," 38, 42.

10. Kleinman et al., "Introduction to 'Bioethics and Beyond,'" vii. Cf. (with further sources) Reubi, "Will to Modernize"; Reubi, "Human Capacity"; and Salter and Salter, "Bioethics and the Global Moral Economy."

11. Bauer and Gaskell, *Biotechnology*; Engelhardt, *Global Bioethics*; R. M. Green et al., *Global Bioethics*.

12. Cooter, "Traffic in Victorian Bodies," 526.

13. The situation was different elsewhere. In Germany, for example, the history of medicine was largely replaced by medical ethics. There, the situation was different

again in that the ethical enterprise was grounded in a different (largely Kantian) philosophical tradition more attentive to language.

14. Roskies, "Neuroethics for the New Millennium"; see also Gazzaniga, *Ethical Brain*; Levy, *Neuroethics*; cf. Chorover, "Who Needs Neuroethics"; and Birnbacher, "Neuroethics and Stem Cell Transplantation."

15. Earlier scholarship, such as that on the Hippocratic Oath by Edelstein (*Ancient Medicine*) in 1967, was never an attempt to historicize medical ethics as a whole.

16. Baker et al., *The Codification of Medical Morality*; Wear et al., *Doctors and Ethics*.

17. See also Waddington, "Development of Medical Ethics," which informed the outlook and approach of his monograph.

18. Cooter, "Resistible Rise of Bioethics," 1995; Cooter, "Ethical Body."

19. A long list of the "pioneering soldiers" is provided in "Who Was Who in the History of Medical Ethics," appended to Baker and McCullough, *World History of Medical Ethics*, 693–720.

20. Armstrong, "Embodiment and Ethics," 878.

21. D. Fox and Klein, "Ethics and Health Policy," 667, maintain that "whatever the achievements of ethicists in other areas, their influence on the development of legislation and regulation in the United Kingdom and the United States has been at best mixed."

22. Messikomer et al., "Presence and Influence of Religion in American Bioethics"; Stevens, *Bioethics in America*, x, 48.

23. Engelhardt et al., "Bioethics: Institutionalization," 281.

24. Stevens, *Bioethics in America*, x–xi, xii.

25. Subjects on the agenda of medical ethics in the United Kingdom in the 1960s, such as mental health, bisexuality, war, nuclear weapons, cannabis use, unemployment, health and poverty, and the welfare state, simply disappeared thereafter. See Whong-Barr, "Clinical Ethics Teaching in Britain."

26. Stevens, *Bioethics in America*, xiii. See also Adams, "Artificial Kidneys and the Emergence of Bioethics."

27. As a structuring, as opposed to an analytical device, this use of "discourse" is not new in the history of medical ethics; Maehle deployed it in 1993 to chart theological and philosophical concepts around vivisection: "Ethical Discourse on Animal Experimentation." Jonsen, referring to public discussion of medical ethics post-1970s, styled one of the chapters in his *Birth of Bioethics* "Bioethics as a Discourse."

28. See Reverby, *Examining Tuskegee*.

29. Baker and McCullough, *World History of Medical Ethics*, 282.

30. Ibid., 283.

31. Cooter and Fissell, "Exploring Natural Knowledge."

32. Baker and McCullough, *World History of Medical Ethics*, 283.

33. As Hacking points out, "If you hold that a discourse consists in the totality of what is said in some domain, then you go beyond reading the intellectual highs of the heroes of science and you sample what is being said everywhere—including not only the annals of public hygiene but also the broadsheets of the day. You inevitably have to consider who is doing what to whom": "Archaeology of Michel Foucault," 80.

34. Faden and Beauchamp, *Informed Consent*, 157–63. Which is not to say that the notion fell from heaven or was merely some individual socio-religious informed prick of consciousness on Beecher's part. In his famous 1966 "ground breaking" article on "Ethics and Clinical Research," he refers to Sir Robert Platt in the United Kingdom pointing to "a general awakening of social consciousness" in medicine, and to a correspondence with the UK "whistle blower" Maurice Pappworth.

35. Reubi, "Ethics Governance, Modernity."

36. This is eloquently laid out in ibid.

37. Titmuss, *Gift Relationship*, discussed in ibid., 46ff.

38. Hacking, "Making Up People."

39. Baker and McCullough, *World History of Medical Ethics*, 485.

40. Foucault, "Questions of Methods," 79.

41. Daston and Galison, *Objectivity*; Cracraft, "Implicit Morality."

42. See, for example, Smail, *On Deep History*; and McGilchrist, *Master and His Emissary*.

43. N. Rose, "Value of Life," 48.

44. Ibid., 40. For an exposé of the spirit of biocapital largely avoided by bioethicists, see that by the cultural anthropologist Fortun, *Promising Genomics*.

45. Fortun, *Promising Genomics*, 40, 46.

Chapter 9. Cracking Biopower

Author's note: The final, definitive version of this essay has been published in *History of the Human Sciences* 23 (2010): 109–28, by Sage Publications Ltd. All rights reserved.

i. Cooter, review of Nadesan, *Governmentality*.

ii. For an overview, see Stein, "Birth of Biopower."

iii. E-mail to Cooter, 21 June 2011.

1. Foucault, *Society Must Be Defended*; Foucault, *Security, Territory, Population*; Foucault, *Birth of Biopolitics*.

2. Among recent works, besides those by Nikolas Rose and Esposito taken up below, see Nadesan, *Governmentality*; Lemke, *Gouvernmentalität*; Muhle, *Eine Genealogie der Biopolitik*; Weiß, *Bios und Zoë*; and Raman and Tutton, "Life, Science,

and Biopower." Pioneering, in many respects, was the brief intervention by Donnelly, "On Foucault's uses of the notion 'biopower.'" More substantial early interventions are: Lemke, "The Birth of Bio-Politics"; and Rabinow and Rose, "Biopower Today," the first version of which ("Thoughts on the Concept of Biopower Today") was a presentation in 2003.

3. Only a fraction of the writings in *Dits et écrits* have been translated into English.

4. Lemke, *Gouvernmentalität und Biopolitic*, 12.

5. Foucault, "Birth of Biopolitics."

6. *Dits et écrits*, vol. 4, 258–64. The word stems from the French adjective "gouvernemental" (concerning the government) not, as often thought, from the compound of gouverner (to rule) and mentalité (way of thinking), which has led many commentators to reduce Foucault's use of it to "thinking about government": Lemke, "An Indigestible Meal"; Lemke, *Gouvernmentalität und Biopolitic*, 12–13; Krasmann and Volkmer, *Foucault's "Geschichte der Gouvernmentalität"*; Purtschert et al., *Gouvernmentalität und Sicherheit*; Schäfer, Review of Purtschert. According to Lemke, Foucault deployed the concept of governmentality "as a 'guideline' for a 'genealogy of the modern state' embracing a period from Ancient Greece up until contemporary forms of neo-liberalism": "An Indigestible Meal," 354. Cf. Nadesan, *Governmentality*, 217, who acknowledges it as "a rather slippery concept that Foucault at times used to describe his method of analyzing the governmentalization of the liberal state, and at other times referred more concretely to historically specific arts of government, or governmentalities, linking the individual to social relations of power."

7. Deleuze, *Foucault*, 25–26.

8. Foucault, "Birth of Biopolitics."

9. Barthes, in 1957, used "governmentality" in his *Mythologies*.

10. Esposito, *Bíos*, 16–18.

11. Lemke, *Gouvernmentalität und Biopolitik*, 13–14.

12. On Roberts, see Hayward, "Biopolitics of Arthur Keith and Morley Roberts."

13. Esposito, *Bíos*, 22, refers to Lynton K. Caldwell's 1964 article "Biopolitics: Science, Ethics, and Public Policy," which was taken from James C. Davies, *Human Nature in Politics*, published the year before. On the use of biopolitics in recent eugenic scholarship, see Levine and Bashford, "Eugenics in the Modern World." For an overview of biopolitics from the 1970s to the 1990s, see Somit and Peterson, "Biopolitics After Three Decades."

14. Foucault, *History of Sexuality*, 143.

15. Campbell, "Introduction," xlii.

16. N. Rose, *Politics of Life*, 5, 51.

17. Ibid., 8, 58; Rabinow and Rose, "Biopower Today," 201.

18. N. Rose, *Politics of Life*, 54.

19. Ibid.

20. Rabinow and Rose, "Biopower Today," 203.

21. Heath et al., "Genetic Citizenship"; Lemke, *Gouvernmentalität und Biopolitik*, 78.

22. Lemke, *Gouvernmentalität und Biopolitik*, 79.

23. Rabinow, "Artificiality and Enlightenment."

24. N. Rose, *Politics of Life*, 113.

25. Ibid., 252, 33.

26. Fuller, "A Strong Distinction Between Humans and Non-Humans," 82–83; see also Fuller, "Putting the Brain at the Heart of General Education."

27. N. Rose, *Politics of Life*, 5.

28. Foucault, *Politics, Philosophy, Culture*, cited in Shildrick, *Leaky Bodies*, 211.

29. Foucault, *Politics, Philosophy, Culture*, 211–24; Eribon, *Foucault*, 289; Eribon, *L'infréquentable Michel Foucault*; Empson, "Political Immunity of Discourse"; Afary and Anderson, *Foucault and the Iranian Revolution*.

30. N. Rose, *Politics of Life*, 42.

31. Muhle, *Eine Genealogie der Biopolitik*, 62–63.

32. Ibid., 63.

33. On Foucault and Canguilhem's different understandings of "norms" and "life," see Muhle, *Eine Genealogie der Biopolitik*, 162–64. See also Macherey, "Für eine Naturgeschichte der Normen," 187, who demonstrates that Foucault's understanding of "norms" is intrinsically historical. In contrast to Canguilhem ("The Concept of Life"), who believed in an inner dynamic of life creating normativities, Foucault was interested in the causation, dynamics, and action of norms.

34. N. Rose, "Was ist Leben," 168.

35. *Dispositif* was Foucault's means to avoid referring to "structure"; it was a way to indicate something that gives shape to "discourse." See Veyne, *Foucault*, 180, n. 17, and 11–28, which is far superior to the discussion in Deleuze, "What Is a Dispositive."

36. N. Rose, "Was ist Leben," 168.

37. N. Rose, *Politics of Life*, 49–50.

38. Knorr-Cetina, "The Rise of a Culture of Life," 77.

39. N. Rose, *Politics of Life*, 3.

40. Lemke, *Gouvernmentalität und Biopolitik*, 91.

41. Campbell, "Introduction," xxxiii.

42. Cf. Fuller, "Why Science Studies Has Never Been Critical of Science"; and Fuller, "A Strong Distinction Between Humans and Non-Humans."

43. Esposito, *Bíos*, 13.

44. Biosocialities has also recently been adopted and elaborated empirically in Gibbon and Novas, *Biosocialities*, which contains an afterword by Rabinow.

45. Cooter, "Biocitizenship."

46. "Epistemological citizenship" has been suggested as a somewhat better term: Jasanoff, *Designs on Nature*, though this, too, retains the inherent contradiction of "citizenship" as democratic and voluntarily participatory.

47. Rabinow and Rose, "Biopower Today," 119.

48. Esposito, *Bíos*, 11.

49. Ibid., 32.

50. Ibid.

51. Ibid., 38.

52. Rabinow and Rose, "Biopower Today," 119; Esposito, *Bíos*, 31.

53. Esposito, *Bíos*, 31.

54. Ibid., 8.

55. Ibid.

56. Ibid., 9.

57. Ibid., 85; Foucault, "Nietzsche, Genealogy, History"; Foucault, "Truth and Power," 133.

58. Esposito, *Bíos*, 47.

59. Ibid., 9.

60. Ibid., 48.

61. Ibid., 111; Bauman, *Modernity and the Holocaust*.

62. Esposito, *Bíos*, 111.

63. Ibid., 148.

64. Ibid., 191.

65. Empson, "Political Immunity of Discourse," 59.

66. Esposito, *Bíos*, 12.

67. Ibid.

68. Empson, "Political Immunity of Discourse," 58, 52.

69. Shaviro, "Biopolitics."

70. Esposito, *Bíos*, 194.

71. Campbell, "Introduction," vii.

72. Foucault, "Truth and Power," 114–15.

Chapter 10. The New Poverty of Theory

1. The conference prospectus stated that its aim was "to provide a purely data-driven perspective on the neural correlates of consciousness," but this was then ethically legitimized in the claim that "until animals have their own storytellers, humans will always

have the most glorious part of the story": http://fcmconference.org/, accessed 18 April 2012. The very notion of a "non-human animal" well illustrates the current posthuman turn.

ii. The faith and aspiration announced by Prime Minister Tony Blair in Lisbon in November 2000 remains unchanged in the current government: "Biotechnology is the next wave of the knowledge economy and I want Britain to become its European hub": quoted in N. Rose, *Politics of Life*, 35. Appropriately, the showcase location for this new initiative was on the site of the money-spinning hopeful, the 2012 London Olympics, where the ideologues of beneficent biotechnology, the Wellcome Trust, have announced a major property investment.

iii. Turney, *Sci-Tech Report*, cited in Fuller, "Why Science Studies Has Never Been Critical of Science," 12.

iv. P. Joyce, "What Is the Social."

v. Fuller, "Why Science Studies Has Never Been Critical of Science." Our text remains faithful to the original, composed in 2010, but in the notes I have made reference to Fuller's essay where appropriate.

vi. Ibid., 5.

vii. Scott, "History-Writing as Critique."

1. Scott, "History-Writing as Critique."

2. Fissell, for example, writing on the new cultural history of medicine found herself having to choose between the politically engaged social history of dead bodies by Ruth Richardson, and what she perceived as the non-politically engaged cultural history of them by Mike Sappol, and opted for the former: "Making Meaning from the Margins," 388–89. More recently Sappol, nostalgically, has spoken of how Foucauldian scholarship around the body (heralded as an agenda for social, political, and cultural change, reform, and revolution) has unfortunately "tended to bear only indirectly on the great issues of our time. Poor body!—it's [now but] a category of displaced political analysis and struggle . . . [mere] productive remains for readers and scholars to pick over": "Introduction: Empires in Bodies," 35.

3. Scott, "History-Writing as Critique," 19.

4. P. Joyce, "What Is the Social," 219.

5. Ibid., 248.

6. Ibid., 219.

7. So, too, for Carlo Ginzburg and, in different ways, for Natalie Zemon Davis and Carolyn Steedman in the 1980s, the struggle was with recovering the historical voice of individual human experience (in the face of the flattening reductive and quantifying practices of *longue durée* studies by the Annales School). For Steedman, as she more recently argues in opposition to the kind of culturalist "identity studies" hitherto celebrated by Joyce, it was without the need to abandon "class" as a material force,

or to perceive it (as Joyce did) merely as a historical construct: *Labours Lost*, see esp. 17ff. Material objects, she argues, are bought into and used in a world in which they contribute to define identity and status, often in terms of class and "the social."

8. E. P. Thompson, "Poverty of Theory," 196. Thompson accused the fashionable followers of Althusser of being a bourgeois *"lumpen-intelligentsia"* and indicted them for "disorganising the constructive intellectual discourse of the Left, and . . . reproducing continually the élitist division between theory and practice" (195).

9. P. Joyce, "What Is the Social," 215 (emphasis added).

10. Ibid., 231.

11. For its take up in other domains, intended partly as a lever for its move into history-writing, see Bennett and Joyce, *Material Powers*.

12. Vernon, *Hunger*, viii–ix (emphasis added). For a very different, richer, and in some ways contesting view of hunger, see Tallis, *Hunger*.

13. In this respect, Joyce is not unlike the Althusserians assailed by Thompson, their main journal being *Theoretical Practice*.

14. P. Joyce, "What Is the Social," 224.

15. Ibid., 223, 224, 247, and 220, respectively.

16. Rigney, "Being an Improper Historian," 158.

17. Ibid., 158.

18. Scott, "History-Writing as Critique"; Butler, "What Is Critique"; Butler, "Critique, Dissent, Disciplinarity"; White, "Afterword." Scott ("History-Writing as Critique," 23) draws in part on the definition of critique provided by the literary theorist Barbara Johnson: "an analysis that focuses on the grounds of the system's possibility" and "reads backwards from what seems natural, obvious, self-evident, or universal in order to show that these things have their history, their reason for being the way they are."

19. P. Joyce, "The Gift of the Past," 93. This expresses many of the same views (and reproduces many of the same sentences) as the article focused on here, the main difference being the latter's emphasis on materiality and Joyce's sketching of his project on a "paper empire" in nineteenth-century India.

20. Ibid., 95. "Critical history," he adds (not very helpfully), offers "a critique of present power by its acts of remembrance" (97).

21. Ibid., 95. This is also the project of the historian Chris Otter (cited by Joyce, ibid., 228–30) in his work on light and vision in Britain, 1800–1910: *Victorian Eye*, 2.

22. The "slipperiness" of Latour, and hence perhaps that of Joyce in drawing on his work, is constitutive of Latour's particular way of describing the world; the language of description is intentionally slippery because "the reality" described is thus conceived, as we will elaborate below. Amsterdamska in her review of Latour's *Science in Action* points not only to his "extraordinary powers of language" and penchant for "verbal

pyrotechnics," but also his often politically pernicious logic: "Surely You Are Joking," 499.

23. Anderson, *The Collectors of Lost Souls*. For an account of Latour as "one of the most prominent guides to our liminal times," see Restivo, "Latour." While there are now cartloads of critical engagements with Latour's writings from various disciplinary stands, there exist no historicized accounts, such as we attempt below. The nearest such is Fuller, "Why Science Studies Has Never Been Critical of Science."

24. On the context (of the 1980s) in which the metaphor of the "network" moved from negative to positive sociological connotation and came to be privileged as an investigative tool, see Boltanski and Chiapello, *New Spirit of Capitalism*, 141–49. It is revealing that Latour, who along with Steve Woolgar is often (wrongly) credited with starting the "science studies revolution," now holds the position of vice-director of the Institut d'études politiques de Paris ("Sciences Po"), a conservative school for management and political elites whose research programs are sustained by collaborations with industry and the state. Woolgar is currently a professor of marketing in Oxford. The appeal and appreciation of Latour among the managerially minded and client dependent is also witnessed in the bestowing of an honorary degree on Latour in July 2010 by Nigel Thrift, the senior manager (vice-chancellor) of Warwick University.

25. The name is "so awkward, so confusing, so meaningless that it deserves to be kept," he writes in *Reassembling the Social*, 9. It was originally elaborated by Latour and Callon as the "sociology of translation" where "translation" was meant to "cover the process whereby one things represents another so well that the voice of the represented is effectively silenced": Fuller, "Why Science Studies Has Never Been Critical of Science," 13. See also, Latour, "Recalling ANT." Fuller is informative on the relation between this formulation and Mitterand's economic policies in 1980s France and the trend to top-down management.

26. See Latour, *Science in Action*; Latour, *Pasteurization of France*—his, as it were, "thick description" of actants.

27. Latour, *We Have Never Been Modern*, 13.

28. See, for example, Ong and Collier, *Global Assemblages*. N. Rose, *Inventing Our Selves*, 38, defines assemblages as the "localization and connecting together of routines, habits and technologies within specific domains of action and value . . . hybrid of flesh, artifact, knowledge, passion, and technique." For him, its use signifies the adoption of non-narrative, not-linear thinking—i.e., the adoption of Foucauldian genealogy.

29. Fleck, *Genesis*; and on Kuhn, Fuller, *Kuhn*.

30. For example, Daston and Galison, *Objectivity*, on the various makings of "objectivity" in the natural sciences (who also avoid using "the social" for its presumed theory); and R. Smith on the common cultural existence of the natural and human sciences: *Being Human*. See also Daston, "Science Studies"; and Pestre, "Thirty Years of Science Studies."

31. Shapin, "Here and Everywhere," 312. Latour's critique of Shapin and Schaffer's *Leviathan and the Air Pump* was made in these terms: on the one hand Latour praised it for the great care it took "to use the expression 'scientific fact' not as a resource but rather as a historical and political invention," but on the other, criticized it for taking "no such precautions where political language itself is concerned." Shapin and Schaffer use the words "power," "interest," and "politics" in all innocence, Latour asserted, "yet who invented these words, with their modern meaning? Hobbes! And thus, as Hobbes accused Boyle, Shapin, and Schaffer are to be accused of 'seeing double' themselves, [of] walking sideways, criticizing science but swallowing politics as the only valid source of explanation. . . . If nature and epistemology are not made up of transhistoric entities, then neither are history and sociology—unless one adopts some authors' asymmetrical posture and agrees to be simultaneously constructivist where nature is concerned and realist where society is concerned": Latour, *We Have Never Been Modern*, 27. Not surprisingly, much of the debate between ANT and SSK has rotated around the meaning of "symmetry" (a concept introduced by the Edinburgh School sociologist of science David Bloor); see, for example, Latour, "For David Bloor."

32. Latour, *Science in Action*, 72; Amsterdamska, "Surely You Are Joking," 499. This is very different from the "non-humanism" of Haraway in *Modest Witness*, whose approach is epistemological and sociological: see Lash, "Objects the Judge" 9, n. 4; see also Gray, *Cyborg Citizen*, which, from the standpoint of an anarchist democrat, is deeply skeptical of the aristocrat Latour.

33. Latour does not himself view facts as socially or epistemologically transcendent; see, for example, note 31, above.

34. As he puts it in *Reassembling the Social*, "The social is not a type of thing either visible or to be postulated. It is visible only by the *traces* it leaves (under trials) when a *new* association is being produced between elements which themselves are in no way 'social'" (8). Latour's thinking on "the social" (as well as that of Nigel Thrift) owes debts to the late nineteenth century anti-Darwinian sociology of Gabriel de Tarde's *Social Law*. See Latour, *Reassembling the Social*, 14–16; Latour, "Gabriel Tarde"; Ruffing, *Latour*, 90; and Candea, *Social After Gabriel Tarde*.

35. More accurately, we should call it "anti-modernist" in that "modernity" (defined in terms of humanism) is regarded by Latour as having obfuscated "the simultaneous birth of "nonhumanity": *We Have Never Been Modern*, 13. The latter's second chapter on the "modern constitution" is structured around the conceit that "modernity is often defined in terms of humanism" (13), whereas "modernity has nothing to do with the invention of humanism" (34).

36. Latour, *We Have Never Been Modern*, 44.

37. Latour, "Why Has Critique Run Out of Steam." Cf. Mallavarapu and Prasad, "Facts, Fetishes, and the Parliament of Things"; Scott, "History-Writing as Critique"; Butler, "What Is Critique"; and Butler, "Critique, Dissent, Disciplinarity."

38. In *Science in Action* Latour posits that "the ideal of explanation is not a desirable goal." Quoted in Amsterdamska, "Surely You Are Joking," 503, who adds that, on that basis, "we can pretty much abandon all responsibility for what we are saying."

39. Latour, *We Have Never Been Modern*, 44, although it should be said that Latour fudges "critique" in the sense that Johnson defines it (quoted above, n. 18) with "criticism" or "denunciation" being allowed to stand in its stead. For Latour on Marxism, as "deluded humanism," see *We Have Never Been Modern*, 36, and on postmodernists as "disappointed rationalists" who should move on to "to empirical studies of the networks that give meaning to the work of purification" (46). Cf. Fuller, who perceives ANT as "extending the postmodern condition from the humanities to the science and engineering faculties" inasmuch as it signifies the reduction of critical inquiry in universities to "a cluster of buildings where representatives of these discourse have chance encounters and set up temporary alliances, subject to the terms set down by buildings' custodians": "Why Science Studies Have Never Been Critical of Science," 18. See also Law and Hassard, *Actor Network Theory and After*.

40. Latour, *Eine neue Soziologie*, 421, quoted in Ruffing, *Latour*, 86.

41. On "flexibility" as the key concept in the transformation of management strategies and economics from the 1970s, see Boltanski and Chiapello, *New Spirit of Capitalism*, 194–95. What business management wants today, write Hammer and Champy in *Reengineering the Corporation* (7), is "companies that are lean, nimble, flexible, responsive, competitive, innovative, efficient, customer-focused, and profitable," quoted in Liu, *Laws of Cool*, 17.

42. Boltanski and Chiapello, *New Spirit of Capitalism*, 110.

43. *Neoliberalism*, 3.

44. Rancière, *Politics of Aesthetics*.

45. Fuller, "Why Science Studies Has Never Been Critical of Science," 16.

46. Arendt, *Eichmann in Jerusalem*.

47. As Fuller points out, the tactic of avoiding uncomfortable realities about human injustices is inherent to ANT: "it involves flattening the ontology of the social world so that structures are replaced by networks, and all parties are presented as exerting their own kind of power over each other, according to the alliances they can form in a given circumstance. Claims that the natives are subjugated or suffering are thus converted into ones about their hidden competences and agency. As a result, the contingency of the natives' condition may not be reduced, but the client's responsibility for it is." As he adds, "the current [ANT-associated] fashion for distributing agency across both people and things merely underscores the value of the masses as means to the ends of other parties, since in many cases nonhumans turn out to be at least as helpful as humans in achieving those ends": "Why Science Studies Has Never Been Critical," 11, 20.

48. P. Joyce, "What Is the Social," 227.

49. Ibid., 225.

50. See, for example, E. Wilson, *Affect and Artificial Intelligence*, which explicitly follows Latour on the equivalence of humans and material things.

51. [See the discussion and sources cited in Chapter 1.] For a thoughtful reflection on how this "new empiricism" limits the ability of historians to engage politically with emergent representations of health and illness performing overtly political work (and serves overall to de-legitimize the study of representations), see Gilman, "Representing Health and Illness."

52. Papoulias and Callard, "Biology's Gift," 33.

53. Ibid., 33. The point has more recently been amplified and particularized in Leys, "Turn to Affect." See also the critique by Korf, "A Neural Turn."

54. Papoulias and Callard, "Biology's Gift," 34. Affect theorist Liza Blackman (who is also the current editor of the journal *Body & Society*) proclaims affect as referring "to those processes that are non-cognitive, that are more bodily and which pass between subjects in ways which are difficult to see, in the conventional methodological sense, but which we feel nonetheless." She confesses to taking strength from contemporary neuroscience: "Biology, Brain Theory and History."

55. Papoulias and Callard, "Biology's Gift," 36.

56. Ibid., 39.

57. P. Joyce, "What Is the Social," 248.

58. Ibid. (emphasis added). This is a good example of the process of "enrolling" that Latour (in *Science in Action*, 1987) cut his teeth on in reference to the production of scientific truth.

59. Cusset, *French Theory*; see also Harvey, *Neoliberalism*, 42, 47, 198.

60. Raphael Samuel was more or less of this mind when, much in the spirit of Thompson's critique of the Althusserians, he wrote "Reading the Signs." But he made the mistake of thinking that "French theory" was motivated by hostility to Marxism.

61. Latour was prominent in the "wars"; Alan Sokal, the physicist who penned the sensational pastiche of postmodernism in *Social Texts* in 1996, came to devote a whole chapter to criticizing Latour a few years later: Sokal and Brimont, *Fashionable Nonsense*.

62. Treichler, "Beyond Cosmo," 272.

63. At the core of Marxism, which was incorporated into social history, was an understanding of human relationships in relation to material conditions. The novelty in Joyce's claim lies, Latour-like, in the displacement of those human relations through a focus on "objects" or "things" alone. Latour treats humans as mere cogs in the wheels of a machine, with machines, rather than humans (as in Marxism) as the ultimate source of value in the world. See Fuller, "Why Science Studies Has Never Been Critical," 21.

64. This is not the place to rehearse the case for, and nature of, the epistemic shift that took place in the 1980s and 1990s. This is now beginning to become clear though work focused on biotechnology, biopower, biocitizenship, biocapital, and the primacy of technology in postmodernity. See, for example, Forman, "Primacy of Science in Modernity."

65. R. Williams, *Key Words*, 75–76, cited in Butler, "What Is Critique," 212.

66. Adorno, "Cultural Criticism and Society," 30, quoted in Butler, "What Is Critique," 213.

67. Foucault, "What Is Enlightenment," 46. See also Foucault, "What Is Critique." Cf. Richard Rorty, "Continuity Between the Enlightenment and 'Postmodernism.'"

68. Butler, "What Is Critique," 213.

69. Ibid., 217.

70. White, "Afterword," 224, 226.

71. This builds upon E. H. Carr's observation that, since the idea of the past exists only in the mind of the historian, the study of the history is always that of the politics of the moment in which it is written. It was for this reason that Carr insisted that the historian of history-writing should attend seriously to the biographies of historians, particularly to their political and institutional affiliations: *What Is History*, chapter 1: "The Historians and His Facts."

72. Butler, "What Is Critique," 212–14, 218; Scott, "History-Writing as Critique," 22.

73. White, "Afterword," 224.

74. Joyce refers to his essay on "The Gift of the Past" as "one of a number of Foucault-influenced notions of 'critical history'" ("What Is the Social," 248, n. 89). Elsewhere (e.g., in the introduction to *Material Powers*) he also acknowledges his debt to Foucault as, of course, he did in his previous turn to discourse.

75. In what can only be regarded as a sleight-of-hand on Joyce's part ("What Is the Social," 248, n. 89), he contrasts his own "theoretical route to critical history" to the "poetic route" of Hayden White in *Manifestos for History*. Thereby, he trivializes White's contribution to thinking on contemporary history, at the same time as appropriating it to, and legitimizing, his own notion of "critical history."

76. Butler, "What Is Critique," 219. Joyce does not depart from the rationalist agendas of conventional historians whose cultural functions have remained unquestioned while, unnoticed (as Ermarth has observed) "the world has moved on" and the historian's mental tools, "derived so substantially from conventional history, are increasingly inadequate": "Closed Space of Choice," 63.

77. Garber, *Academic Instincts*, cited in Gilman, "Representing Health and Illness," 297.

78. Spivak, "Can the Subaltern Speak."

79. P. Joyce, "What Is the Social," 247.

BIBLIOGRAPHY

Abdy, Jane. *The French Poster: Cheret to Cappiello*. New York: C. N. Potter, 1969.

Ackerknecht, E. H. "Anticontagionism Between 1821 and 1867." *Bulletin of the History of Medicine* 22 (1948): 562–93.

Adams, D. "Artificial Kidneys and the Emergence of Bioethics: The History of 'Outsiders' in the Allocation of Hemodialysis." *Social History of Medicine* 24 (2011): 461–77.

Adelson, Naomi. "Visible/Human/Project: Visibility and Invisibility at the Next Anatomical Frontier." In *Figuring It Out: Science, Gender, and Visual Culture*, ed. Ann B. Shteir and Bernard Lightman, 358–76. Hanover, NH: Dartmouth College Press, 2006.

Adorno, Theodor. "Cultural Criticism and Society." In Adorno, *Prisms*, trans. Samuel and Shierry Weber, 17–34. Cambridge, MA: MIT Press, 1981.

Adorno, Theodor, and Max Horkheimer. "The Culture Industry: Enlightenment as Mass Deception [1969]." In *The Culture Studies Reader*, ed. Simon During, 31–42. London: Routledge, 1993.

———. *Towards a New Manifesto*, trans. Rodney Livingstone. London: Verso, 2011.

Afrary, Janet, and Kevin B. Anderson. *Foucault and the Iranian Revolution: Gender and the Seductions of Islam*. Chicago: University of Chicago Press, 2005.

Agamben, Giorgio. *Homo Sacer: Sovereign Power and Bare Life*, trans. Daniel Heller-Roazen. Stanford, CA: Stanford University Press, 1995.

Agar, Nicholas. *Humanity's End: Why We Should Reject Radical Enhancement*. Cambridge, MA: MIT Press, 2010.

———. *Liberal Eugenics: In Defence of Human Enhancement*. Oxford: Blackwell, 2004.

"AIDS Plakate International Bildsammlung, 1985–1997." Edition Braus. CD-ROM. Produced by Stiftung NeoCortex for Medizinische Fakultät der Universität, Basel (n.d.).

Aisenberg, Andrew. *Contagion: Disease, Government, and the "Social Question" in Nineteenth-Century France*. Stanford, CA: Stanford University Press, 1999.

"Alpha." "Remarks on the Cholera Morbus." *Lancet* 1 (1831): 108.
Altman, Dennis. "Globalization and the AIDS Industry." *Contemporary Politics* 4 (1998): 233–45.
Amsterdamska, Olga. "Surely You Are Joking, Monsieur Latour." *Science, Technology and Human Values* 15 (1990): 495–504.
Anderson, Warwick. *The Collectors of Lost Souls: Turning Kuru Scientists into Whitemen*. Baltimore: Johns Hopkins University Press, 2008.
———. "From Subjugated Knowledge to Conjugated Subjects: Science and Globalisation, or Postcolonial Studies of Science?" *Postcolonial Studies* 12 (2009): 389–400.
Ankersmit, Frank. "History and Post-Modernism." *History and Theory* 28 (1989): 137–53.
———. " 'Presence' and Myth." *History and Theory* 45 (2006): 328–36.
Appadurai, Arjun. "Grassroots Globalization and the Research Imagination." In *Globalization*, ed. Arjun Appadurai, 1–21. Durham, NC: Duke University Press, 2001.
Appleby, Joyce. *Economic Thought and Ideology in Seventeenth-Century England*. Princeton, NJ: Princeton University Press, 1978.
Appleby, Joyce, Lynn Hunt, and Margaret Jacob. *Telling the Truth About History*. New York: Norton, 1994.
Arditti, Rita, Pat Brennan, and Steve Cavrak, eds. *Science and Liberation*. Boston: South End Press, 1980.
Arendt, Hannah. *Eichmann in Jerusalem: A Report on the Banality of Evil*. London: Faber, 1963.
———. "What Is Authority?" In *Between Past and Future: Eight Exercises in Political Thought*, 91–141. London: Penguin, 1977.
Armstrong, David. "Embodiment and Ethics: Constructing Medicine's Two Bodies." *Sociology of Health and Illness* 28 (2006): 866–81.
———. *A New History of Identity: A Sociology of Medical Knowledge*. Houndsmills, UK: Palgrave, 2002.
———. *Political Anatomy of the Body*. Cambridge: Cambridge University Press, 1983.
Aronowitz, Robert A. *Making Sense of Illness: Science, Society and Illness*. Cambridge: Cambridge University Press, 1998.
Ashcroft, R. E. "Bioethics and Conflicts of Interests." *Studies in History and Philosophy of Science*, Part C: *Studies in History and Philosophy of Biological and Biomedical Science* 35 (2004): 155–65.

Attridge, Derek, G. Bennington, and R. Young, eds. *Poststructuralism and the Question of History*. Cambridge: Cambridge University Press, 1987.

Aulich, Jim, and John Hewitt. *Seduction or Instruction? First World War Posters in Britain and Europe*. Manchester: Manchester University Press, 2007.

Aynsley, Jeremy. *Graphic Design in Germany, 1890–1945*. London: Thames and Hudson, 2000.

Baker, Robert, and L. B. McCullough, eds. *The Cambridge World History of Medical Ethics*. Cambridge: Cambridge University Press, 2009.

Baker, Robert, D. Porter, and Roy Porter, eds. *The Codification of Medical Morality*, vol. 1: *Medical Ethics and Etiquette in the Eighteenth Century*. Dordrecht: Kluwer Academic, 1993.

Baldwin, Peter. *Contagion and the State in Europe, 1830–1930*. Cambridge: Cambridge University Press, 1999.

———. *Disease and Democracy: The Industrialized World Faces AIDS*. Berkeley: University of California Press, 2005.

Barham, Peter. *Forgotten Lunatics of the Great War*. New Haven: Yale University Press, 2004.

Barnes, Barry. "Elusive Memories of Technoscience." *Perspectives on Science* 13 (2005): 142–65.

Barnes, Barry, and Steven Shapin, eds. *Natural Order: Historical Studies of Scientific Culture*. Beverly Hills, CA: Sage, 1978.

Barnicoat, John. *A Concise History of Posters*. London: Thames and Hudson, 1972.

———. "Poster." In *The Dictionary of Art*, vol. 25, ed. Jane Turner, 345–55. London: Macmillan, 1996.

Barron, Stephanie, ed. *"Degenerate Art": The Fate of the Avant-Garde in Nazi Germany*. New York: Harry Abrams, 1991.

Barrow, Logie. "Socialism in Eternity: Plebeian Spiritualists, 1853–1913." *History Workshop Journal* 9 (1980): 37–69.

Barthes, Roland. *Mythologies*. Paris: Éditions du Seuil, 1957.

Bastos, Cristiana. *Global Reponses to AIDS: Science in Emergency*. Bloomington: Indiana University Press, 1999.

Baudrillard, Jean. *In the Shadow of the Silent Majorities . . . or the End of the Social, and Other Essays*. New York: Semiotext(e), 1983.

Bauer, M. W., and G. Gaskell. *Biotechnology: The Making of a Global Controversy*. Cambridge: Cambridge University Press, 2002.

Baum, Michael. "Evidence-Based Art." *Journal of the Royal Society of Medicine* 94 (June 2001): 306–7.

Bauman, Zygmunt. *Modernity and the Holocaust*. Cambridge: Polity, 1989.

Bayly, Christopher. *The Birth of the Modern World, 1780–1914*. Oxford: Blackwell, 2004.

Beck, Ulrich. "How Modern Is Modern Society." *Theory, Culture and Society* 9 (1992): 163–69.

———. "The Reinvention of Politics: Towards a Theory of Reflexive Modernization." In *Reflexive Modernization: Politics, Tradition and Aesthetics in the Modern Social Order*, ed. U. Beck, A. Giddens, and S. Lash, 1–55. Cambridge: Polity, 1994.

———. *Risk Society: Towards a New Modernity*, trans. M. Ritter. London: Sage, 1992.

———. *Was ist Globalisierung? Irrtümer des Globalismus—Antworten auf Globalisierung*. Frankfurt: Suhrkamp, 2007.

Beckett, F. "Protest Politics." *AIDS Matters* 8 (1992): 5–6.

Beecher, Henry. "Ethics and Clinical Research." *New England Journal of Medicine* 274 (1966): 1354–60.

Beer, Gillian. *Darwin's Plots: Evolutionary Narrative in Darwin, George Eliot and Nineteenth-Century Fiction*. London: Routledge, 1983.

Beevor, Antony. "Eyes on the Prize." *Intelligent Life, The Quarterly from The Economist* 4 (2010): 79.

Belkin, G. S. Review of *Bioethics in Social Context*. *Journal of the History of Medicine and Allied Sciences* 57 (2002): 373.

Bell, Daniel. *The End of Ideology: On the Exhaustion of Political Ideas in the Fifties*. New York: Free Press, 1960.

Bencard, Adam. "History in the Flesh: Investigating the Historicized Body." PhD diss., University of Copenhagen, 2007.

———. "Life Beyond Information—Contesting Life and the Body in History and Molecular Biology." In *Contested Categories: Life Sciences in Society*, ed. Susanne Bauer and Ayo Wahlberg, 135–54. Farnham: Ashgate, 2009.

Bennett, Tony. *The Birth of the Museum: History, Theory, Politics*. London: Routledge, 1995.

———. "Texts in History: The Determinations of Readings and Their Texts." In *Post-Structuralism and the Question of History*, ed. Derek Attridge, G. Bennington, and Robert Young, 63–81. Cambridge: Cambridge University Press, 1987.

Bennett, Tony, and Patrick Joyce, eds. *Material Powers: Cultural Studies, History and the Material Turn*. London: Routledge, 2010.

Bentham, Jeremy. *Works of Jeremy Bentham*, ed. J. Bowring, vol. 11. New York: Russell and Russell, 1962.

Bentley, Michael. "Past and 'Presence': Revisiting Historical Ontology." *History and Theory* 45 (2006): 349–61.

Berridge, Virginia. *AIDS in the UK: The Making of Policy, 1981–1994*. Oxford: Oxford University Press, 1996.

Bess, Michael. "Icarus 2.0: A Historian's Perspective on Human Biological Enhancement." *Technology and Culture* 49 (2008): 114–26.

Best, Steven. *The Politics of Historical Vision: Marx, Foucault, Habermas*. New York: Guilford, 1995.

Biehl, Joao. *The Will to Live: AIDS Therapy and the Politics of Survival (In-Formation)*. Princeton, NJ: Princeton University Press, 2007.

Biernoff, Suzannah. *Sight and Embodiment in the Middle Ages*. London: Palgrave Macmillan, 2002.

Billig, Michael. *Banal Nationalism*. London: Sage, 1995.

Bilson, G. *A Darkened House: Cholera in Nineteenth-Century Canada*. Toronto: University of Toronto Press, 1980.

Birnbacher, D. "Neuroethics and Stem Cell Transplantation." *Medicine Studies* 1 (2009): 67–76.

Blackman, Lisa. "Biology, Brain Theory and History: What, If Anything, Can Historians Learn from Biology?" Discussion paper, Senate House, University of London, 11 November 2011.

Blaut, James M., et al. *1492: The Debate on Colonialism, Eurocentrism, and History*. Trenton, NJ: Africa World Press, 1992.

Bloch, Iwan. *Der Ursprung der Syphilis: Eine medizinische und kulturgeschichtliche Untersuchung*, vol. 1: *Ursprung der Syphilis*. Jena: Gustav Fischer, 1901; vol. 2: *Kritik der Lehre von der Altertumssyphilis*. Jena: Gustav Fischer, 1911.

Bloch, Marc. *The Historian's Craft*. Manchester: Manchester University Press, 1992.

———. *The Royal Touch: Sacred Monarchy and Scrofula in England and France*, trans. J. E. Anderson. London: Routledge, 1973.

Boggs, Carl. "How Can We End the End of Politics?" *Democracy and Nature* 7 (March 2001): 205–8.

Boltanski, Luc, and Eve Chiapello. *The New Spirit of Capitalism*, trans. Gregory Elliott. London: Verso, 2005.

Bonah, C. " 'Experimental Rage': The Development of Medical Ethics and the Genesis of Scientific Facts. Ludwik Fleck: An Answer to the Crisis of Modern Medicine in Interwar Germany?" *Social History of Medicine* 15 (2002): 187–207.

Bonfield, Lloyd, R. S. Smith, and K. Wrightson, eds. *The World We Have Gained: Essays in Honour of Peter Laslett*. Oxford: Blackwell, 1986.

Bonnell, Victoria E., and Lynn Hunt, eds. *Beyond the Cultural Turn*. Berkeley: University of California Press, 1999.

Borneman, John. "AIDS in the Two Berlins." *October* 43 (1987): 223–36.

Bosk, C. L., ed. *What Would You Do? Juggling Bioethics and Ethnography*. Chicago: University of Chicago Press, 2008.

Bourdieu, Pierre. *Homo Academicus*, trans. Peter Collier. Cambridge: Polity, 1988.

———. *The Logic of Practice*. Cambridge: Polity, 1992.

Bowring, J., ed. *Works of Jeremy Bentham*, vol. 11. New York: Russell and Russell, 1962.

Brandt, Allan M. "Emerging Themes in the History of Medicine." *Milbank Quarterly* 69 (1991): 199–214.

———. *No Magic Bullet: A Social History of Venereal Disease in the United States Since 1880*. New York: Oxford University Press, 1985.

Braverman, Harry. *Labour and Monopoly Capital: The Degradation of Work in the Twentieth Century*. New York: Monthly Review, 1974.

Brewer, John. "The Error of Our Ways: Historians and the Birth of Consumer Society." Working Paper no. 12 (ESRCC and HEC, 2005).

Brinckmann, Justus. *Katalog der Plakat-Ausstellung: Hamburg 1896 Museum für Kunst und Gewerbe*. Hamburg: Lütcke und Wulff E. H. Senatsbuchdruckerei, 1896.

Brock, Adrian C., ed. *Internationalizing the History of Psychology*. New York: New York University Press, 2006.

Brock, Richard. "An 'Onerous Citizenship': Globalization, Cultural Flows and the HIV/AIDS Pandemic in Hari Kunzru's *Transmission*." *Journal of Postcolonial Writing* 44 (2008): 379–90.

Brown, E. Richard. *Rockefeller Medicine Men: Medicine and Capitalism in America*. Berkeley: University of California Press, 1979.

Brown, Nik. "Shifting Tenses: Reconnecting Regimes of Truth and Hope." *Configurations* 13 (2007): 313–55.

Browne, Lord John. "Securing a Sustainable Future for Higher Education: An Independent Review of Higher Education and Student Finance." 12 October 2010, available at www.independent.gov.uk/browne-report.

Bryson, Norman. "The Neural Interface." Introduction to *Blow Up Photography, Cinema and the Brain*, ed. Warren Neidlich. New York: Distributed Arts, 2003.

Burchell, G., Colin Gordon, and Peter Miller, eds. *The Foucault Effect: Studies in Governmental Rationality*. Hemel Hempstead, UK: Harvester/Wheatsheaf, 1991.

Burke, Seán. *The Death and Return of the Author: Criticism and Subjectivity in Barthes, Foucault and Derrida*. Edinburgh: Edinburgh University Press, 1992.
Burney, Ian. *Bodies of Evidence: Medicine and the Politics of the English Inquest, 1830–1926*. Baltimore: Johns Hopkins University Press, 2000.
———. "Roundtable Discussion on *The Making of a Social Body*." *Journal of Victorian Culture* 4 (1999): 104–16.
Burrow, John. *A History of Histories: Epics, Chronicles, Romance and Inquiries from Herodotus and Thucydides to the Twentieth Century*. London: Allen Lane, 2007.
Burton, Antoinette. "Thinking Beyond the Boundaries: Empire, Feminism and the Domains of History." *Social History* 26 (2001): 60–71.
Bury, M. R. "Social Constructivism and the Development of Medical Sociology." *Sociology of Health and Illness* 8 (1986): 137–69.
Butler, Judith. *Bodies That Matter: On the Discursive Limits of Sex*. New York: Routledge, 1993.
———. "Critique, Dissent, Disciplinarity." *Critical Inquiry* 35 (2009): 773–95.
———. "What Is Critique? An Essay on Foucault's Virtue." In *The Political*, ed. David Ingram, 212–26. Oxford: Blackwell, 2002.
Butterfield, Herbert. *History and Human Relations*. London: Longman, 1951.
———. *The Whig Interpretation of History* [1931]. Harmondsworth, UK: Penguin Books, 1973.
Bynum, Caroline Walker. "Why All the Fuss About the Body? A Medievalist's Perspective." *Critical Inquiry* 22 (1995): 1–33.
Campbell, Timothy. Introduction to Roberto Esposito, *Bíos: Biopolitics and Philosophy*, trans. Timothy Campbell, vii–xlii. Minneapolis: University of Minnesota Press, 2008.
Candea, Matei, ed. *The Social After Gabriel Tarde: Debate and Assessment*. Abingdon, UK: Routledge, 2010.
Canguilhem, Georges. "The Concept of Life [1973]." In *A Vital Rationalist: Selected Writings from Georges Canguilhem*, ed. F. Delaporte, trans. A. Goldhammer, and intro. Paul Rabinow, 303–19. New York: Zone Books, 1994.
———. *The Normal and the Pathological*. Boston: Zone Books, 1991.
Cantor, G. N. "A Critique of Shapin's Social Interpretation of the Edinburgh Phrenology Debate." *Annals of Science* 32 (1975): 245–56.
Carr, E. H. *What Is History?* London: Penguin, 1961.
Carrette, Jeremy, and Richard King. *Selling Spirituality: The Silent Takeover of Religion*. London: Routledge, 2005.

Cartwright, Lisa. *Screening the Body: Tracing Medicine's Visual Culture*. Minneapolis: University of Minnesota Press, 1995.

Casey, Catherine. *Work, Self and Society*. London: Routledge, 1995.

Chandler, James. "Introduction: Doctrines, Disciplines, Discourses, Departments." *Critical Inquiry* 35 (2009): 739–40.

Chandler, James, Arnold I. Davidson, and Harry Harootunian, eds. *Questions of Evidence: Proof, Practice, and Persuasion Across the Disciplines*. Chicago: University of Chicago Press, 1991.

Chorover, S. "Who Needs Neuroethics?" *Lancet* 365 (2005): 2081–82.

Clarke, Adele E., et al. "Biomedicalization: A Theoretical and Substantive Introduction." In *Biomedicalization: Technoscience, Health, and Illness in the U.S.*, ed. Adele Clarke et al., 1–44. Durham, NC: Duke University Press, 2010.

———, eds. *Biomedicalization: Technoscience, Health, and Illness in the U.S.* Durham, NC: Duke University Press, 2010.

Clarke, Paul Barry. "Deconstruction." In *Dictionary of Ethics, Theology and Society*, ed. P. B. Clarke and A. Linzey, 216–23. London: Routledge, 1996.

Classen, Constance, and David Howes. "The Museum as Sensescape: Western Sensibilities and Indigenous Artifacts." In *Sensible Objects: Colonialism, Museums and Material Culture*, ed. Elizabeth Edwards, C. Gosden, and R. Phillips, 199–222. Oxford: Berg, 2006.

Coleman, William. *Yellow Fever in the North: The Methods of Early Epidemiology*. Madison: University of Wisconsin Press, 1987.

Collingwood, R. G. *The Idea of History*. Oxford: Oxford University Press, 1961.

Collini, Stefan. "Impact on Humanities Researchers Must Take a Stand Now or Be Judged and Rewarded as Salesmen." *Times Literary Supplement*, 13 November 2009, 18.

———. "[Lord] Browne's Gamble." *London Review of Books* 32 (4 November 2010): 23–25.

Collins, Harry, and Trevor Pinch. *Frames of Meaning: The Social Construction of Extraordinary Science*. London: Routledge, 1982.

Conrad, Peter. *The Medicalization of Society: On the Transformation of Human Conditions into Treatable Disorders*. Baltimore: Johns Hopkins University Press, 2007.

Cook, Harold J. "Borderlands: A Historian's Perspective on Medical Humanities in the US and the UK." *Journal of Medical Ethics; Medical Humanities Online*, 7 June 2010, 1–2.

———. *Matters of Exchange: Commerce, Medicine, and Science in the Dutch Golden Age*. New Haven: Yale University Press, 2007.
Cook, Matt. "From Gay Reform to Gaydar, 1967–2006." In *A Gay History of Britain: Love and Sex Between Men Since the Middle Ages*, ed. Matt Cook et al., 179–214. Westport, CT: Greenwood, 2007.
Cooper, Melenda. *Life as Surplus: Biotechnology and Capitalism in the Neoliberal Era*. Seattle: University of Washington Press, 2008.
Cooter, Roger. "After Death/After-'Life': The Social History of Medicine in Post-Postmodernity." *Social History of Medicine* 20 (2007): 441–64.
———. "Biocitizenship." *Lancet* 372 (15 November 2008): 1725.
———. "Biology, Brain Theory and History: What, If Anything, Can Historians Learn from Biology?" Discussion paper, Senate House, University of London, 11 November 2011.
———. "Crisis." *Lancet* 373 (14 March 2009): 887.
———. *The Cultural Meaning of Popular Science: Phrenology and the Organization of Consent in Nineteenth-Century Britain*. Cambridge: Cambridge University Press, 1984.
———. "The Dead Body." In *Companion to Medicine in the Twentieth Century*, ed. Roger Cooter and John Pickstone, 469–85. Amsterdam: Harwood Academic, 2000; repr. London: Routledge, 2003.
———. "Deploying 'Pseudoscience': Then and Now." In *Science, Pseudo-Science and Society*, ed. M. Hanen, M. Osler, and R. G. Weyant, 237–72. Waterloo, ON: Laurier University Press, 1980.
———. 'The Disabled Body." In *Companion to Medicine in the Twentieth Century*, ed. Roger Cooter and John Pickstone, 367–83. Amsterdam: Harwood Academic, 2000; repr. London: Routledge, 2003.
———. "The Ethical Body." In *Companion to Medicine in the Twentieth Century*, ed. Roger Cooter and John Pickstone, 451–68. Amsterdam: Harwood Academic, 2000; repr. London: Routledge, 2003.
———. "La médecine dans la pensée historique contemporaine." In *Histoire de la pensée médicale contemporaine*, ed. Bernardino Fantini. Paris: Éditions du Seuil (forthcoming).
———. "Medicine and Modernity." In *Handbook to History of Medicine*, ed. Mark Jackson, 100–116. Oxford: Oxford University Press, 2011.
———. "NeuroPatients in Historyland." In *The Neurological Patient in History*, ed. Stephen Jacyna and Stephen Casper, 215–22. Rochester, NY: University of Rochester Press, 2012.

———. "Phrenology and British Alienists, c. 1825–1845." *Medical History* 20 (1976): 1–21, 135–51, repr. in *Madhouses, Mad-Doctors and Madmen: The Social History of Psychiatry in the Victorian Era*, ed. Andrew Scull, 58–104. Philadelphia: University of Pennsylvania Press, 1981.

———. "The Power of the Body: The Early Nineteenth Century." In *Natural Order: Historical Studies of Scientific Culture*, ed. Barry Barnes and Steven Shapin, 73–92. London: Sage, 1978.

———. "Preisgabe der Demokratie Wie man die Geschichts- und Geisteswissenschaften in die Naturwissenschaft einwickelt." In *Wissenschaft und Demokratie*, ed. Michael Hagner, 88–111, 260–67. Zurich: Suhrkamp, 2013.

———. "The Resistible Rise of Medical Ethics." *Social History of Medicine* 8 (1995): 275–88.

———. Review of Anderson, *Collectors of Lost Souls*. *Isis* 100 (2009): 941–42.

———. Review of Nadesan, *Governmentality*. *Medical History* 54 (2010): 124–26.

———. *Surgery and Society in Peace and War: Orthopaedics and the Organization of Modern Medicine, 1880–1948*. Manchester: Manchester University Press, 1993.

———. "Teamwork." *Lancet* 363 (10 April 2004): 1245.

———. "The Traffic in Victorian Bodies: Medicine Literature and History." *Victorian Studies* 45 (2003): 513–27.

———. *When Paddy Met Geordie: The Irish in County Durham and Newcastle, 1840–1880*, with a foreword by Donald M. MacRaild. Sunderland, UK: University of Sunderland Press, 2005.

Cooter, Roger, and Mary Fissell. "Exploring Natural Knowledge: Science and the Popular in the Eighteenth Century." In *Cambridge History of Science*, vol. 4: *Science in the Eighteenth Century*, ed. Roy Porter, 145–79. Cambridge: Cambridge University Press, 2003.

Cooter, Roger, and John Pickstone, eds. *Medicine in the Twentieth Century*. Amsterdam: Harwood Academic, 2000; repr. London: Routledge, 2003.

Cooter, Roger, and Stephen Pumphrey. "Separate Spheres and Public Places: Reflections on the History of Science Popularisation and Science in Popular Culture." *History of Science* 32 (1994): 237–67.

Cooter, Roger, and Claudia Stein. "Can Historians Change Philosophy?" Paper presented to the Centro de Ciencias Humanas y Sociales—CSIC, Madrid, 16 June 2011.

———. "Positioning the Image of Aids." *Endeavour* 34 (2010): 12–15.

———. "Protect Yourself." In *Public Health Campaigns: Getting the Message Across*, ed. World Health Organization, 66–88. Geneva: World Health Organization, 2009.

———. "Visual Imagery and Epidemics in the Twentieth Century." In *Imagining Illness: Public Health and Visual Culture*, ed. David Serlin, 169–92. Minneapolis: University of Minnesota Press, 2010.

Cooter, Roger, and Steve Sturdy. "Of War, Medicine and Modernity: Introduction." In *War, Medicine and Modernity*, ed. Roger Cooter, Mark Harrison, and Steve Sturdy, 1–21. Stroud, UK: Sutton, 1998.

———. "Science, Scientific Management and the Transformation of Medicine in Britain, c. 1870–1950." *History of Science* 36 (1998): 421–66.

Cordle, Daniel. *Postmodern Postures: Literature, Science and the Two Cultures Debate*. Aldershot, UK: Ashgate, 1999.

Corning, Peter. *The Fair Society: The Science of Human Nature and the Pursuit of Social Justice*. Chicago: University of Chicago Press, 2011.

Cosgrove, Denis. "Landscape and Landschaft. Lecture Delivered at the 'Spatial Turn in History' Symposium German Historical Institute, February 19, 2004." *German Historical Institute Bulletin* 35 (2004): 57–71.

Cracraft, James. "Implicit Morality." *History and Theory* 43 (2004): 31–42.

Crary, Jonathan. *Suspensions of Perception: Attention, Spectacle, and Modern Culture*. Cambridge, MA: MIT Press, 2000.

Crawford, Matthew. "The Limits of Neuro-Talk." *New Atlantis* (Winter 2008): 65–78.

Crimp, Douglas. *Melancholia and Moralism: Essays on AIDS and Queer Politics*. Cambridge, MA: MIT Press, 2002.

———. "Portraits of People with AIDS [1992]." In *Melancholia and Moralism: Essays on AIDS and Queer Politics*. Cambridge, MA: MIT Press, 2002.

Crimp, Douglas, with Adam Rolston. *AIDS Demo Graphics*. Seattle: Bay Press, 1990.

Cromby, John, Time Newton, and Simon J. Williams, eds. "Neuroscience and Subjectivity." *Subjectivity*, special issue, 4, no. 3 (2011).

Cubitt, Geoffrey. *History and Memory*. Manchester: Manchester University Press, 2007.

Cunningham, Andrew. "Transforming Plague: The Laboratory and the Identity of Infectious Disease." In *The Laboratory Revolution in Medicine*, ed. Andrew Cunningham and Perry Williams, 209–44. Cambridge: Cambridge University Press, 1992.

Cuno, James. *Whose Muse? Art Museums and the Public Trust*. Princeton, NJ: Princeton University Press, 2004.

Cusset, François. *French Theory: How Foucault, Derrida, Deleuze, & Co. Transformed the Intellectual Life of the United States*, trans. Jeff Fort. Minneapolis: University of Minnesota Press, 2008.

Daston, Lorraine. "Baconian Facts, Academic Civility, and the Prehistory of Objectivity." In *Rethinking Objectivity*, ed. Allan Megill, 37–63. Durham, NC: Duke University Press, 1994.

———. "Historical Epistemology." In *Questions of Evidence: Proof, Practice, and Persuasion Across the Disciplines*, ed. James Chandler, Arnold I. Davidson, and Harry Harootunian, 282–89. Chicago: University of Chicago Press, 1991.

———. "Science Studies and the History of Science." *Critical Inquiry* 35 (2009): 798–815.

———, ed. *Biographies of Scientific Objects*. Chicago: University of Chicago Press, 2000.

Daston, Lorraine, and Peter Galison. *Objectivity*. New York: Zone Books, 2007.

Davidson, Arnold. "On Epistemology and Archaeology: From Canguilhem to Foucault." In *The Emergence of Sexuality: Historical Epistemology and the Formation of Concepts*, ed. Arnold Davidson, 192–206. Cambridge, MA: Harvard University Press, 2001.

Davies, James C. *Human Nature in Politics: The Dynamics of Political Behavior*. New York: Wiley, 1918.

Davis, Mike. *City of Quartz. Excavating the Future in Los Angeles*. London: Verso, 1990.

Davis, Natalie Zemon. "A Life of Learning." The Charles Homer Haskins lecture for 1997, American Council of Learned Societies. *Occasional Papers*, no. 39, 1997.

Dean, Mitchell. *Critical and Effective Histories: Foucault's Methods and Historical Sociology*. London: Routledge, 1994.

De Beauvoir, Simone. *The Second Sex* [1949]. New York: Knopf, 1989.

De Certeau, Michel. *The Practice of Everyday Life*, trans. Steven Rendall. Berkeley: University of California Press, 1984.

Delaporte, François. *Disease and Civilization: The Cholera in Paris*, trans. Arthur Goldhammer, with a foreword by Paul Rabinow. Cambridge, MA: MIT Press, 1986.

———. "The History of Medicine According to Foucault." In *Foucault and the Writing of History*, ed. Jan Goldstein, 137–49. Oxford: Blackwell, 1994.

Deleuze, Gilles. *Foucault*, trans. S. Hand. London: Athlone, 1988.

———. "What Is a Dispositif?" In *Michel Foucault, Philosopher*, trans. T. Armstrong, 159–68. Hemel Hempstead, UK: Harvester/Wheatsheaf, 1992.

Derrida, Jacques. *Of Grammatology*, trans. G. C. Spivak. Baltimore: Johns Hopkins University Press, 1974.

Desmond, Adrian, and Jim Moore. Response to their *Darwin. Journal of Victorian Culture* 3 (1998): 152.

Desrosières, Alain. *The Politics of Large Numbers: A History of Statistical Reasoning*, trans. Camille Naish. Cambridge, MA: Harvard University Press, 1998.

De Tarde, Gabriel. *Social Law: An Outline of Sociology*, trans. H. C. Warren. New York: Macmillan, 1899.

DeVries, R., and J. Subedi, eds. *Bioethics and Society: Constructing the Ethical Enterprise*. Upper Saddle River, NJ: Prentice Hall, 1998.

Diamond, S. "Anthropology in Question." In *Reinventing Anthropology*, ed. D. Hymes, 401–29. New York: Vintage, 1974.

Dickson, D. "Science and Political Hegemony in the Seventeenth Century." *Radical Science Journal* 8 (1979): 7–37.

Dikovitskaya, Margaret. *Visual Culture: The Study of the Visual After the Cultural Turn*. Cambridge, MA: MIT Press, 2006.

Directions to Plain People as a Guide for Their Conduct in the Cholera. London: n.p., 1831.

Dirlik, Aruf. "Confounding Metaphors, Inventions of the World: What Is World History For?" In *Writing World History, 1800–2000*, ed. Benedikt Stuchtey and Eckhardt Fuchs, 91–133. Oxford: Oxford University Press, 2003.

Dittberner, Job. *The End of Ideology and American Social Thought, 1930–1960*. Ann Arbor: UMI Research Press, 1979.

Dolan, Brian. "Second Opinions: History, Medical Humanities, and Medical Education." *Social History of Medicine* 23 (2010): 393–405.

Dolan, Brian, and Allison Tillack. "Pixels, Patterns and Problems of Vision: The Adaptation of Computer-Aided Diagnosis for Mammography in Radiological Practice in the U.S." *History of Science* 48 (2010): 27–49.

Dommann, Monika. "Vom Bild zum Wissen: Eine Bestandsaufnahme wissenschaftshistorischer Bildforschung." *Gesnerus* 61 (2004): 77–89.

Donnelly, M. "On Foucault's Uses of the Notion 'Biopower.'" In *Michel Foucault, Philosopher*, trans. T. Armstrong, 199–203. Hemel Hempstead, UK: Harvester/Wheatsheaf, 1992.

Donzelot, Jacques. *L'invention du social*. Paris: Fayard, 1984.

Döring, Jürgen, ed. *Gefühlsecht: Graphikdesign der 90er Jahre*. Hamburg: Museum für Kunst und Gewerbe, 1996.

Douglas, Mary. *Implicit Meanings: Essays in Anthropology*. London: Routledge and Kegan Paul, 1975.

———. *Natural Symbols: Explorations in Cosmology*. New York: Vintage, 1973.

———. *Purity and Danger: An Analysis of Concepts of Pollution and Taboo*. London: Routledge and Kegan Paul, 1966.

———. "The Social Preconditions of Radical Skepticism." In *Power, Action and Belief: A New Sociology of Knowledge?*, ed. John Law, 68–87. London: Routledge and Kegan Paul, 1986.

Downey, Garry Lee, and Joseph Dumit, eds. *Cyborgs and Citadels: Anthropological Interventions in Emerging Sciences and Technologies*. Santa Fe, NM: School of American Research Press, 1997.

Dreyfus, Hubert, and Paul Rabinow, eds. *Michel Foucault: Beyond Structuralism and Hermeneutics*. Brighton, UK: Harvester, 1982.

Driscoll, Lawrence. *Reconsidering Drugs: Mapping Victorian and Modern Drug Discourses*. New York: Palgrave, 2000.

Drucker, Peter. *Managing in Turbulent Times*. New York: Harper and Row, 1980.

Duden, Barbara. *The Woman Beneath the Skin: A Doctor's Patients in Eighteenth-Century Germany*, trans. T. Dunlop. Cambridge, MA: Harvard University Press, 1991.

Durey, Michael. "Bodysnatchers and Benthamites: The Implications of the Dead Body Bill for the London Schools of Anatomy, 1820–42." *London Journal* 2 (1976): 200–225.

———. *The Return of the Plague: British Society and the Cholera, 1831–2*. Dublin: Gill and Macmillan, 1979.

During, Simon. *Foucault and Literature: Towards a Genealogy of Writing*. London: Routledge, 1992.

———, ed. *The Culture Studies Reader*. London: Routledge, 1993.

Dutton, Dianna B. *Worse Than the Disease: Pitfalls of Medical Progress*. Cambridge: Cambridge University Press, 1988.

Eagleman, David. *Incognito: The Sacred Lives of the Brain*. Edinburgh: Canongate, 2011.

Eagleton, Terry. *The Illusions of Postmodernism*. Oxford: Blackwell, 1996.

Eagleton, Terry, and Matthew Beaumont. *The Task of the Critic: Terry Eagleton in Dialogue*. London: Verso, 2009.

Easlea, Brian. *Liberation and the Aims of Science*. Edinburgh: Scottish Academic Press, 1980.

Easton, L. D., and K. H. Guddat, eds. *Writings of the Young Marx on Philosophy and Society*. New York: Doubleday, 1967.

Edelstein, L. *Ancient Medicine: Selected Papers of Ludwig Edelstein*, ed. O. Temkin and C. L. Temkin. Baltimore: Johns Hopkins University Press, 1967.

Edwards, Adrian, et al. "Presenting Risk Information: A Review of the Effects of 'Framing' and Other Manipulations on Patient Outcomes." *Journal of Health Communication* 6 (2001): 61–82.

Edwards, Martin. "Put Out Your Tongue! The Role of Clinical Insight in the Study of the History of Medicine." *Medical History* 55 (2011): 301–306.

Eley, Geoff. *A Crooked Line: From Cultural History to the History of Society*. Ann Arbor: University of Michigan Press, 2006.

———. "Is All the World a Text? From Social History to the History of Society Two Decades Later." In *The Historical Turn in the Human Sciences*, ed. Terrence J. McDonald, 193–243. Ann Arbor: University of Michigan Press, 1996.

Eley, Geoff, and Keith Nield. "Farewell to the Working Class?" *International Labour and Working-Class History* 57 (2000): 1–30.

———. "Reply: Class and the Politics of History." *International Labour and Working-Class History* 57 (2000): 76–87.

———. "Why Does Social History Ignore Politics?" *Social History* 5 (1980): 249–71.

Elkins, James. *Visual Studies: A Skeptical Introduction*. New York: Routledge, 2003.

Elliott, Carl. *A Philosophical Disease: Bioethics, Culture and Identity*. New York: Routledge, 1999.

———. "Six Problems with Pharma-Funded Bioethics." *Studies in History and Philosophy of Science*, Part C: *Studies in History and Philosophy of Biological and Biomedical Sciences* 35 (2004): 125–29.

———. "When Pharma Goes to the Laundry: Public Relations and the Business of Medical Education." *Hastings Center Report* 34 (2004): 18–23.

Elliott, Stuart. "Another Furor over a Benetton Ad." *New York Times*, 29 January 1992, section D, 17.

Empson, E. "The Political Immunity of Discourse [a review of R. Esposito, *Bíos*]." *Mute* 2 (2009): 46–59.

Engelhardt, H. T., ed. *Global Bioethics: The Collapse of Consensus*. Salem, MA: M & M Scrivener, 2006.

Engelhardt, H. T., A. S. Iltis, and F. Herand. 2003. "Bioethics: Institutionalization of." In *Nature Encyclopedia of the Human Genome*, ed. David Cooper, vol. 1, 281–85. London: Macmillan, 2003.

Epstein, Jim. "Signs of the Social." *Journal of British Studies* 36 (1997): 473–84.

Epstein, Steven. *Impure Science: AIDS, Activism, and the Politics of Knowledge*. Berkeley: University of California Press, 1996.

Eribon, Didier, ed. *L'infréquentable Michel Foucault: Renouveaux de la pensée critique*. Paris: EPEL, 2001.

———. *Michel Foucault*, trans. B. Wing. London: Faber, 1991.

Ermarth, Elizabeth Deeds. "The Closed Space of Choice: A Manifesto on the Future of History." In *Manifestos for History*, ed. Keith Jenkins, Sue Morgan, and Alun Munslow, 50–66. London: Routledge, 2007.

Ernst, Waltraud. "The Normal and the Abnormal: Reflections on Norms and Normativity." In *Histories of the Normal and the Abnormal: Social and Cultural Histories of Norms and Normativity*, ed. Waltraud Ernst, 1–25. London: Routledge, 2006.

Esposito, Roberto. *Bíos: Biopolitics and Philosophy*, trans. and intro. Timothy Campbell. Minneapolis: University of Minnesota Press, 2008.

———. *Immunitas: The Protection and Negations of Life* [2002], trans. Zakiya Hanafi. Cambridge: Polity, 2011.

Evans, Richard. *Death in Hamburg: Society and Politics in the Cholera Years, 1830–1910*. Oxford: Clarendon, 1987.

———. *In Defence of History*. London: Granta Books, 1997.

———. "The Wonderfulness of Us (The Tory Interpretation of History)." *London Review of Books* 33 (17 March 2011): 9–12.

Faden, R., and T. Beauchamp. *A History and Theory of Informed Consent*. New York: Oxford University Press, 1986.

Featherstone, Mike. *Consumer Culture and Postmodernism*, 2nd ed. London: Sage, 2007.

Featherstone, Mike, Mike Hepworth, and Bryan Turner, eds. *The Body: Social Process and Cultural Theory*. London: Sage, 1991.

Fee, Elizabeth, and Theodore M. Brown. *Making Medical History: The Life and Times of Henry E. Sigerist*. Baltimore: Johns Hopkins University Press, 1997.

Feher, Michael, ed. *Fragments for a History of the Human Body*. 3 vols. New York: Zone Books, 1989.

Ferber, Sarah. *Bioethics in Historical Perspective: Medicine and Culture*. London: Palgrave, 2013.

Fernández-Armesto, Felipe. "Global History, Methods and Objectives." Paper presented to the Polyphonic History Seminar, Madrid, 22 January 2008.

Feudtner, Chris. " 'Minds the Dead Have Ravished': Shell Shock, History, and the Ecology of Disease Systems." *History of Science* 31 (1993): 377–420.

Field, Becky, and Kaye Wellings. *Stopping AIDS: AIDS/HIV Public Education and the Mass Media in Europe*. London: Longman, 1996.

Field, Becky, et al. *Promoting Safer Sex: A History of the Health Education Authority's Mass Media Campaigns on HIV, AIDS and Sexual Health, 1987–1996*. London: Health Education Authority, 1997.

Figlio, Karl. "Chlorosis and Chronic Disease in Nineteenth-Century Britain: The Social Construction of Somatic Illness in a Capitalist Society." *Social History* 3 (1978): 167–97.

———. "Second Thoughts on 'Sinister Medicine.'" *Radical Science Journal* 10 (1980): 159–66.

———. "Sinister Medicine? A Critique of Left Approaches to Medicine." *Radical Science Journal* 9 (1979): 14–68.

Finnegan, Diarmid. "The Spatial Turn: Geographical Approaches in the History of Science." *Journal of the History of Biology* 41 (2008): 369–88.

Fishman, Jennifer R. "The Making of Viagra: The Biomedicalization of Sexual Dysfunction." In *Biomedicalization: Technoscience, Health, and Illness in the U.S.*, ed. Adele E. Clarke et al., 289–306. Durham, NC: Duke University Press, 2010.

Fissell, Mary. "Making Meaning from the Margins: The New Cultural History of Medicine." In *Locating Medical History: The Stories and their Meanings*, ed. Frank Huisman and John Harley Warner, 364–89. Baltimore: Johns Hopkins University Press, 2004.

Fleck, Ludwik. *Genesis and Development of a Scientific Fact* [1935], trans. F. Bradley. Chicago: University of Chicago Press, 1979.

Flynn, Thomas R. *Existentialism: A Very Short History*. Oxford: Oxford University Press, 2006.

Forman, Paul. "The Primacy of Science in Modernity, of Technology in Postmodernity, and of Ideology in the History of Technology." *History and Technology* 23 (2007): 1–152.

———. "(Re)cognizing Postmodernity: Helps for Historians—of Science Especially." *Berichte zur Wissenschaftsgeschichte* 33 (2010): 1–19.

Forrester, John. *The Seductions of Psychoanalysis: Freud, Lacan, and Derrida*. Cambridge: Cambridge University Press, 1990.

Forster, Thomas Furley. *Brief Inquiry into the Causes and Mitigation of Pestilential Fever, and into the Opinion of the Ancients Respecting Epidemical Diseases* [*The Pamphleteer*, vol. 24, no. 48, Pamphlet 5], 2nd ed. London, 1824.

Forster, Thomas I. M. *Facts and Enquiries Respecting the Source of Epidemia*, 3rd ed. London: Keating and Brown, 1832.

———. "On the Atmospherical and Terrestrial Commotions Which Have Accompanied the Cholera Morbus of the Present Period." *Lancet* (22 October 1831): 113.

Fortun, M. *Promising Genomics: Iceland and deCODE Genetics in a World of Speculation*. Berkeley: University of California Press, 2008.

Fotopoulos, Takis. "The Myth of Postmodernity." *Democracy and Nature* 7 (March 2001): 27–75.

Foucault, Michel. *The Archaeology of Knowledge* [1969], trans. A. M. Sheridan Smith. London: Tavistock, 1972.

———. *The Birth of Biopolitics: Lectures at the Collège de France, 1978–1979*, ed. M. Senellart, F. Ewald, and A. Fontana, English series ed. A. Davidson, trans. G. Burchell. Houndsmill, UK: Palgrave Macmillan, 2008.

———. "The Birth of Bio-Power." In *The Essential Foucault*, ed. Paul Rabinow and Nikolas Rose, 202–7. New York: New Press, 2003.

———. "The Confession of the Flesh." In *Power/Knowledge: Selected Interviews and Other Writings, 1972–1977*, ed. C. Gordon, trans. C. Gordon et al., 194–228. Brighton, UK: Harvester/Wheatsheaf, 1980.

———. *Discipline and Punish: The Birth of the Prison* [1975], trans. A. Sheridan. New York: Random, 1977.

———. *Dits et écrits: Schriften in vier Bänden*, ed. D. Defert and F. Ewald. 4 vols. Frankfurt: Suhrkamp, 2001–5.

———. "An Ethics of Pleasure [1983]." In *Foucault Live: Collected Interviews, 1961–84*, trans. Lysa Hochroth and John Johnston, ed. S. Lotringer, 371–81. New York: Semiotext(e), 1989.

———. *The History of Sexuality*, vol. 1: *An Introduction*, trans. R. Hurley. Harmondsworth, UK: Penguin, 1978.

———. *Michel Foucault, Philosopher*, trans. T. Armstrong. Hemel Hempstead, UK: Harvester/Wheatsheaf, 1992.

———. "Nietzsche, Genealogy, History [1971]." In *Language, Counter-Memory, Practice*, ed. Donald Bouchard, 139–64. Oxford: Blackwell, 1977.

———. "The Politics of Health in the Eighteenth Century." In Michel Foucault, *Power/Knowledge: Selected Interviews and Other Writings, 1972–1977*, ed. Colin Gordon, 166–82. Brighton, UK: Harvester/Wheatsheaf, 1980.

———. *Politics, Philosophy, Culture: Interviews and Other Writings, 1977–1984*, ed. L. D. Kritzman. New York: Routledge, 1988.

———. *Power/Knowledge: Selected Interviews and Other Writings, 1972–1977*, ed. Colin Gordon. Brighton, UK: Harvester/Wheatsheaf, 1980.

———. "Questions of Methods." In *The Foucault Effect: Studies in Governmentality*, ed. G. Burchell, C. Gordon, and P. Miller, 73–86. Chicago: University of Chicago Press, 1991.

———. "Questions on Geography." In *Power/Knowledge: Selected Interviews and Other Writings, 1972–1977*, ed. Colin Gordon, 63–77. Brighton, UK: Harvester/Wheatsheaf, 1980.

———. *Security, Territory, Population: Lectures at the Collège de France, 1977–78*. Basingstoke, UK: Palgrave Macmillan, 2007.

———. *"Society Must Be Defended": Lectures at the Collège de France, 1975–76*, trans. D. Macey. New York: Picador, 2003.

———. "Truth and Power." In *Power/Knowledge: Selected Interviews and Other Writings, 1972–1977*, ed. C. Gordon, trans. C. Gordon et al., 109–33. Brighton: Harvester/Wheatsheaf, 1980.

———. "What Is Critique?" In *The Political*, ed. David Ingram, 191–211. Oxford: Blackwell, 2002.

———. "What Is Enlightenment?" trans. Catherine Porter. In *The Foucault Reader*, ed. Paul Rabinow, 32–50. New York: Pantheon, 1984.

Fox, Daniel, and Rudolf Klein. "Ethics and Health Policy in the United Kingdom and the United States: Legislation and Regulation." In *Cambridge World History of Medical Ethics*, ed. R. Baker and L. McCullough, 667–77. Cambridge: Cambridge University Press, 2009.

Fox, Nick J. "Derrida, Meaning and the Frame." In *Beyond Health: Postmodernism and Embodiment*, 134–35. London: Free Association, 1999.

Fox, Renée, and Judith Swazey. *Observing Bioethics*. Oxford: Oxford University Press, 2008.

Fox-Genovese, Elizabeth, and Elizabeth Lasch-Quinn, eds. *Reconstructing History: The Emergence of a New Historical Society*. New York: Routledge, 1999.

Frank, Manuel. *Freedom from History and Other Untimely Essays*. New York: New York University Press, 1971.

Freedberg, David. "Memory in Art: History and the Neuroscience of Response." In *The Memory Process: Neuroscientific and Humanistic Perspectives*, ed. S. Nalbantian, P. M. Matthews, and J. L. McClelland, 337–58. Cambridge, MA: MIT Press, 2011.

Frow, John. "The Literary Frame." *Journal of Aesthetic Education* 16 (1982): 25–30.

Fujimura, Joan H., and Danny Y. Chou. "Dissent in Science: Styles of Scientific Practice and the Controversy over the Cause of AIDS." *Sociology of Science and Medicine* 38 (1994): 1017–36.

Fukuyama, Francis. *The Origins of Political Order: From Prehuman Times to the French Revolution*. London: Profile Books, 2011.

Fulbrook, Mary. *Historical Theory: Ways of Imaging the Past*. London: Routledge, 2002.

Fuller, Steve. "Disciplinary Boundaries and the Rhetoric of the Social Sciences." *Poetics Today* 12 (1991): 302.

———. *New Humanist*, 12 September 2008. Available at http://biog.newhumanist.org.uk/2008/09/more-from-great-grayling-fuller-debate.html.

———. "Putting the Brain at the Heart of General Education in the Twenty-First Century: A Proposal." *Interdisciplinary Science Review* 36 (2011): 359–72.

———. Review of Smail's *On Deep History*. *Interdisciplinary Science Review* 34 (2009): 389–92.

———. "A Strong Distinction Between Humans and Non-Humans Is No Longer Required for Research Purposes: A Debate Between Bruno Latour and Steve Fuller," ed. Colin Barron. *History of the Human Sciences* 16 (2003): 77–99.

———. *Thomas Kuhn: A Philosophical History for Our Times*. Chicago: University of Chicago Press, 2000.

———. "Warwick 'Human Futures' Seminar on Chris Renwick's 'Biology and the Making of British Sociology.'" Audio lecture, 21 October 2010. Available at http://www2.warwick.ac.uk/fac/soc/sociology/staff/academicstaff/sfuller/fullers_index/audio.

———. "Why Science Studies Has Never Been Critical of Science: Some Recent Lessons on How to Be a Helpful Nuisance and a Harmless Radical." *Philosophy of the Social Sciences* 30 (2000): 5–32.

Gallagher, Catherine. *The Body Economic: Life, Death, and Sensation in Political Economy and the Victorian Novel*. Princeton, NJ: Princeton University Press, 2006.

Gallagher, Catherine, and Thomas Laqueur, eds. *The Making of the Modern Body: Sexuality and Society in the Nineteenth Century*. Berkeley: University of California Press, 1987.

Gallo, Max. *Geschichte der Plakate*. Herrsching: Pawlet, 1975.

Garber, Marjorie. *Academic Instincts*. Princeton, NJ: Princeton University Press, 2001.

Garfield, Simon. *The End of Innocence: Britain in the Time of AIDS*. London: Faber, 1994

Garoian, Charles R. "Art Education and the Aesthetics of Health in the Age of AIDS." *Studies in Art Education* 39 (1997): 6–23.

Garrett, Laurie. *Betrayal of Trust: The Collapse of Global Public Health*. New York: Hyperion, 2000.

Gatter, Philip. *Identity and Sexuality: AIDS in Britain in the 1990s*. London: Cassell, 1999.

Gazzaniga, M. S. *The Ethical Brain*. New York: Dana Press, 2005.

Geertz, Clifford. "Thick Description: Towards an Interpretive Theory of Culture." In *The Interpretation of Cultures*, ed. Clifford Geertz, 3–33. New York: Basic Books, 1973.

Geison, Gerald. *The Private Science of Louis Pasteur*. Princeton, NJ: Princeton University Press, 1995.

Gibbon, S., and C. Novas, eds. *Biosocialities, Genetics and the Social Sciences*. London: Routledge, 2008.

Gibbons, Joan. *Art and Advertising*. London: I. B. Tauris, 2005.

Gills, Barry, and William R. Thompson, eds. *Globalization and Global History*. New York: Routledge, 2006.

Gilman, Sander. "The Beautiful Body and AIDS." In *Picturing Health and Illness: Images of Identity and Difference*, 115–72. Baltimore: Johns Hopkins University Press, 1995.

———. *Difference and Pathology: Stereotypes of Sexuality, Race, and Madness*. Ithaca, NY: Cornell University Press, 1985.

———. *Disease and Representation: Images of Illness from Madness to AIDS*. Ithaca, NY: Cornell University Press, 1988.

———. "How and Why Do Historians of Medicine Use or Ignore Images in Writing Their Histories?" In *Picturing Health and Illness: Images of Identity and Difference*, 9–32. Baltimore: Johns Hopkins University Press, 1995.

———. *Inscribing the Other*. Lincoln: Nebraska University Press, 1991.

———. *The Jew's Body*. London: Routledge, 1991.

———. *Making the Body Beautiful: A Cultural History of Aesthetic Surgery*. Princeton, NJ: Princeton University Press, 1999.

———. *Picturing Health and Illness: Images of Identity and Difference*. Baltimore: Johns Hopkins University Press, 1995.

———. "Representing Health and Illness: Thoughts for the Twenty-First Century." *Medical History* 55 (2011): 259–300.

Giroux, Henry A. "Schooling and the Culture of Positivism: Notes on the Death of History." *Educational Theory* 29 (1979): 263–84.

Goffman, Erving. *Frame Analysis: An Essay on the Organization of Experience*. New York: Harper and Row, 1974.

———. *The Presentation of Self in Everyday Life*. New York: Doubleday, 1959.

Golinski, Jan. *Making Natural Knowledge: Constructivism and the History of Science*. Cambridge: Cambridge University Press, 1998.

[Gooch, Robert]. "Contagion and Quarantine." *Quarterly Review* 27 (1822): 524–53.

Gooch, Robert. "Plague, a Contagious Disease." *Quarterly Review* 33 (1826): 241–42.

Gordin, Michael. "Abgrenzung und Demokratie: Die politischen Valenzen der Wissenschaftsgrenze." In *Wissenschaft und Demokratie*, ed. Michael Hagner, 70–87. Zurich: Suhrkamp, 2013.

Grafton, Anthony. "Britain: The Disgrace of the Universities." *New York Review of Books*, 8 April 2010, 32.

———. "Can the Colleges Be Saved?" *New York Review of Books*, 6 June 2012, 22–24.

———. "History Under Attack." *Perspectives on History* 49 (6 January 2011). Available at http://www.historians.org/perspectives/issues/2011/1101/1101pre1.cfm.

Grant, Mariel. *Propaganda and the Role of the State in Inter-War Britain*. Oxford: Clarendon, 1994.

Granville, Augustus Bozzi. *Autobiography*, ed. P. B. Granville. London: King, 1874.

Gray, Chris Hables. *Cyborg Citizen: Politics in the Posthuman Age*. New York: Routledge, 2001.

Green, Anna, and Kathleen Troup. "The Challenge of Poststructuralism/Postmodernism." In *The Houses of History: A Critical Reader in Twentieth-Century History and Theory*, ed. Green and Troup, 297–307. Manchester: Manchester University Press, 1999.

Green, Monica. "Letting the Genome Out of the Bottle: On Creating Alliances Between Medical History and Historicist Sciences." Paper presented at the conference on "The Future of the History of Medicine," London, July 2010.

Green, R. M., A. Donovan, and S. A. Jauss, eds. *Global Bioethics: Issues of Conscience for the Twenty-First Century*. Oxford: Clarendon, 2008.

Gregg, Jessica L. *Virtually Virgins: Sexual Strategies and Cervical Cancer in Recife, Brazil*. Stanford, CA: Stanford University Press, 2003.

Gregory, John. *Observations on the Duties and Offices of a Physician, and On the Method of Prosecuting Enquiries in Philosophy*. London: W. Strahan and T. Cadell, 1770.

Grosz, Elizabeth. *Volatile Bodies: Toward a Corporeal Feminism*. Bloomington: Indiana University Press, 1994.

Grosz, Elizabeth, and Elspeth Probyn, eds. *Sexy Bodies: The Strange Carnalities of Feminism*. New York: Routledge, 1995.

Guha, Ranajit. *History at the Limit of World History*. New York: Columbia University Press, 2002.

Gumbrecht, Hans Ulrich. "Presence Achieved in Language (with Special Attention Given to the Presence of the Past)." *History and Theory* 45 (2006): 317–27.

———. *Production of Presence: What Meaning Cannot Convey*. Stanford, CA: Stanford University Press, 2004.

Habermas, Jürgen. *The New Conservatism: Cultural Criticism and the Historians' Debate*, trans. and ed. Sherry W. Nicholsen. Cambridge: Polity, 1989.

Hacking, Ian. "The Archaeology of Michel Foucault [1981]." In *Historical Ontology*, ed. Ian Hacking, 73–86. Cambridge, MA: Harvard University Press, 2002.

———. "Bio-Power and the Avalanche of Printed Numbers." *Humanities in Society* 5 (1982): 279–95.
———. *The Emergence of Probability*. Cambridge: Cambridge University Press, 1975.
———. *Historical Ontology*. Cambridge, MA: Harvard University Press, 2002.
———. "Making Up People [1986]." In *Historical Ontology*, ed. Ian Hacking, 99–114. Cambridge, MA: Harvard University Press, 2002.
———. *The Taming of Chance*. New York: Cambridge University Press, 1990.
Hall, John H. "Cultures of Inquiry and the Re-Thinking of Disciplines." In *Social in Question: New Bearings in History and the Social Sciences*, ed. Patrick Joyce, 191–210. London: Routledge, 2002.
Hall, Stuart, ed. *Representation: Cultural Representations and Signifying Practices*. London: Sage and Open University, 1997.
Halttunen, Karen. "Cultural History and the Challenge of Narrativity." In *Beyond the Cultural Turn: New Directions in the Study of Society and Culture*, ed. Victoria E. Bonnell and Lynn Hunt, 165–81. Berkeley: University of California Press, 1999.
Hamlin, Christopher. *Cholera: The Biography*. Oxford: Oxford University Press, 2009.
———. "Predisposing Causes and Public Health in Early Nineteenth-Century Medical Thought." *Social History of Medicine* 5 (1992): 43–70.
Hammer, Michael, and James Champy. *Reengineering the Corporation*. New York: HarperCollins, 1993.
Hancock, Philip, et al., eds. *The Body, Culture, and Society*. Buckingham, UK: Open University, 2000.
Hanen, M., M. Osler, and R. G. Weyant, eds. *Science, Pseudo-science and Society*. Waterloo, ON: Laurier University Press, 1980.
Hansen, Bert. "Medical History for the Masses: How American Comic Books Celebrated Heroes of Medicine in the 1940s." *Bulletin of the History of Medicine* 78 (2004): 148–91.
———. *Picturing Medical Progress from Pasteur to Polio: A History of Mass Media Images and Popular Attitudes in America*. New Brunswick, NJ: Rutgers University Press, 2009.
Haraway, Donna. *Modest Witness@Second_Millennium.FemaleMan©_Meets_Oncomouse™: Feminism and Technoscience*. London: Routledge, 1997.
———. "Situated Knowledges: The Science Question in Feminism and the Privilege of Partial Perspective." *Feminist Studies* 14 (1988): 575–99. Reprinted in *Simians, Cyborgs and Women: The Reinvention of Nature*, 183–201. London: Routledge, 1991.

Hardie, Martin, and A. K. Sabin, eds. *War Posters Issued by Belligerent and Neutral Nations, 1914–1919*. London: A. and C. Black, 1920.

Harley, David. "Rhetoric and the Social Construction of Sickness and Healing." *Social History of Medicine* 12 (1999): 407–35.

Harvey, David. *A Brief History of Neoliberalism*. Oxford: Oxford University Press, 2005.

Hatemi, Peter, and Rose McDermott, eds. *Man Is by Nature a Political Animal: Evolution, Biology, and Politics*. Chicago: University of Chicago Press, 2011.

Hayek, Friedrich von. *The Sensory Order: An Inquiry into the Foundations of Theoretical Psychology*. Chicago: University of Chicago Press, 1952.

Hayward, Rhodri. "The Biopolitics of Arthur Keith and Morley Roberts." In *Regenerating England: Science, Medicine and Culture in Inter-war Britain*, ed. C. Lawrence and A-K. Mayer, 251–74. Amsterdam: Rodopi, 2000.

Head, Simon. "The Grim Threat to British Universities." *New York Review of Books*, 13 January 2011.

Heath, D., R. Rapp, and K.-S. Taussig. "Genetic Citizenship." In *A Companion to the Anthropology of Politics*, ed. D. Nugent and J. Vincent, 152–67. Oxford: Blackwell, 2004.

Helfand, William H. *Quack, Quack, Quack: The Sellers of Nostrums in Prints, Posters, Ephemera and Books*. New York: Grolier Club, 2002.

Hess, Volker. "Standardizing Body Temperature: Quantification in Hospitals and Daily Life, 1850–1900." In *Body Counts: Medical Quantification in Historical and Sociological Perspective*, ed. Gérard Jorland, A. Opinel, and G. Weisz, 109–26. Montreal: McGill-Queen's University Press, 2005.

Hewison, Robert. *The Heritage Industry: Britain in a Climate of Decline*. London: Methuen, 1987.

Hezig, R. "On Performance, Productivity, and Vocabularies of Motive on Recent Studies of Science." *Feminist Theory* 5 (2004): 127–47.

Hillier, Bevis. *Plakate*. Hamburg: Hoffmann und Campe, 1969.

Hobsbawm, Eric. *On History*. London: Abacus, 1998.

Hofer, Hans-Georg, and Lutz Sauerteig. "Perspektiven einer Kulturgeschichte der Medizin." *Medizinhistorisches Journal* 42 (2007): 105–41.

Hoffmann, Detlef. "The German Art Museum and the History of the Nation." In *Museum Culture: Histories, Discourses, Spectacles*, ed. Daniel J. Sherman and Irit Rogoff, 3–21. Minneapolis: University of Minnesota Press, 1994.

Hoffmaster, B. *Bioethics in Social Context*. Philadelphia: Temple University Press, 2001.

———. "Can Ethnography Save the Life of Medical Ethics?" *Social Science and Medicine* 35 (1992): 1421–32.

Holt, Thomas C. "Experience and the Politics of Intellectual Inquiry." In *Questions of Evidence: Proof, Practice, and Persuasion Across the Disciplines*, ed. James Chandler, Arnold I. Davidson, and Harry Harootunian, 388–96. Chicago: University of Chicago Press, 1991.

Hopkins, M., P. Martin, P. Nightingale, A. Kraft, and S. Mahdi. (2007). "The Myth of the Biotech Revolution: An Assessment of Technological, Clinical and Organisational Change." *Research Policy* 36 (2007): 566–89.

Hopwood, Nick. "Pictures of Evolution and Charges of Fraud: Ernst Haeckel's Embryological Illustrations." *Isis* 97 (2006): 260–301.

Horkheimer, Max. *The Eclipse of Reason*. Oxford: Oxford University Press, 1947.

Howell, Keith. *Broadcasting It: An Encyclopedia in Film, Radio and TV in the UK, 1923–1993*. London: Cassell, 1993.

Huber, Valeska. "The Unification of the Globe by Disease? The International Sanitary Conferences on Cholera, 1851–1894." *Historical Journal* 49 (2006): 454–74.

Hunt, Lynn. "Does History Need Defending?" *History Workshop Journal* 46 (1998): 241–49.

———, ed. *The New Cultural History*. Berkeley: University of California Press, 1989.

Hüppauf, Bernd, and Peter Weingart, eds. *Science Images and Popular Images of the Sciences*. London: Routledge, 2008.

Hutchinson, Harold. *The Poster: An Illustrated History from 1860*. London: Studio Vista, 1968.

Iggers, Georg. *The German Conception of History: The National Tradition of Historical Thought from Herder to the Present*. Middletown, CT: Wesleyan University Press, 1983.

———. *Historiography in the Twentieth Century: From Scientific Objectivity to the Postmodern Challenge*. Middletown, CT: Wesleyan University Press, 1997.

Iggers, Georg G., and Q. Edward Wang. *A Global History of Modern Historiography*. London: Pearson/Longman, 2008.

Illich, Ivan. *Medical Nemesis: The Expropriation of Health*. London: Marian Boyars, 1976.

Inkster, Ian. "Marginal Men: Aspects of the Social Role of the Medical Community in Sheffield, 1790–1850." In *Health Care and Popular Medicine in Nineteenth-Century England*, ed. John Woodward and D. Richards, 128–63. London: Croom Helm, 1977.

Jackson, Mark. *Asthma: The Biography*. Oxford: Oxford University Press, 2009.

Jacobs, Margaret C. *The Cultural Meaning of the Scientific Revolution*. New York: Knopf, 1988.
Jameson, Fredric. *Postmodernism; or, The Cultural Logic of Late Capitalism*. London: Verso, 1991.
———. *The Prison-House of Language: A Critical Account of Structuralism and Russian Formalism*. Princeton, NJ: Princeton University Press, 1972.
Jardine, Nick, and Marina Frasca-Spada. "Splendours and Miseries of the Science Wars." *Studies in History and Philosophy of Science* 28 (1997): 219–35.
Jasanoff, Sheila. *Designs on Nature: Science and Democracy in Europe and the United States*. Princeton, NJ: Princeton University Press, 2005.
———, ed. *States of Knowledge: The Co-Production of Science and Social Order*. London: Routledge, 2004.
Jay, Martin. *The Dialectical Imagination: A History of the Frankfurt School and the Institute of Social Research, 1923–50*. London: Heinemann, 1973.
———. *Downcast Eyes: The Denigration of Vision in Twentieth-Century French Thought*. Berkeley: University of California Press, 1993.
Jenkins, Keith. *Re-Thinking History*. London: Routledge, 1991.
Jenkins, Keith, Sue Morgan, and Alun Munslow, ed. *Manifestos for History*. London: Routledge, 2007.
Jenner, Mark. "Body, Image, Text in Early Modern Europe." *Social History of Medicine* 12 (1999): 143–54.
Jenner, Mark, and Bertrand Taithe. "The Historiographical Body." In *Medicine in the Twentieth Century*, ed. Roger Cooter and John Pickstone, 187–200. Amsterdam: Harwood Academic, 2000.
Jewson, Nicholas. "The Disappearance of the Sick-Man from Medical Cosmology, 1770–1870." *Sociology* 10 (1976): 225–44.
John, S. *Bioethics*. Cambridge: Cambridge University Press, 2002.
Jones, Anne Hudson. "Narrative-Based Medicine: Narrative in Medical Ethics." *British Medical Journal* 318 (1999): 255.
Jones, Caroline A., and Peter Galison, eds. *Picturing Science, Producing Art*. New York: Routledge, 1998.
Jones, Colin, and Roy Porter, eds. *Reassessing Foucault: Power, Medicine and the Body*. London: Routledge, 1994.
Jones, Colin, and D. Wahrman, eds. *A Cultural Revolution? England and France, 1750–1820*. Berkeley: University of California Press, 2002.
Jones, Gareth Stedman. *Languages of Class: Studies in English Working Class History, 1832–1982*. Cambridge: Cambridge University Press, 1983.

Jonsen, A. R. *The Birth of Bioethics*. New York: Oxford University Press, 1998.

Jordan, Tim. *Activism! Direct Action, Hacktivism and the Future of Society*. London: Reaktion Books, 2002.

Jordanova, Ludmilla. "Has the Social History of Medicine Come of Age?" *Historical Journal* 36 (1993): 437–49.

———. *History in Practice*. London: Arnold, 2000.

———. "Medicine and Visual Culture." *Social History of Medicine* 3 (1990): 89–99.

———. *Nature Displayed: Gender, Science and Medicine, 1760–1820*. London: Longman, 1999.

———. "The Sense of a Past in Eighteenth-Century Medicine." The Stenton Lecture 1997, University of Reading, 1999.

———. "The Social Construction of Medical Knowledge." *Social History of Medicine* 8 (1995): 361–81.

Joseph, Jonathan. "Derrida's Spectres of Ideology." *Journal of Political Ideologies* 6 (2001): 95–115.

Joyce, Kelly A. "The Body as Image: An Examination of the Economic and Political Dynamics of Magnetic Resonance Imaging and the Construction of Difference." In *Biomedicalization: Technoscience, Health, and Illness in the U.S.*, ed. Adele Clarke et al., 197–217. Durham, NC: Duke University Press, 2010.

———. "From Numbers to Pictures. The Development of Magnetic Resonance Imaging and the Visual Turn in Medicine." *Science as Culture* 15 (2006): 1–22.

Joyce, Patrick. "The End of Social History?" *Social History* 20 (1995): 73–91.

———. "The Gift of the Past: Towards a Critical History." In *Manifestos for History*, ed. Keith Jenkins, Sue Morgan, and Alun Munslow, 88–97. London: Routledge, 2007.

———. "The Return of History: Postmodernism and the Politics of Academic History in Britain." *Past and Present* 158 (1998): 207–35.

———, ed. *Social in Question: New Bearings in History and the Social Sciences*. London: Routledge, 2002.

———. "What Is the Social in Social History?" *Past and Present* 206 (2010): 213–48.

Judovitz, Dalia. *The Culture of the Body: Genealogies of Modernity*. Ann Arbor: University of Michigan Press, 2001.

Kay, Carolyn. *Art and the German Bourgeoisie: Alfred Lichtwark and Modern Painting in Hamburg, 1886–1914*. Toronto: University of Toronto Press, 2002.

Kay, Lily E. *Who Wrote the Book of Life? A History of the Genetic Code*. Stanford, CA: Stanford University Press, 2000.

Kemp, Martin. *Seen/Unseen: Art, Science, and Intuition from Leonardo to the Hubble Telescope*. Oxford: Oxford University Press, 2006.

Kendall, Gavin, and Gary Wicham. "Health and the Social Body." In *Private Risks and Public Dangers*, ed. Sue Scott et al., 8–18. Aldershot, UK: Avebury, 1992.

Kent, Christopher. "Victorian Social History: Post-Thompson, Post-Foucault, Postmodern." *Victorian Studies* 40 (1996): 97–133.

Kevles, Bettyann H. *Naked to the Bone: Medical Imaging in the Twentieth Century*. New Brunswick, NJ: Rutgers University Press, 1997.

Kinner, David. "Groundhog Day? The Strange Case of Sociology, Race and 'Science.'" *Sociology* 41 (2007): 931–43.

Kiple, Kenneth, ed. *The Cambridge World History of Human Diseases*. Cambridge: Cambridge University Press, 1993.

Kirk, Robert, and Mick Worboys. "Medicine and Species: One Medicine, One History?" In *The Oxford Handbook to the History of Medicine*, ed. Mark Jackson, 561–77. Oxford: Oxford University Press, 2011.

Klein, Naomi. *No Logo: Taking Aim at the Brand Bullies*. London: Flamingo/HarperCollins, 2000.

———. *The Shock Doctrine: The Rise of Disaster Capitalism*. London: Allan Lane, 2007.

Klein, Rudolf. "The Crises of the Welfare States." In *Medicine in the Twentieth Century*, ed. Roger Cooter and John Pickstone, 155–170. Amsterdam: Harwood Academic, 2000.

Kleinman, A., R. Fox, and A. Brandt. Introduction to special issue: "Bioethics and Beyond." *Daedalus* 128 (1999): vii–x.

Klusacek, Allan, and Ken Morrison, eds. *A Leap in the Dark: AIDS, Art and Contemporary Cultures*. Montreal: Véhicule, 1992.

Knorr-Cetina, K. "The Rise of a Culture of Life." *EMBO Reports* 6 (2005): 76–80.

Konner, Melvin. *The Evolution of Childhood: Relationships, Emotion, Mind*. Cambridge, MA: Harvard University Press, 2010.

Koopman, Colin. "Foucault Across the Disciplines: Introductory Notes on Contingency in Critical Inquiry." *History of the Human Sciences* 24 (2011): 1–12.

———. *Genealogy as Critique: Foucault and the Problems of Morality*. Bloomington: Indiana University Press, 2013.

Korf, Benedikt. "A Neural Turn? On the Ontology of the Geographical Subject." *Environment and Planning* 40 (2008): 715–32.

Koven, Seth. "The Whitechapel Picture Exhibition and the Politics of Seeing." In *Museum Culture: Histories, Discourses, Spectacles*, ed. Daniel J. Sherman and Irit Rogoff, 22–48. Minneapolis: University of Minnesota Press, 1994.

Krasmann, S., and M. Volkmer, eds. *Michel Foucault's "Geschichte der Gouvernementalität" in den Sozialwissenschaften*. Bielefeld: Transcript, 2008.

Krüger, Lorenz. "Does a Science Need Knowledge of Its History?" In *Why Does History Matter to Philosophy and the Sciences? Selected Essays of Lorenz Krüger*, ed. T. Sturm, W. Carl, and L. Daston, 221–31. Berlin: Walter de Gruyter, 2005.

Kuklick, Henrika. "Professional Status and the Moral Order." In *Disciplinarity at the Fin de Siècle*, ed. Amanda Anderson and Joseph Valente, 126–52. Princeton, NJ: Princeton University Press, 2002.

Kuo, James S. "Swimming with the Sharks—the MD, MBA." *Lancet* 350 (1997): 828.

Kurzweill, Edith. "Michel Foucault: Ending the Era of Man." *Theory and Society* 4 (1977): 295–420.

Labinger, Jay A., and Harry Collins, eds. *The One Culture? A Conversation About Science*. Chicago: University of Chicago Press, 2001.

Labisch, Alfons. "Von Sprengels 'Pragmatischer Medizingeschichte' zu Kocks 'Psychischen Apiori' in Geschichte der Medizin und Geschichte in der Medizin." In *Die Institutionalisierung der Medizinhistoriographie*, ed. Andreas Frewer and Volker Roelcke, 235–54. Stuttgart: Franz Steiner, 2001.

Laclau, Ernesto. *New Reflections on the Revolution of Our Time*. London: Verso, 1990.

Laqueur, Thomas. *Making Sex: Gender and Sex from the Greeks to Freud*. Cambridge, MA: Harvard University Press, 1990.

Lash, Scott. "Objects the Judge: Latour's Parliament of Things." *eipcp.net*, 2006. Available at http://translate.eipcp.net/transversal/0107/lash/en cached 17 December 2010.

Laslett, Peter. *The World We Have Lost*. London: Methuen, 1965.

Lasn, Kalle. *Culture Jam: How to Reverse America's Suicidal Consumer Binge—and Why We Must*. New York: Quill, 2000.

Latour, Bruno. *Eine neue Soziologie für eine neue Gesellschaft*. Frankfurt: Suhrkamp, 2008.

———. "For David Bloor . . . and Beyond: A Reply to Bloor's 'Anti-Latour.'" *Studies in the History and Philosophy of Science* 30 (1999): 113–29.

———. "Gabriel Tarde and the End of the Social." In *Social in Question: New Bearings in History and the Social Sciences*, ed. Patrick Joyce, 117–32. London: Routledge, 2002.

———. "How to Talk About the Body? The Normative Dimension of Science Studies." *Body and Society* 10 (2004): 205–29.

———. "On Recalling ANT." In *Actor Network Theory and After*, ed. John Law and John Hassard, 15–25. Oxford: Blackwell, 1999.

———. "On the Partial Existence of Existing *and* Nonexisting Objects." In *Biographies of Scientific Objects*, ed. Lorraine Daston, 247–69. Chicago: University of Chicago Press, 2000.

———. *The Pasteurization of France*, trans. A. Sheridan and John Law. Cambridge, MA: Harvard University Press, 1988.

———. *Reassembling the Social: An Introduction to Actor-Network Theory*. Oxford: Oxford University Press, 2005.

———. *Science in Action: How to Follow Scientists and Engineers Through Society*. Cambridge, MA: Harvard University Press, 1987.

———. "Tarde's Idea of Quantification." In *The Social After Gabriel Tarde*, ed. Matei Candea, 145–62. Abingdon, UK: Routledge, 2010.

———. *We Have Never Been Modern*, trans. Catherine Porter. New York: Harvester Wheatsheaf, 1993.

———. "Why Has Critique Run Out of Steam? From Matters of Fact to Matters of Concern." *Critical Inquiry* 30 (2004): 225–48.

Latour, Bruno, and Peter Weibel, eds. *Iconoclash: Beyond the Image Wars in Science, Religion, and Art*. Cambridge, MA: MIT Press, 2002.

Law, John, ed. *Power, Action and Belief: A New Sociology of Knowledge?* London: Routledge & Kegan Paul, 1986.

Law, John, and John Hassard, eds. *Actor Network Theory and After*. Oxford: Blackwell, 1999.

Lawrence, Christopher. "The Nervous System and Society in the Scottish Enlightenment." In *Natural Order: Historical Studies of Scientific Culture*, ed. Barry Barnes and Steven Shapin, 19–40. London: Sage, 1979.

Lemke, Thomas. "'The Birth of Bio-Politics'—Michel Foucault's Lecture at the Collège de France on Neo-Liberal Governmentality." *Economy and Society* 30 (2001): 190–207.

———. *Gouvernmentalität und Biopolitik*, 2nd ed. Wiesbaden: VS Verlag für Sozialwissenschaften, 2008.

———. "An Indigestible Meal? Foucault, Governmentality and State Theory." *Distinktion: Scandinavian Journal of Social Theory* 15 (2007). Available at http://www.thomaslemkeweb.de/publikationen/IndigestibleMealfinal5.pdf.

Levine, George, ed. *One Culture: Essays in Science and Literature*. Madison: University of Wisconsin Press, 1987.

Levine, Phillipa, and Alison Bashford. "Eugenics in the Modern World." In *The Oxford Handbook of the History of Eugenics*, ed. P. Levine and A. Bashford, 3–25. Oxford: Oxford University Press, 2010.

Levy, N. *Neuroethics: Challenges for the Twenty-First Century*. Cambridge: Cambridge University Press, 2007.

Lewens, T. "The Commercial Exploitation of Ethics." *Studies in History and Philosophy of Science*, Part C: *Studies in History and Philosophy of Biological and Biomedical Sciences* 35 (2004): 145–53.

Leys, Ruth. "The Turn to Affect: A Critique." *Critical Inquiry* 37 (2011): 434–72.

Lightman, Bernard. "The Visual Theology of Victorian Popularizers of Science: From Reverent Eye to Chemical Retina." *Isis* 91 (2000): 651–80.

Lilford, R. J., et al. "Medical Practice: Where Next?" *Journal of the Royal Society of Medicine* 94 (2001): 559–62.

Linkner, Beth. "Resuscitating the 'Great Doctor': The Career of Biography in Medical History." In *The History and Poetics of Scientific Biography*, ed. Thomas Söderqvist, 221–39. Aldershot, UK: Ashgate, 2007.

Liu, Alun. *The Laws of Cool: Knowledge Work and the Culture of Information*. Chicago: University of Chicago Press, 2004.

Lock, Margaret, and Vinh-Kim Nguyen. *An Anthropology of Biomedicine*. Oxford: Wiley-Blackwell, 2010.

Long, Liza. *Rehabilitating Bodies: Health, History, and the American Civil War*. Philadelphia: University of Pennsylvania Press, 2004.

Longmate, Norman. *King Cholera: The Biography of a Disease*. London: Hamish Hamilton, 1966.

Lorimer, Hayden. "Cultural Geography: The Busyness of Being 'More-Than-Representational.'" *Progress in Human Geography* 29 (2005): 83–94.

Lowenthal, David. *The Heritage Crusade and the Spoils of History*. New York: Free Press, 1996.

Luckin, Bill. "The Crisis, the Humanities and Medical History." *Medical History* 55 (2011): 283–87.

———. Review of Pelling, *Cholera, Fever and English Medicine*. *Social History* 4 (1979): 566.

Lupton, Deborah. *The Imperative of Health: Public Health and the Regulated Body*. London: Sage, 1995.

———. *Medicine as Culture: Illness, Disease and the Body in Western Societies*. London: Sage, 1994.

Lynch, Michael. "Living with Kaposi's Sarcoma and AIDS." *Body Politic* 88 (1982): 31–37.

Lyotard, Jean-François. *The Postmodern Condition: A Report on Knowledge* [1979], trans. Geoff Bennington and Brian Massumi. Manchester: Manchester University Press, 1984.

Macherey, P. "Für eine Naturgeschichte der Normen." In *Spiele der Wahrheit: Michel Foucaults Denken*, ed. F. Ewalt and B. Waldenfels, 171–92. Frankfurt: Suhrkamp, 1991.

MacIntyre, A. "Does Applied Ethics Rest on a Mistake?" *Monist* 67 (1984): 498–513.

MacLaren, Angus. "Bourgeois Ideology and Victorian Philanthropy: The Contradictions of Cholera." In *Social Class in Scotland: Past and Present*, 36–54. Edinburgh: Donald, 1976.

Maclean, Charles. *Summary of Facts and Inferences Respecting the Causes, Proper and Adventitious, of Plague, and Other Pestilential Diseases, with Proofs of the Non-Existence of Contagion in These Maladies*. London, 1820 [extracted from *Pamphleteer*, vol. 16, 153–92].

Macmillan, Margaret. *The Uses and Abuses of History*. London: Profile Books, 2009.

Maehle, A.-H. "The Ethical Discourse on Animal Experimentation, 1650–1900." In *Doctors and Ethics: The Earlier Historical Setting of Professional Ethics*, ed. Andrew Wear, J. Geyer-Kordesch, and R. French, 203–51. Amsterdam: Rodopi, 1993.

Malcolm, Barnard. *Approaches to Understanding Visual Culture*. Houndsmill, UK: Palgrave, 2001.

———. *Art, Design and Visual Culture: An Introduction*. Basingstoke: Macmillan, 1998.

Mallavarapu, Srikanth, and Amit Prasad. "Facts, Fetishes, and the Parliament of Things: Is There Any Space for Critique?" *Social Epistemology* 20 (2006): 185–99.

Malthus, Thomas. *An Essay on the Principle of Population* (1798), ed. and intro. Geoffrey Gilbert. Oxford: Oxford University Press, 1993.

Marks, Lara. "What Is Biotechnology?" Available at http://www.whatisbiotechnology.org/login/f2ba49e314.

Marwick, Arthur. "All Quiet on the Postmodern Front: The 'Return to Events' in Historical Study." *Times Literary Supplement*, 3 February 2001, 13–14.

———. *The Nature of History*, 3rd ed. Basingstoke, UK: Macmillan, 1989.

———. *The New Nature of History: Knowledge, Evidence, Language*. Basingstoke, UK: Palgrave, 2001.

———. "Two Approaches to Historical Study: The Metaphysical (Including 'Postmodernism') and the Historical." *Journal of Contemporary History* 30 (1995): 4–35.

Marx, Karl. "The Eighteenth Brumaire of Louis Bonaparte [1858]." In *Pelican Marx Library: Political Writings*, vol. 2: *Surveys From Exile*, ed. and intro. David Fernbach, 146–249. Harmondsworth, UK: Penguin, 1973.

Marx, Karl, and F. Engels. *The German Ideology* [1845–46], ed. C. J. Arthur. New York: International Publishers, 1970.

Massey, Doreen. *For Space*. Thousand Oaks, CA: Sage, 2005.
Mayfield, David, and Susan Thorne. "Social History and Its Discontents: Gareth Stedman Jones and the Politics of Language." *Social History* 17 (1992): 165–88.
McCarthy, T. *The Critical Theory of Jürgen Habermas*. Cambridge, MA: MIT Press, 1978.
McDonald, J. C. "The History of Quarantine in Britain During the Nineteenth Century." *Bulletin of the History of Medicine* 25 (1951): 22–44.
McDonald, Rónón. *The Death of the Critic*. London: Continuum, 2007.
McGilchrist, Iain. *The Master and His Emissary: The Divided Brain and the Making of the Western World*. New Haven: Yale University Press, 2009.
McGrath, Roberta. "Health, Education and Authority: Difference and Deviance." In *Pleasure Principles: Politics, Sexuality and Ethics*, ed. Victoria Harwood et al., 157–83. London: Lawrence and Wishart, 1993.
McKay, Richard. "Imagining 'Patient Zero': Sexuality, Blame, and the Origins of the North American AIDS Epidemic." DPhil diss., Oxford University, 2010.
McKendrick, Neil, John Brewer, and J. H. Plumb. *The Birth of a Consumer Society: The Commercialization of Eighteenth-Century England*. London: Europa, 1982.
McMillan, J. "Is Corporate Money Bad for Bioethics?" *Studies in History and Philosophy of Science*, Part C: *Studies in History and Philosophy of Biological and Biomedical Sciences* 35 (2004): 167–75.
McMillen, Liz. "The Science Wars Flare at the Institute for Advanced Study." *Chronicle of Higher Education*, 16 May 1997.
McNeill, William. *Plagues and Peoples*. Oxford: Blackwell, 1977.
———. "The Rise of the West After Twenty-Five Years." *Journal of World History* 1 (1990): 1–21.
Megill, Allan. "Fragmentation and the Future of Historiography." *American Historical Review* 96 (1991): 693–98.
Mendelsohn, Everet. "The Social Construction of Scientific Knowledge." In *The Social Production of Scientific Knowledge*, ed. E. Mendelsohn, P. Weingart, and R. Whitley, 3–26. Boston: Reidel, 1977.
Merleau-Ponty, Maurice. *Phenomenology of Perception* [1945], trans. Colin Smith. London: Routledge and Kegan Paul, 1962.
———. *The Primacy of Perception and Other Essays*, ed. J. M. Edie. Evanston, IL: Northwestern University Press, 1964.
Messikomer, C. M., R. Fox, and J. P. Swazey. "The Presence and Influence of Religion in American Bioethics." *Perspectives in Biology and Medicine* 44 (2001): 485–508.

"Microarrays: Chipping Away at the Mysteries of Science and Medicine." National Center for Biotechnology Information, 2007. Available at http://www.ncbi.nlm.nih.gov/About/primer/microarrays.html.

Mildenberger, Florian. "Die Geburt der Rezeption: Michel Foucault und Werner Liebbrand." *Sudhoffs Archiv* 90 (2006): 97–105.

Miller, James, ed. *Fluid Exchanges: Artists and Critics in the AIDS Crisis*. Toronto: University of Toronto Press, 1992.

Mintz, David. "What's in a Word: The Distancing Function of Language in Medicine." *Journal of Medical Humanities* 13 (1992): 223–33.

Mirzoeff, Nicholas. *An Introduction to Visual Culture*. London: Routledge, 1999.

———, ed. *The Visual Culture Reader*. London: Routledge, 1998.

Mitchell, William J. T. "Art, Fate, and the Disciplines: Some Indicators." *Critical Inquiry* 35 (2009): 1026–27.

———. " 'Critical Inquiry' and the Ideology of Pluralism." *Critical Inquiry* 8 (1982): 609–18.

———. "The Pictorial Turn [1992]." In Mitchell, *Picture Theory*, 11–34. Chicago: University of Chicago Press, 1994.

Mold, Alex. "Patient Groups and the Construction of the Patient-Consumer in Britain: An Historical Overview." *Journal of Social Policy* 39 (2010): 505–21.

Mommsen, Wolfgang J. "Moral Commitment and Scholarly Detachment: The Social Function of the Historian." In *Historians and Social Values*, ed. J. T. Leessen and Ann Rigney, 45–55. Amsterdam: Amsterdam University Press, 2000.

Morgan, Lynn M. *Icons of Life: A Cultural History of Human Embryos*. Berkeley: University of California Press, 2009.

Morris, R. J. *Cholera 1832: The Social Response to an Epidemic*. London: Croom Helm, 1976.

Muhle, M. *Eine Genealogie der Biopolitik: Zum Begriff des Lebens bei Foucault und Canguilhem*. Bielefeld: Transcript, 2008.

Mukherjee, Siddhartha. *The Emperor of All Maladies: A Biography of Cancer*. London: Fourth Estate, 2011.

Myerson, Jeremy, and Graham Vickers. *Rewind: Forty Years of Design and Advertising*. London: Phaidon, 2002.

Nadesan, M. H. *Governmentality, Biopower, and Everyday Life*. London: Routledge, 2008.

Nelkin, Dorothy. "Promotional Metaphors and Their Popular Appeal." *Public Understanding of Science* 3 (1994): 25–31.

Nesbitt, G. L. *Benthamite Reviewing: The First Twelve Years of the Westminster Review, 1824–1836*. New York: Columbia University Press, 1936.

Newton, T. "Truly Embodied Sociology: Marrying the Social and the Biological?" *Sociological Review* 51 (2003): 20–42.

Nieto-Galan, Agustí. "Antonio Gramsci Revisited: Historians of Science, Intellectuals, and Struggle for Hegemony." *History of Science* 49 (2011): 453–78.

Nietzsche, Friedrich. *Vom Natzen und Nachteil der Historie für das Leben* [1874], trans. Peter Preuss. Indianapolis, IN: Hackett, 1980.

Nikolow, Sybilla, and Lars Bluma. "Science Images: Between Scientific Fields and the Public Sphere." In *Science Images and Popular Images of the Sciences*, ed. Bernd Hüppauf and Peter Weingart, 33–51. London: Routledge, 2008.

Norris, Christopher. *Against Relativism: Philosophy of Science, Deconstruction and Critical Theory*. Oxford: Blackwell, 1997.

Novick, Peter. *That Noble Dream: The "Objectivity Question" and the American Historical Profession*. Cambridge: Cambridge University Press, 1988.

Nowotny, Helga, et al. *The Public Nature of Science Under Assault: Politics, Markets, Science and the Law*. Heidelberg: Springer, 2005.

Nussbaum, Martha. *Not for Sale: Why Democracy Needs the Humanities*. Princeton, NJ: Princeton University Press, 2010.

———. "Skills for Life: Why Cuts in the Humanities Pose a Threat to Democracy Itself." *Times Literary Supplement*, 30 April 2010, 13.

Oakes, Guy. *Weber and Rickert: Concept Formation in the Cultural Sciences*. Cambridge, MA: MIT Press, 1988.

O'Connor, Erin. *Raw Material: Producing Pathology in Victorian Culture*. Durham, NC: Duke University Press, 2000.

Oddy, Derek, et al., eds. *The Rise of Obesity in Europe*. Farnham, UK: Ashgate, 2009.

Ogilvie, Brian. *The Science of Describing: Natural History in Renaissance Europe*. Chicago: Chicago University Press, 2006.

O'Manique, Colleen. *Neoliberalism and AIDS Crisis in Sub-Saharan Africa: Globalization's Pandemic*. London: Palgrave, 2004.

Ong, Aihwa, and Stephen J. Collier, eds. *Global Assemblages: Technology, Politics, and Ethics as Anthropological Problems*. Oxford: Blackwell, 2006.

Onians, John. *Neuroarthistory: From Aristotle and Pliny to Baxandall and Zeki*. New Haven: Yale University Press, 2007.

Orwell, George. "Inside the Whale [1940]." In *The Collected Essays, Journalism and Letters of George Orwell*, vol. 1: *An Age Like This, 1920–1940*, ed. Sonia Orwell and Ian Angus, 540–78. London: Penguin, 1970.

Osborne, Thomas. "History, Theory, Disciplinarity." In *Social in Question: New Bearings in History and the Social Sciences*, ed. Patrick Joyce, 175–90. London: Routledge, 2002.

Otter, Chris. *The Victorian Eye: A Political History of Light and Vision in Britain*. Chicago: University of Chicago Press, 2008.

Overy, Richard. "The Historical Present." *Times Higher Education*, 29 April 2010, 34.

Oxford Dictionary of National Biography, ed. Lawrence Goldman. Oxford: Oxford University Press, 2004.

Packard, Vance. *The Hidden Persuaders*. Harmondsworth, UK: Penguin, 1957.

Palladino, Paolo. "And the Answer Is . . . 42." *Social History of Medicine* 13 (2000): 142–51.

———. "Medicine Yesterday, Today, and Tomorrow." *Social History of Medicine* 14 (2001): 539–51.

Palmer, Bryan. *Descent into Discourse: The Reification of Language and the Writing of Social History*. Philadelphia: Temple University Press, 1990.

———. "Is There Now, or Has There Ever Been, a Working Class?' " In *After the End of History*, ed. Alan Ryan, 97–102. London: Collins and Brown, 1992.

Pang, Alex Soojung-Kim. "Visual Representation and Post-Constructivist History of Science." *Historical Studies in the Physical and Biological Sciences* 28 (1997): 139–71.

Papoulias, Constantina, and Felicity Callard. "Biology's Gift: Interrogating the Turn to Affect." *Body and Society* 16 (2010): 29–56.

Parallax. "To Jean François Lyotard." Special issue, 6, no. 4 (2000): 1–145.

Parsons, G. P. "The British Medical Profession and Contagion Theory: Puerperal Fever as a Case Study, 1830–1860." *Medical History* 22 (1978): 138–50.

Parsons, Keith, ed. *The Science Wars*. Amherst, NY: Prometheus Books, 2003.

Paul, Gerhard. *Visual History: Ein Studienbuch*. Berlin: Vandenhoeck und Ruprecht, 2006.

Pauwels, Luc, ed. *Visual Cultures of Science: Rethinking Representational Practices in Knowledge Building and Science Communication*. Hanover, NH: University Press of New England, 2006.

Pelling, Margaret. *Cholera, Fever and English Medicine, 1825–1865*. Oxford: Oxford University Press, 1978.

———. "The Meaning of Contagion: Reproduction, Medicine and Metaphor." In *Contagion: Historical and Cultural Studies*, ed. Alison Bashford and Claire Hooker, 15–38. London: Routledge, 2001.

———. Review of R. J. Morris, *Cholera 1832*. *Annals of Science* 36 (1979): 203.

Pence, G. *Brave New Bioethics*. Lanham, MD: Rowman and Littlefield, 2002.

Pestre, Dominique. *Science, argent et politique: Un essai d'intersection*. Paris: INRA, 2003.

———. "Thirty Years of Science Studies: Knowledge, Society and the Political." *History and Technology* 20 (2004): 351–69.

Peters, Rik. "Actes de Presence: Presence in Fascist Political Culture." *History and Theory* 45 (2006): 362–74.

Petersen, Alan, and Robin Bunton, eds. *Foucault: Health and Medicine*. London: Routledge, 1997.

Petryna, Adriana. *When Experiments Travel: Clinical Trials and the Global Search for Human Subjects*. Princeton, NJ: Princeton University Press, 2009.

Piaget, Jean. *Main Trends in Inter-Disciplinary Research*. New York: Harper and Row, 1970.

Pick, Daniel. *War Machine: The Rationalisation of Slaughter in the Modern Age*. New Haven: Yale University Press, 1993.

Pickering, Andrew. *The Mangle of Practice*. Chicago: Chicago University Press, 1995.

Pickstone, John. "A Brief History of Medical History." In *Making History: The Changing Face of the Profession in Britain*, 15 January 2012. Available at http://www.history.ac.uk/makinghistory/resources/articles/history_of_medicine.html.

———. "Bureaucracy, Liberalism and the Body in Post-Revolutionary France: Bichat's Physiology and the Paris School of Medicine." *History of Science* 14 (1981): 115–42.

———. "The Development and Present State of History of Medicine in Britain." *Dynamis* 19 (1999): 457–86.

———. "Medical Botany (Self-Help Medicine in Victorian England)." *Memoirs of the Literary and Philosophical Society of Manchester* 119 (1976–77): 85–95.

———. "Medicine, Society, and the State." In *Cambridge Illustrated History of Medicine*, ed. Roy Porter, 304–41. Cambridge: Cambridge University Press, 1996.

———. "Production, Community and Consumption: The Political Economy of Twentieth-Century Medicine." In *Medicine in the Twentieth Century*, ed. Roger Cooter and John Pickstone, 1–19. Amsterdam: Harwood Academic, 2000.

———. "The Rule of Ignorance: A Polemic on Medicine, English Health Service Policy, and History." *British Medical Journal*, 3 March 2011, 997.

———. *Ways of Knowing: A New History of Science, Technology and Medicine*. Manchester: Manchester University Press, 2000.

———. "Working Knowledges Before and After Circa 1800: Practices and Disciplines in the History of Science, Technology and Medicine." *Isis* 98 (2007): 489–516.

Pink, Sarah. *The Future of Visual Anthropology: Engaging the Senses*. Abingdon, UK: Routledge, 2006.

Pisani, Elizabeth. *The Wisdom of Whores: Bureaucrats, Brothels and the Business of AIDS*. London: Granta, 2008.

Poovey, Mary. *A History of the Modern Fact: Problems of Knowledge in the Sciences of Wealth and Society*. Chicago: University of Chicago Press, 1998.

———. *Making a Social Body: British Cultural Formation, 1830–1864*. Chicago: University of Chicago Press, 1995.

Porter, Carolyn. "History and Literature: 'After the New Historicism.'" *Social History* 21 (1990): 253–72.

Porter, Dorothy. "The Mission of Social History of Medicine: An Historical View." *Social History of Medicine* 8 (1995): 345–59.

Porter, Roy. *Bodies Politic: Disease, Death and Doctors in Britain, 1650–1900*. London: Reaktion Books, 2001.

———. *Gibbon: Making History*. London: Weidenfeld and Nicolson, 1988.

———. *Health for Sale: Quackery in England, 1660–1850*. Manchester: Manchester University Press, 1989.

———. "History of the Body." In *New Perspectives on Historical Writing*, ed. Peter Burke, 206–32. Cambridge: Polity, 1991.

———. "History of the Body Reconsidered." In *New Perspectives on Historical Writing*, ed. Peter Burke, 232–60. Cambridge: Polity, 2001.

———. Interview. *Lancet* 350 (1997): 1410.

———. "The Patient's View: Doing Medical History from Below." *Theory and Society* 14 (1985): 175–98.

———. Review of Petersen and Bunton, *Foucault*. *Social History of Medicine* 12 (1999): 177–78.

Porter, T. M. *Trust in Numbers: The Pursuit of Objectivity in Science and Public Life*. Princeton, NJ: Princeton University Press, 1989.

Post, Robert. "Debating Disciplinarity." *Critical Inquiry* 35 (2009): 749–70.

Power, Michael. *The Audit Society: Rituals of Verification*. New York: Oxford University Press, 1997.

Powers, Richard. *Galatea 2.2*. London: Abacus, 1995.

Poynter, F. L. N. "Thomas Southwood Smith—the Man (1788–1861)." *Proceedings of the Royal Society of Medicine* 55 (1962): 381–92.

Purtschert, P., K. Meyer, and Y. Winter, eds. *Gouvernmentalität und Sicherheit: Zeitdiagnostische Beiträge im Anschluss an Foucault*. Bielefeld: Transcript, 2008.

Putnam, Hilary. *Reason, Truth and History*. Cambridge: Cambridge University Press, 1981.

Putnam, Robert D. *Making Democracy Work*. Princeton, NJ: Princeton University Press, 1993.

Rabinow, Paul. "Artificiality and Enlightenment: From Sociobiology to Biosociality." In *Zone Books 6: Incorporations*, ed. Jonathan Crary and Sanford Kwinter, 234–52. New York: Zone Books, 1992.

Rabinow, Paul, and Nikolas Rose. "Biopower Today." *Biosocieties* 1 (2006): 195–217.

———, eds. *The Essential Foucault*. New York: New Press, 2003.

Rajan, Kaushik Sunder. *Biocapital: The Constitution of Postgenomic Life*. Durham, NC: Duke University Press, 2006.

Ramachandran, Vilayanur. *The Tell-Tale Brain: Unlocking the Mystery of Human Nature*. London: Heinemann, 2011.

Raman, Sujatha, and Richard Tutton. "Life, Science, and Biopower." *Science, Technology and Human Values* 35 (2010): 711–34.

Rancière, Jacques. *Aesthetics and Its Discontents*, trans. Steven Corcoran. Cambridge: Polity, 2009.

———. *On the Shores of Politics*, trans. Liz Heron. London: Verso, 1995.

———. *The Politics of Aesthetics: The Distribution of the Sensible*, trans. and intro. Gabriel Rockhill. London: Continuum, 2006.

Rawls, Walton. *Wake Up America! World War I and the American Poster*. New York: Abbeville, 1988.

Renaud, Lisa, with Caroline Bouchard. *La Santé s'affiche au Québec: Plus de 100 ans d'histoire*. Quebec: Presses de l'Université du Québec, 2005.

Restivo, Sal. "Bruno Latour: The Once and Future Philosopher." In *The New Blackwell Companion to Major Social Theorists*, ed. George Ritzer and Jeffrey Stepinsky. Available at http://www.salrestivo.org/LatourFinal.10.pdf.

Reubi, David. "Ethics Governance, Modernity and Human Beings' Capacity to Reflect and Decide: A Genealogy of Medical Research Ethics in the UK and Singapore." PhD diss., London School of Economics, 2009.

———. "The Human Capacity to Reflect and Decide: Bioethics and the Reconfiguration of the Research Subject in the British Biomedical Sciences." *Social Studies of Science* 42 (2012): 348–68.

———. "The Will to Modernize: A Genealogy of Biomedical Research Ethics in Singapore." *International Political Sociology* 4 (2010): 142–58.

Reverby, Susan. *Examining Tuskegee: The Infamous Syphilis Study and Its Legacy*. Chapel Hill: University of North Carolina Press, 2009.

Rheinberger, Hans-Jörg. "Beyond Nature and Culture: Modes of Reasoning in the Age of Molecular Biology and Medicine." In *Living and Working with the New Medical Technologies*, ed. Margaret M. Lock, Allan Young, and Alberto Cambrosio, 19–30. Cambridge: Cambridge University Press, 2000.

Richard, Hollis. *Graphic Design: A Concise History*. London: Thames and Hudson, 1994.

Richardson, Ruth. *Death, Dissection and the Destitute*. London: Routledge, 1987.

Rickards, Maurice. *Posters of Protest and Revolution*. Bath: Adams and Dart, 1970.

———. *Posters of the First World War*. London: Evelyn, Adams and Mackay, 1968.

———. *The Rise and Fall of the Poster*. Newton Abbot: David and Charles, 1971.

Rigby, Hugh, and Susan Leibtag. *HardWare: The Art of Prevention*. Edmonton, AB: Quon Editions, 1994.

Rigney, Ann. "Being an Improper Historian." In *Manifestos for History*, ed. Keith Jenkins, Sue Morgan, and Alun Munslow, 149–59. London: Routledge, 2007.

Robert-Sterkendries, Marine. *Posters of Health*. Brussels: Therabel, 1996.

Robson, David. "Disorderly Genius: How Chaos Drives the Brain." *New Scientist* 2714 (29 June 2009): 34–37.

Rogoff, Irit. "Studying Visual Culture." In *The Visual Culture Reader*, ed. Nicholas Mirzoeff, 14–26. London: Routledge, 1998.

Rooney, Elen. *Seductive Reasoning: Pluralism as the Problematic of Contemporary Literary Theory*. Ithaca, NY: Cornell University Press, 1989.

Rorty, Richard. "The Continuity Between the Enlightenment and 'Postmodernism.'" In *What's Left of Enlightenment? A Postmodern Question*, ed. K. M. Baker and P. H. Reill, 19–36. Stanford, CA: Stanford University Press, 2001.

Rose, Hilary, and Steven Rose, eds. *The Radicalisation of Science*. London: Macmillan, 1976.

Rose, Nikolas. "Beyond Medicalisation." *Lancet* 369 (24 February 2007): 700–702.

———. *Governing the Soul: The Shaping of the Private Self*. London: Routledge, 1989.

———. "The Human Sciences in a Biological Age." *Institute for Culture and Society Occasional Paper Series* 3 (February 2012).

———. *Inventing Our Selves: Psychology, Power, and Personhood*. Cambridge: Cambridge University Press, 1996.

———. "Medicine, History and the Present." In *Reassessing Foucault: Power, Medicine and the Body*, ed. Colin Jones and Roy Porter, 48–72. London: Routledge, 1994.

———. "Neurochemical Selves." *Society* 41 (2003): 46–59.

———. *The Politics of Life Itself: Biomedicine, Power, and Subjectivity in the Twenty-First Century*. Princeton, NJ: Princeton University Press, 2007.

———. "'Screen and Intervene': Governing Risky Brains." *History of the Human Sciences* 23 (2010): 79–105.

———. "The Value of Life: Somatic Ethics and the Spirit of Biocapital." *Daedalus* 137 (2008): 36–48.

———. "Was ist Leben—Versuch einer Wiederbelebung." In *Bios and Zoë: Die menschliche Natur im Zeitalter ihrer technischen Reproduzierbarkeit*, ed. M. Weiß, 152–78. Frankfurt: Suhrkamp, 2009.

Rose, Nikolas, and Joelle Abi-Rached. "The Birth of the Neuromolecular Gaze." *History of the Human Sciences* 23 (2010): 11–36.

Rose, Nikolas, and Peter Miller. "Political Power Beyond the State: Problematics of Government." *British Journal of Sociology* 43 (1992): 173–205.

Rose, Nikolas, and Carlos Novas. "Biological Citizenship." In *Global Assemblages: Technology, Politics, and Ethics as Anthropological Problems*, ed. Aihwa Ong and Stephen J. Collier, 439–63. Oxford: Blackwell, 2006.

Rose, Nikolas, and Paul Rabinow. "Biopower Today." *Biosocieties* 1 (2006): 195–218.

Rosenberg, Charles. "Anticipated Consequences: Historians, History, and Health Policy." In *Our Present Complaint: American Medicine, Then and Now*, 185–205. Baltimore: Johns Hopkins University Press, 2007.

———. *The Cholera Years: The United States in 1832, 1849, and 1866*. Chicago: University of Chicago Press, 1962.

———. "Disease and Social Order in America: Perceptions and Expectations." *Milbank Quarterly* 64, suppl. 1 (1986): 34–55. Reprinted in *AIDS: The Burdens of History*, ed. Elizabeth Fee and Daniel Fox, 12–32. Berkeley: University of California Press, 1988.

———. "Disease in History: Frames and Framers." *Milbank Quarterly* 67 (1989): 1–15.

———. "Erwin H. Ackerknecht, Social Medicine, and the History of Medicine." *Bulletin of the History of Medicine* 81 (2007): 511–32.

———. "Florence Nightingale on Contagion: The Hospital as Moral Universe." In *Healing and History: Essays for George Rosen*, ed. Charles Rosenberg, 116–36. New York: N. Watson, 1979.

———. "Holism in Twentieth-Century Medicine." In *Greater Than the Parts: Holism in Biomedicine, 1920–1950*, ed. C. Lawrence and G. Weisz, 335–55. Oxford: Oxford University Press, 1998.

———. "Introduction: Framing Disease: Illness, Society, and History." In *Framing Disease: Studies in Cultural History*, ed. Charles Rosenberg and Janet Golden, xiii–xxvi. New Brunswick, NJ: Rutgers University Press, 1992.

———. "Meanings, Policies, and Medicine: On the Bioethical Enterprise and History." *Daedalus* 128 (1999): 27–46.

———. "Mechanism and Morality: On Bioethics in Context." In *Our Present Complaint: American Medicine, Then and Now*, 166–84. Baltimore: Johns Hopkins University Press, 2007.

———. "What Is Disease? In Memory of Owsei Temkin." *Bulletin of the History of Medicine* 77 (2003): 491–505.

Rosenstone, Robert. "Space for the Bird to Fly." In *Manifestos for History*, ed. Keith Jenkins, Sue Morgan, and Alun Munslow, 11–18. London: Routledge, 2007.

Roskies, A. "Neuroethics for the New Millennium." *Neuron* 35 (2002): 21–23.

Ross, Andrew. *Science Wars*. Durham, NC: Duke University Press, 1996.

Rousseau, George S. *Nervous Acts: Essays on Literature, Culture and Sensibility*. Basingstoke, UK: Palgrave, 2004.

Rowson, Jonathan. "Transforming Behaviour Change: Beyond Nudge and Neuromania." Royal Society of Arts report, 2011. Available at http://www.thersa.org/projects/social-brain.

Ruffing, Reiner. *Bruno Latour*. Stuttgart: Fink, 2009.

Runia, Eelco. "Presence." *History and Theory* 45 (2006): 1–29.

Rüsen, Jaürn. "Historical Objectivity as a Matter of Social Values." In *Historians and Social Values*, ed. J. T. Leessen and Ann Rigney, 57–68. Amsterdam: Amsterdam University Press, 2000.

Rütten, Thomas. "Karl Sudhoff and 'The Fall' of German Medical History." In *Locating Medical History: The Stories and Their Meanings*, ed. Frank Huisman and John Harley Warner, 95–114. Baltimore: Johns Hopkins University Press, 2004.

Safranski, Rudiger. *Nietzsche: A Philosophical Biography*, trans. Shelley Frisch. London: Granta, 2002.

Said, Edward. *Culture and Imperialism*. New York: Knopf, 1993.

———. *Orientalism*. New York: Pantheon, 1978.

Salter, Brian, and C. Salter. "Bioethics and the Global Moral Economy: The Cultural Politics of Human Embryonic Stem Cell Science." *Science, Technology and Human Values* 32 (2007): 554–81.

Salvemini, Lorella Pagnucco. *United Colors: The Benetton Campaigns*. London: Scriptum Editions, 2002.

Samuel, Raphael. "On the Methods of *History Workshop*: A Reply." *History Workshop Journal* 9 (1980): 162–76.

———. "Reading the Signs: Fact Grubbers and Mind Readers." *History Workshop Journal* 32 (1991): 88–109; and 33 (1992): 220–51.

Sappol, Michael. "Introduction: Empires in Bodies: Bodies in Empires." In *A Cultural History of the Human Body in the Age of Empire*, ed. Michael Sappol and Stephen P. Rice, 1–35, 261–68. Oxford: Berg, 2010.

———. *A Traffic in Dead Bodies: Anatomy and Embodied Social Identity in Nineteenth-Century America*, Princeton, NJ: Princeton University Press, 2002.

Sardar, Ziauddin. *Thomas Kuhn and the Science Wars*. Duxford, UK: Icon, 2000.

Sauer, Michael. "Hinweg damit!' Plakate als historische Quellen zur Politik-und Mentalitätsgeschichte." In *Visual History ein Studienbuch*, ed. Paul Gerhard, 37–56. Berlin: Vandenhoeck und Ruprecht, 2006.

Saul, John Ralston. *The Collapse of Globalism and the Reinvention of the World*. London: Atlantic Books, 2005.

Sawday, Jonathan. *The Body Emblazoned: Dissection and the Human Body in Renaissance Culture*. London: Routledge, 1995.

Schäfer, H. Review of Purtschert et al., *Gouvernmentalität und Sicherheit. Foucault Studies*, no. 7 (2009): 170–77.

Schaffer, Simon. "Instruments as Cargo in the China Trade." *History of Science* 44 (2006): 217–46.

———. "A Social History of Plausibility: Country, City and Calculation in Augustan Britain." In *Rethinking Social History: English Society, 1570–1920, and Its Interpretation*, ed. Adrian Wilson, 128–57. Manchester: Manchester University Press, 1993.

Scheper-Hughes, Nancy. "The Last Commodity: Post-Human Ethics and the Global Traffic in 'Fresh' Organs." In *Global Assemblages: Technology, Politics, and Ethics as Anthropological Problems*, ed. Aihwa Ong and Stephen J. Collier, 145–67. Oxford: Blackwell, 2006.

Schiller, Herbert. *Culture Inc.: The Corporate Takeover of Public Expression*. New York: Oxford University Press, 1989.

Schnädelbach, Herbert. *Philosophy in Germany, 1831–1933*, trans. Eric Matthews. Cambridge: Cambridge University Press, 1983.

Scott, Joan W. "The Evidence of Experience." *Critical Enquiry* 17 (1991): 773–97.

———. "History-Writing as Critique." In *Manifestos for History*, ed. Keith Jenkins, Sue Morgan, and Alun Munslow, 19–38. London: Routledge, 2007.

Scull, Andrew. *Hysteria: The Biography*. Oxford: Oxford University Press, 2009.

———. *Museums of Madness: The Social Organization of Insanity in Nineteenth-Century England*. London: Allen Lane, 1979.

Sevecke, Torsten. *Wettbewerbsrecht und Kommunikationsgrundrechte: Zur rechtlichen Bewertung gesellschaftskritischer Aufmerksamkeitswerbung in der Presse und auf Plakaten am Beispiel der Benetton-Kampagnen*. Baden-Baden: Nomos, 1997.

Shapin, Steven. "Citation for Mary Douglas, 1994 Bernal Prize Recipient [Society for Social Studies of Science]." *Science, Technology, and Human Values* 20 (1995): 259–261.

———. "Here and Everywhere: Sociology of Scientific Knowledge." *Annual Reviews in Sociology* 21 (1995): 289–321.

———. *The Scientific Life: A Moral History of a Late Modern Vocation*. Chicago: University of Chicago Press, 2008.

Shapin, Steven, and Simon Schaffer. *Leviathan and the Air Pump: Hobbes, Boyle, and the Experimental Life*. Princeton, NJ: Princeton University Press, 1985.

Shaviro, S. "Biopolitics, the Pinocchio Theory [a review of Esposito, *Bíos*]," 17 August 2009. Available at http://www.shaviro.com/Blog/?=695.

Sherman, Daniel J. "Quatremère/Benjamin/Marx: Art Museums, Aura, and Commodity Fetishism." In *Museum Culture: Histories, Discourses, Spectacles*, ed. Daniel J. Sherman and Irit Rogoff, 124–43. Minneapolis: University of Minnesota Press, 1994.

Shildrick, Margrit. *Leaky Bodies and Boundaries: Feminism, Postmodernism and (Bio)ethics*. London: Routledge, 1997.

Shilling, Chris. *The Body and Social Theory*, 2nd ed. London: Sage, 2003.

Shortt, Samuel. "Clinical Practice and the Social History of Medicine: A Theoretical Accord." *Bulletin of the History of Medicine* 55 (1981): 533–42.

Shteir, Ann, and Bernard Lightman, eds. *Figuring It Out: Science, Gender, and Visual Culture*. Hanover, NH: Dartmouth College Press, 2006.

Sim, Stuart. *Derrida and the End of History*. Trumpington, UK: Icon, 1999.

Sinding, Christiane. "The Power of Norms: Georges Canguilhem, Michel Foucault, and the History of Medicine." In *Locating Medical History: The Stories and Their Meanings*, ed. Frank Huisman and John Harley Warner, 262–84. Baltimore: Johns Hopkins University Press, 2004.

Sloterdijk, Peter. *Critique of Cynical Reason*. Minneapolis: University of Minnesota Press, 1987.

Smail, Daniel Lord. "An Essay on Neurobiology." Version 1.4, 8 June 2010, available at http://creative commons.org/licenses/by/3.0.

———. *On Deep History and the Brain*. Berkeley: University of California Press, 2008.

Smart, Barry. *Foucault, Marxism and Critique*. London: Routledge, 1983.

———. *Michel Foucault*, rev. ed. London: Routledge, 2002.

Smith, F. B. *The People's Health, 1830–1910*. London: Croom Helm, 1979.

Smith, Roger. *Being Human: Historical Knowledge and the Creation of Human Nature*. Manchester: Manchester University Press, 2007.

———. *Inhibition: History and Meaning in the Sciences of Mind and Brain*. Berkeley: University of California Press, 1992.

Smith, Southwood. "Anatomy." *Westminster Review* 10 (1829): 116–48.

———. "Contagion and Sanitary Laws." *Westminster Review* 3 (1825): 134–67.

———. "Education." *Westminster Review* 1 (1824): 43–79.

———. "Plague—Typhus Fever—Quarantine." *Westminster Review* 3 (1825): 134–67, 499–530.

———. "Spasmodic Cholera." *Westminster Review* 15 (1831): 457–90.

———. *Treatise on Fevers*. London: Longman, Rees, Orme, Brown, and Green, 1830.

———. "Use of the Dead to the Living." *Westminster Review* 2 (1824): 59–97.

Smith, Wesley J. *Culture of Death: The Assault on Medical Ethics in America*. San Francisco: Encounter Books, 2000.

Söderqvist, Thomas. "To Give Global Genomes a Local Habitation and a Name: The Genechip as a Visualisation Device." Paper presented to Society for the Social History of Medicine Annual Conference, Warwick University, 2006.

Sokal, Alan. "Transgressing the Boundaries: Towards a Transformative Hermeneutics of Quantum Gravity." *Social Text* 46/47 (1996): 217–52.

Sokal, Alan, and Jean Brimont. *Fashionable Nonsense: Postmodern Intellectual's Abuse of Science*. New York: Picador, 1998.

Somit, A., and S. Peterson. "Biopolitics After Three Decades—a Balance Sheet." *British Journal of Political Science* 28 (1998): 559–71.

Sontag, Susan. *AIDS and Its Metaphors* [1989]. Reprinted in Sontag, *Illness as Metaphor* and *AIDS and Its Metaphors*. London: Penguin, 1991.

———. *Illness as Metaphor* [1978]. Reprinted in Sontag, *Illness as Metaphor* and *AIDS and Its Metaphors* [1989]. London: Penguin, 1991.

———. "Posters: Advertisement, Art, Political Artefact, Commodity. Introductory Essay." In *The Art of Revolution: 96 Posters from Cuba*, ed. Dugald Stermer. New York: McGraw-Hill, 1970.

Southgate, Beverley. " 'Humani nil alienum': The Quest for 'Human Nature.' " In *Manifestos for History*, ed. Keith Jenkins, Sue Morgan, and Alun Munslow, 67–76. London: Routledge, 2007.

Sparks, Tabitha. *The Doctor in the Victorian Novel: Family Practices*. Farnham, UK: Ashgate, 2009.

Spielmann, Heinz. *Justus Brinckmann*. Hamburg: Ellert und Richter, 2002.

Spivak, Gayatri. "Can the Subaltern Speak?" In *Marxism and the Interpretation of Culture*, ed. Cary Nelson and Lawrence Grossberg, 271–313. Urbana: University of Illinois Press, 1988.

Stadler, Max. "The Neuromance of Cerebral History." In *Critical Neuroscience: A Handbook of the Social and Cultural Contexts of Neuroscience*, ed. S. Choudhury and J. Slaby, 135–58. Oxford: Wiley-Blackwell, 2011.

Stafford, Barbara Marie. *Body Criticism: Imaging the Unseen in Enlightenment Art and Medicine*. Cambridge, MA: MIT Press, 1991.

———. *Echo Objects: The Cognitive Work of Images*. Chicago: University of Chicago Press, 2007.

———. *Visual Analogy: Consciousness as the Art of Connecting*. Cambridge, MA: MIT Press, 2001.

Stedman Jones, Gareth. "The Determinist Fix." *History Workshop Journal* 42 (1996): 19–35.

Steedman, Carolyn. *Dust*. Manchester: Manchester University Press, 2001.

———. *Labours Lost: Domestic Service and the Making of Modern England*. Cambridge: Cambridge University Press, 2007.

———. *Landscape for a Good Woman*. London: Virago, 1986.

———. *Past Tenses: Essays on Writing Autobiography and History*. London: Rivers Oram Press, 1992.

Stein, Claudia. "The Birth of Biopower in Eighteenth-Century Germany." *Medical History* 55 (2011): 331–37.

———. "Divining and Knowing: The Historical Method of Karl Sudhoff." *Bulletin of the History of Medicine* (forthcoming).

———. *Negotiating the French Pox in Early-Modern Germany*. Aldershot, UK: Ashgate, 2009.

Stern, Alexandra Minna, and Howard Markel. "Commentary: Disease Etiology and Political Ideology: Revisiting Erwin H. Ackerknecht's Classic 1948 Essay, 'Anticontagionism Between 1821 and 1867.'" *International Journal of Epidemiology* 38 (2009): 31–33.

Stevens, M. L. T. *Bioethics in America: Origins and Cultural Politics*. Baltimore: Johns Hopkins University Press, 2000.

Stewart, Thomas A. *Intellectual Capital: The New Wealth of Organization*. London: Nicholas Brealey, 1997.

Stolberg, Michael. "'Abhorreas pinguedinem': Fat and Obesity in Early Modern Medicine (c. 1500–1750)." *Studies in History and Philosophy of Science*, Part C: *Studies in History and Philosophy of Biological and Biomedical Sciences* 43 (2011): 370–78.

Studinka, Felix. "Foreword." In *Poster Collection. Visual Strategies Against AIDS, International AIDS Prevention Posters*. Zurich: Museum für Gestaltung, Lars Muller, 2002.

Sturken, Marita, and Lisa Cartwright. *Practices of Looking: An Introduction to Visual Culture*. Oxford: Oxford University Press, 2001.

Styhre, Alaexander, and Mats Sundger. *Venturing into the Bioeconomy: Professions, Innovation, Identity*. London: Palgrave Macmillan, 2011.

Sudhoff, Karl. "The Aims and Value of Medical History in the Self-Development and Professional Life of the Physician." In Sudoff, *Essays in the History of Medicine*, ed. Fielding H. Garrison, 45–54. New York: Medical Life Press, 1926.

———. "On the Origin of Syphilis." In Sudoff, *Essays in the History of Medicine*, ed. Fielding H. Garrison, 259–74. New York: Medical Life Press, 1926.

Sweeney, Sean, and Ian Hodder, eds. *The Body*. Cambridge: Cambridge University Press, 2002.

Tallis, Raymond. *Aping Mankind: Neuromania, Darwinitis and the Misrepresentation of Humanity*. Durham, UK: Acumen, 2011.

———. *Hunger, the Art of Living*. Durham, UK: Acumen, 2008.

———. *In Defence of Realism*. Lincoln: Nebraska University Press, 1998.

Tannen, T. "Media Giant and Foundation Team Up to Fight HIV/AIDS." *Lancet* 361 (26 April 2003): 1440–41.

Tattersall, Robert. *Diabetes: The Biography*. Oxford: Oxford University Press, 2009.

Teich, Mikulas, and Robert M. Young, eds. *Changing Perspectives in the History of Science: Essays in Honour of Joseph Needham*. London: Heinemann, 1973.

Temkin, Owsei. "An Historical Analysis of the Concept of Infection." In *The Double Face of Janus*, 456–71. Baltimore: Johns Hopkins University Press, 1977.

Theroux, Paul. *Ghost Train to the Eastern Star*. Boston: Houghton Mifflin, 2008.

Thompson, E. P. *The Making of the English Working Class*. London: Gollanz, 1963.

———. "The Poverty of Theory: or, An Orrery of Errors." In *The Poverty of Theory and Other Essays*, 193–397. London: Merlin, 1978.

Thompson, J. "Pictorial Lies? Posters and Politics in Britain, c. 1880–1914." *Past and Present* 197 (2007): 177–210.

Thrift, Nigel. *Non-Representational Theory: Space, Politics, Affect*. London: Routledge, 2008.

———. "Pass It On: Towards a Political Economy of Propensity." *Emotions, Space and Society* 1 (2008): 83–91.

Timmers, Margaret, ed. *The Power of the Poster*. London: V&A Publications, 1998.

Titmuss, Richard. *The Gift Relationship: From Human Blood to Social Policy*, original ed. [1970], with new chapters by Ann Oakley and John Ashton. London: LSE Books, 1997.

Toon, Elizabeth. "Managing the Conduct of Individual Life: Public Health Education and American Public Health, 1910–1940." PhD diss., University of Pennsylvania, 1998.

Tormey, Simon. "Post-Marxism." *Democracy and Nature* 7 (March 2001): 119–34.

Toscani, Oliviero. *Die Werbung ist ein lächelndes Aas*. Frankfurt: Fischer Taschenbuch, 1997.

Tosh, John. *Why History Matters*. London: Palgrave Macmillan, 2008.

Toulmin, Steven. "How Medicine Saved the Life of Ethics." *Perspectives in Biology and Medicine* 25 (1981–82): 736–50.

Treichler, Paula. "AIDS, Africa and Cultural Theory." Reprinted in Treichler, *How to Have Theory in an Epidemic: Cultural Chronicles of AIDS*, 205–234. Durham, NC: Duke University Press, 1999.

———. "AIDS, Gender, and Biomedical Discourse: Current Contests for Meaning." In *AIDS: The Burdens of History*, ed. Elizabeth Fee and Daniel M. Fox, 190–234. Berkeley: University of California Press, 1988.

———. "AIDS, Homophobia, and Biomedical Discourse: An Epidemic of Signification." *Cultural Studies* 1 (1987): 263–305. Reprinted in Treichler, *How to Have Theory in an Epidemic: Cultural Chronicles of AIDS*, 11–41. Durham, NC: Duke University Press, 1999.

———. "Beyond Cosmo: AIDS, Identity, and Inscription of Gender [1992]." Reprinted in Treichler, *How to Have Theory in an Epidemic: Cultural Chronicles of AIDS*, 235–77. Durham, NC: Duke University Press, 1999.

———. "How to Have Theory in an Epidemic." In Treichler, *How to Have Theory in an Epidemic: Cultural Chronicles of AIDS*, 278–314. Durham, NC: Duke University Press, 1999.

———. *How to Have Theory in an Epidemic: Cultural Chronicles of AIDS*. Durham, NC: Duke University Press, 1999.

Treichler, Paula, and Kelly Gates. "'When Pirates Feast . . . Who Pays': The Pirate Figure in Trojan Brand Condom Advertisements, 1926–1932." Paper presented at the American Association for the History of Medicine, 83rd Conference, Rochester, Minnesota, 30 April 2010.

Treichler, Paula, Lisa Cartwright, and Penley Constance, eds. *The Visible Women: Imagining Technologies, Gender, and Science*. New York: New York University Press, 1998.

Turner, Bryan S. *The Body and Society*. London: Sage, 1996.

———. *Medical Power and Social Knowledge*, 2nd ed. London: Sage, 1995.

Turner, Leigh. "Medical Ethics in a Multicultural Society." *Journal of the Royal Society of Medicine* 94 (2001): 592–94.

Turney, Jon, ed. *Sci-Tech Report: Everything You Need to Know About Science and Technology in the 80s*. New York: Pantheon, 1984.

Ueyama, Takahiro. *Health in the Marketplace: Professionalism, Therapeutic Desires, and Medical Commodification in Late-Victorian London*. Palo Alto, CA: Society for the Promotion of Science and Scholarship, 2010.

UNAIDS. "The Joint United Nations Programme on HIV/AIDS (UNAIDS)." Available at http://www.uncsd2012.org.

Usborne, Cornelie, and Willem de Blecourt. "Pains of the Past: Recent Research in the Social History of Medicine in Germany." *Bulletin of the German Historical Institute London* 21 (1999): 5–21.

Valentine, Gill. "Whatever Happened to the Social? Reflections on the 'Cultural Turn' in British Human Geography." *Norwegian Journal of Geography* 55 (2001): 166–72.

Vann, Richard T. "Historians and Moral Evaluations." *History and Theory* 43 (2004): 3–30.

Veeser, H. Aram, ed. *The New Historicism*. London: Routledge, 1989.

Vernon, James. "The End of the Public University in England." Available at http://stormbreaking.blogspot.com/2010/11/end-of-public-university-in-england.html.

———. *Hunger: A Modern History*. Cambridge, MA: Harvard University Press, 2007.

Veyne, Paul. *Foucault: Der Philosoph als Samurai*. Stuttgart: Reclam, 2008.

———. *Writing History: Essay on Epistemology*, trans. M. Moore-Rinvolucri. Middletown, CT: Wesleyan University Press, 1971.

Vitellone, Nicole. *Object Matters: Condoms, Adolescence and Time*. Manchester: Manchester University Press, 2008.

Vrecko, Scott, ed. "Neuroscience, Power, Culture." Special issue of *History of the Human Sciences*, 23 (2010): 1–126.

Vrettos, Athena. *Somatic Fictions*. Stanford, CA: Stanford University Press, 1995.

Waddington, Ivan. "The Development of Medical Ethics: A Sociological Analysis." *Medical History* 19 (1975): 36–51.

Wade, Nicholas. "From 'End of History' Author, a Look at the Beginning and Middle." *New York Times*, 7 March 2011.

Wakley, Thomas. "Editorial." *Lancet* 1 (11 February 1832): 706.

———. "Editorial." *Lancet* 1 (25 February 1832): 774.

Waldby, Catherine. *The Visible Human Project: Informatic Bodies and Posthuman Medicine*. London: Routledge, 2000.
Walker, J. A., and S. Chaplin. *Visual Culture: An Introduction*. Manchester: Manchester University Press, 1997.
Walkowitz, Judith. *City of Dreadful Delight: Narratives of Sexual Danger in Late-Victorian London*. London: Virago, 1992.
Wallace, Mike. "Hijacking History: Ronald Reagan and the Statue of Liberty." *Radical History Review* 37 (1987): 119–130.
Wallis, R., ed. *On the Margins of Science: The Social Construction of Rejected Knowledge*. Keele, UK: University of Keele, 1979.
Warner, John Harley. "The History of Science and the Sciences of Medicine." *Osiris* 10 (1995): 164–93.
Warner, Michael. *Publics and Counterpublics*. New York: Zone Books, 2005.
Watney, Simon. *Policing Desire: Pornography, AIDS and the Media* [1987], 3rd ed. London: Cassell, 1997.
Watney, Simon, and Sunil Gupta. "The Rhetoric of AIDS: A Dossier Compiled by Simon Watney, with Photographs by Sunil Gupta." *Screen* 27 (1986): 72–85.
Wear, Andrew. "Introduction." In *Medicine in Society*, ed. Andrew Wear, 1–14. Cambridge: Cambridge University Press, 1992.
Wear, Andrew, J. Geyer-Kordesch, and R. French, eds. *Doctors and Ethics: The Earlier Historical Setting of Professional Ethics*. Amsterdam: Rodopi, 1993.
Weber, Max. "Science as a Vocation." Available at http://www.ne.jp/asahi/moriyuki/abukuma/weber/lecture/science_frame.html.
Webster, Charles. "The Historiography of Medicine." In *Information Sources in the History of Science and Medicine*, ed. Pietro Corsi and Paul Weindling, 29–43. London: Butterworth Scientific, 1983.
Weiß, M., ed. *Bios und Zoë: Die menschliche Natur im Zeitalter ihrer technischen Reproduzierbarkeit*. Frankfurt: Surkamp, 2009.
Werskey, Gary. "The Marxist Critique of Capitalist Science: A History of Three Movements?" *Science as Culture* 16 (2007): 397–461.
———. *The Visible College: A Collective Biography of British Scientists and Socialists of the 1930s*. London: Allen Lane, 1978.
White, Hayden. "Afterword: Manifesto Time." In *Manifestos for History*, ed. Keith Jenkins, Sue Morgan, and Alun Munslow, 220–31. London: Routledge, 2007.
———. *Metahistory: The Historical Imagination in Nineteenth-Century Europe*. Baltimore: Johns Hopkins University Press, 1972.

———. "Postmodernism and Textual Anxieties." In *The Postmodern Challenge: Perspectives East and West*, ed. Nina Witoszek and Bo Strath, 27–45. London: Sage, 1999.

———. "The Public Relevance of Historical Studies: A Reply to Dirk Moses." *History and Theory* 44 (2005): 333–38.

———. "Response to Arthur Marwick." *Journal of Contemporary History* 30 (1995): 233–46.

Whitfield, Nicholas. "A Genealogy of the Gift: Blood Donation in London, 1921–1946." PhD diss., University of Cambridge, 2011.

Whong-Barr, Michael. "Clinical Ethics Teaching in Britain: A History of the London Medical Group." *New Review of Bioethics* 1 (2003): 73–84.

Wiener, Martin. "Treating 'Historical' Sources as Literary Texts: Literary Historicism and Modern British History." *Journal of Modern History* 70 (1998): 619–38.

Wiesemann, Claudia. "Defining Brain Death: The German Debate in Historical Perspective." In *Coping with Sickness*, ed. John Woodward and Robert Jütte, 149–69. Sheffield: European Association for the History of Medicine and Health Publications, 2000.

Williams, Raymond. *Key Words: A Vocabulary of Culture and Society*. Glasgow: Fontana/Croom Helm, 1976.

———. *Marxism and Literature*. Oxford: Oxford University Press, 1977.

Williams, Simon, and Michael Calnan. "The 'Limits' of Medicalization? Modern Medicine and the Lay Populace in 'Late' Modernity." *Social Science and Medicine* 42 (1996): 1609–20.

Wilson, Adrian. "On the History of Disease-Concepts: The Case of Pleurisy." *History of Science* 38 (2000): 271–319.

Wilson, Elizabeth A. *Affect and Artificial Intelligence*. Seattle: University of Washington Press, 2010.

Winterbottom, Thomas Masterman. "Thoughts on Quarantine and Contagion." *Edinburgh Medical and Surgical Journal* 30 (1828): 62–102.

Wolin, Richard. "Introduction." Jürgen Habermas, *The New Conservatism: Cultural Criticism and the Historians' Debate*, trans. and ed. Sherry W. Nicholsen, vii–xxxiii. Cambridge: Polity, 1989.

Worboys, Michael. *Spreading Germs: Disease Theories and Medical Practice in Britain, 1865–1900*. Cambridge: Cambridge University Press, 2000.

World Health Organization. "World Health Organization Launches Public Information Effort to Increase Global Awareness of AIDS." *WHO Press*, Press release WHO/15, 27 May 1987.

Wright, Peter, and Andrew Treacher, eds. *The Problem of Medical Knowledge: Examining the Social Construction of Medicine*. Edinburgh: Edinburgh University Press, 1982.

Young, Robert M. "Can We Really Distinguish Fact from Value in Science." *Times Higher Education Supplement*, 23 September 1977, 6.

———. *Darwin's Metaphor: Nature's Place in Victorian Culture*. Cambridge: Cambridge University Press, 1985.

———. "Getting Started on Lysenkoism." *Radical Science Journal* 6/7 (1978): 81–105.

———. "Marxism and the History of Science." In *Companion to the History of Modern Science*, ed. R. C. Olby, G. N. Cantor, J. R. R. Christie, and M. J. S. Hodge, 77–86. London: Routledge, 1990.

———. "Science *Is* Social Relations." *Radical Science Journal* 5 (1977): 65–129.

———. "Why Are Figures So Significant? The Role and the Critique of Quantification." In *Demystifying Social Statistics*, ed. J. Irving, I. Miles, and J. Evans, 63–74. London: Pluto, 1979.

Zammito, John. *A Nice Derangement of Epistemes: Post-Postivism in the Study of Science from Quine to Latour*. Chicago: University of Chicago Press, 2004.

Zeller, Thomas. "The Spatial Turn in History." *German Historical Institute Bulletin* 35 (2004): 123–24.

Zinser, Hans. *Rats, Lice and History: Being a Study in Biography*. Boston: Atlantic Monthly, 1935.

Žižek, Slavoj. "A Plea for Leninist Intolerance." *Critical Inquiry* 28 (2002): 542–66.

INDEX

Abel-Smith, Brian, 20
Ackerknecht, Erwin, 42, 43, 67, 68, 82; "Anticontagionism," 49–52
actants, 216, 218
activism, 96, 107, 235n67
Actor-Network Theory (ANT), 206–8, 217–19
ACT UP (American AIDS Coalition to Unleash Power), 125, 127
Adorno, Theodor, 226
aesthetics, 95, 147, 151, 153; politics of, 113, 139, 140, 143, 156–59
affectivity, 254n36
affect theory, 6, 8, 31, 206, 218, 222, 225
Agamben, Giorgio, 188, 189, 198, 201, 204
Agar, Jon, 65
agency, 84, 90, 212, 217, 219, 282n47
*a*historicization, 18. *See also* historians
AIDS/HIV, 167, 225; Africa, 133; "Against Aids" exhibition, 142, 143–44, 145, 151–52; "age of," 114; biological reality, 83, 100; "Don't die of ignorance" campaign, 122; epidemic, 126; global problem, 153, 154; identity politics, 142, 147; medicine, 147, 153; prevention, 135; as representation, 152; social perception, 12; and somatic society, 95; stereotypical images, 133. *See also* posters; UNAIDS
air, idea of, 58–59, 61
Almodóvar, Pedro, 30
Althusser, Louis, 193, 212, 279n8

Amsterdamska, Olga, 279n22
Anatomy Act (1832), 60
animal-centricity, 233n46
animals, "non-human," 205
Ankersmit, Frank, 260n35
Annales School, 65, 278n7
ANT. *See* Actor-Network Theory
Appleby, Joyce, 77
Arbor Ciencia, 93
"archaeological method," 44, 255n52
Arendt, Hannah, 25, 221
Armstrong, David, 47, 103, 176, 256n57
Arnold, John, 69
art, 152; degenerate, 268n29; modernist, 146. *See also* posters
art history, 105, 116
Asimov, Isaac, 209
"assemblages," 216, 280n28
Atwood, Margaret, 205
"audit culture," 33
"authenticity," 106
"autonomy," 37, 174, 180–81, 242n153; patient, 22, 175

Bacon, Francis, 57, 230n14
Balzac, Honoré, 259n13
Barham, Peter, 240n124
Barnes, Barry, 34, 46
Barry, Jonathan, 46
Barthes, Roland, 80
base/superstructure, 221
Bataille, Georges, 254n36
Baudrillard, Jean, 78, 80

337

Bauman, Zygmunt, 200
Beauvoir, Simone de, 105, 107
Beck, Ulrich, 75
Beecher, Henry, 180
Beevor, Antony, 33
Bencard, Adam, 92, 93, 261n44
Benetton, 125, 128
Benoist, Alain de, 254n36
Bentham, Jeremy, 179
Berlant, Jeffrey, 175
Bernal, J. D., 25, 54
Berridge, Virginia, 152
Bess, Michael, 1
Beyond the Cultural Turn, 88
Bichat, Xavier, 59
binaries, 24, 53, 214, 221, 225. *See also* base/superstructure; biology/society; culture/science; dualisms; fact/fiction; fact/value; history/philosophy; history/social context; meanings/materiality; nature/culture; nature/politics; normal/pathological; objectivity/subjectivity; persons/patients; representationalism/authenticity; representationalism/presentationalism; representations/materiality; science/context; science/humanities; science/ideology; science/pseudoscience; science/scientism; science/society; society/culture; society/economy; structure/agency; structure/culture; subject/object; text/context; theory/practice
bio-archaeology, 166
biocapital, 182, 284n64
biocitizenship, 1, 196, 284n64
bio-complex, 190
bio-consumerism, 196
bio-creep, 160
bioeconomics, 190, 230n12
bioethicists, 35, 170, 177
bioethics, 22, 39, 74; birth of, 176; historicization, 177, 180–82;
opportunism, 172–73; tarnished image, 173, 175
Bioethics and Society, 173
Bioethics in Social Context, 173
bio-humanism, 9
biology: "age of," 2, 3; ~/society, 84; destiny, 189; historicization, 6; internalized, 9; and law, 198; molecular, 109; naturalized, 36; non-linear, 222; in sociology, 36
biomedicine, 1, 174
biopolitics 183, 284n64; affirmative, 189, 201, 202, 203, 204; as bioeconomics, 230n12; genealogy of, 203
biopower: black box, 189, 195, 201, 202; Nazi, 204; *vs.* governmentality, 186; semantics, 188, 199. *See also* Foucault, Michel; truth
"Biopower Today," 183, 192, 284n64, 236n78
bios, 188
biosociability, 191, 196
"biosocial context," 230n16
Biosocieties, 183
biotechnology, 1, 30, 95, 177, 196, 206, 284n64
Bivins, Roberta, 65
Blackman, Liza, 283n54
Blair, Tony, 278n*ii*
Bloch, Marc, 101, 265n75
Bloor, David, 46
body: "flesh," 103; Foucault scholarship on, 278n2; fragmented, 2, 102; historicization, 91, 92, 94–97, 101, 105, 108, 261; individualized, 96; knowledge, 126; and modernity, 96, 97, 110; site of experiment with self, 95; as situated knowledge, 68; studies, 81
Body and Society, 96
Bourdieu, Pierre, 236n72
Brandt, Allan, 115, 117–20, 121, 127, 128, 130, 134
Brave New Bioethics, 173

Brinkmann, Justus, 144, 145, 157
Bruckner, Pascal, 254n36
Burke, Peter, 93, 94
Burton, Antoinette, 88
Bush, George W., 77, 153
Butler, Judith, 104–5, 214, 225; *Bodies That Matter*, 81
Butterfield, Herbert, 13
Bynum, Caroline Walker, 104

Callard, Felicity, 222, 223
Calvinism, 57, 58
Cambridge: History and Philosophy of Science Department, 24; Wellcome Unit, 45
Cambridge World History of Medical Ethics, 172, 174–75, 177–82
Campbell, Timothy, 189, 203
Canguilhem, Georges, 94, 194, 195, 203
Canning, George, 246n18
Cantor, David, 65
capitalism, 52, 79, 99, 120, 217, 223
Carr, E. H., 11, 284n71
Casey, Catherine, 88
causality, 87, 161, 219
Centre for the History of Science, Technology and Medicine, Manchester, 28, 65
Chadwick, Edwin, 51
"change," 20
"chimera of origins," 17
cholera, 56, 163, 166, 167
Chomsky, Noam, 28
citizenship: biological, 190; epistemological, 277n46; global, 154, 157; national, 157; social, 1, 196
civil rights movement, 21
class, 79, 80; consciousness, 212; material force, 278n7; struggle, 220
"close reading," 42, 81, 128
"cognitive turn," 109
Cold War, 8, 157
Collingwood, R. G., 11

commercialization, 34, 100, 142
commoditization, 238n99
"complexification," 6, 17, 194
condoms, 144, 146, 149, 153, 266n6, 267n17
consciousness, 99, 158, 212, 221, 222, 227
constructivism, 6, 41, 47, 65, 68. *See also* Latour, Bruno
consumerism, 22, 190, 197
contagium animatum, 55
contextualism, 13, 48, 62, 65, 66, 235n62
Cook, Harold, 139
Corsi, Pietro, 46
Crick, Francis, 205
Crimp, Douglas, 127, 131, 224
"crisis," 21, 219, 225
critical theory. *See* Marxism; postmodernism; theory
critique, 157; historicity of, 66; of modern medicine, 21; notion of, 225, 279n18; of psychiatric practice, 21; as self-awareness, 214; *vs.* criticism, 226
"cultural," 88; ~/science, 24; "do-it-yourself," 76; "jamming," 264n37; redefined, 130–31; theory, 89; "turn," 30, 210–13, 224, 225; *vs.* "the social," 100; "wars," 72, 224
cultural history, 223, 225; "materialized," 211; new, 30, 101, 103
Cunningham, Andrew, 168
Cusset, François, 207, 224

Dalhousie University, 44
Darnton, Robert, 80
Darwinism, 191, 202
Daston, Lorraine, 16–18
Davis, Natalie Zemon, 278n7
death, 128; author, 20; brain, 75; of poststructuralism, 168
decision-making, 74, 140
de-historicization, 171
Delaporte, François, 43

Deleuze, Gilles, 78; *Pure Immanence*, 201
Derrida, Jacques, 78, 79, 80, 82, 85, 98, 254n36, 261n44
"descriptive turn," 211, 219
design technology, 146
determinacy, 84
diagnosis, retrospective, 161, 164–66
"dialectic," 203
Dirlik, Aruf, 159
"disablement," 87
disciplinarity: attack on, 31–32; boundaries, 90; as counter-hegemonic strategy, 38; dissolution, 32; "envy," 227; as epistemic virtue, 241n138; faith in, 38; normalization, 76; as salvation, 40. *See also* interdisciplinarity
"discontinuity," 165
discourse analysis, 64; of appearances, 263n17; use, 177–79; *vs.* ideology, 115
discursivity: *vs.* historicity, 87; *vs.* "reality," 88
disease: concepts, 164, 167; framing, 160, 162; metaphors, 121; postmodern epistemology, 133; signs and symptoms, 167; as social constructs, 83; social perception, 123; stigma, 120. *See also* essentialism; Foucault, Michel
disorders, psychosomatic, 164
dispositifs, 194, 198
"dissensus," 220
DNA analysis, 165
doctor-patient relations, 74, 181
"documentary fallacy," 119
Donzelot, Jacques, 88
Döring, Jürgen, 145, 146
Douglas, Mary, 27, 56, 87, 101
Driscoll, Lawrence, 89
dualisms, 52, 213. *See also* binaries
Duden, Barbara, 103–4, 106, 256n57
Durkheim, Emile, 79
"dynamic equilibriums," 61, 62

Eagleton, Terry, 18, 254n34
eclecticism, 213
ecology, 76, 218, 223
economism, 43, 54, 55, 214, 221
Edgerton, David, 65
"Edinburgh School," 217, 46
Edwards, Martin, 162–63
Eichmann, Adolf, 221
Eley, Geoff, 77
Emerson, Ralph Waldo, 163, 169
empiricism, 4, 31, 114, 165, 173, 219, 222, 194, 195
Enlightenment, 5, 10, 14, 77, 97, 218, 226; narrative, 96, 177, 259n19
epidemiology, 51, 163, 166
epistemes, 12, 16, 234n56
epistemic virtues, 15–16, 17, 214
epistemology: of artifacts, 156; and citizenship, 277n46; of disease, 133; of perception, 137; poststructuralist, 126; prison-house, 13; of signification, 133; translated into science, 15
Epstein, Jim, 89
"equality," 219
Esposito, Roberto, 197–203
essentialisms, 219; categories, 82; biological, 6, 102, 110, 111; and disease, 168; fallacy of, 87; historical, 111; neo-, 92; political, 111; positivist, 161; *vs.* "experience," 104, 108. *See also* history-writing
ethicists. *See* bioethicists
ethics, 2, 14
European Association of the History of Medicine, 112
Evans, Richard, 77
"evidence," 53, 105
evidence-based medicine, 23, 74, 75, 243n164, 252n14
evidence-based policy making, 182
"experience," 103, 105
"experiment," 4, 235n66

facticity, 32, 42, 53, 234n59
fact, 5, 217; as artifact, 234n59; concept of, 167, 246n12; ~/fiction, 12, 33; ~/value, 217, 24, 52, 63, 83; Goethe on, 53; and historians, 239n112; historical, 53; of history, 13; scientific, 165, 281n31; as theory, 53; unsensuous, 54
"false consciousness," 99, 121, 221, 166
Faye, Pierre, 253n34
Fee, Elizabeth, 113
Feher, Michael, 101
feminism, 21, 76, 218
Feyerabend, Paul, 84
Figlio, Karl, 24, 45, 47
Fissell, Mary, 29, 65, 178, 278n2
Fleck, Ludwik, 24, 42, 168, 217, 234n59
"flesh," philosophical concept, 201
flexibility, 219, 222, 282n41
Forman, Paul, 17, 35, 267n13
Forrester, John, 240n124
Foucault, Michel, 24, 27, 73, 81, 84, 254n36; analytics of power, 186, 227; anti-essentialism, 91–92; archaeological method, 44; *Archaeology of Knowledge*, 43; biopower, 98–100, 127, 186, 187, 193, 197, 198, 199, 204; *Birth of the Clinic*, 43; body/knowledge/power, 17, 91, 100, 126, 127, 134; and "critique," 226; *Discipline and Punish*, 44, 79, 94; and disease, 162; *Dits et écrits*, 186; *dispositifs*, 194; end of history, 243n161; epistemic virtues, 16; genealogical method, 17, 44, 204, 280n28; *Histoire de la sexualité*, 79, 199; historicization of, 193; historicization of culture, 66; and history, 203; "history of the present," 5, 30, 193; and historians, 44; in history of science and medicine, 43–44; *Madness and Civilization*, 28; and Marxism, 79, 100, 193; and "medicalization," 99; medical language, 76; political activism, 193, 224; politics of "life," 188, 194, 198;

representationalism, 5; rules of truth, 132; self-governance project, 25; technology of life, 127. *See also* biopower; body; disease; Joyce, Patrick
Fox, Renée, 176
fragmentation, 74, 133, 152, 191, 219, 224
Fragments for a History of the Human Body, 101
"frame": descriptive category, 85; illusion of, 86; substitute for "the social," 83, 84, 86
Francis Crick Memorial Conference, 205
Frankfurt School of Critical Theory, 27, 41, 122, 123, 147, 221
"free actors," 80
"free choice," 22
Freedberg, David, 261n49
free-market ideologues, 76
Freud, Sigmund, 84, 201
Fukuyama, Francis, 80
Fulbrook, Mary, 13
Fuller, Steve, 35–36, 110, 192, 208, 241n138
"future," as a representation, 5

Galison, Peter, 16–18
Gallagher, Catherine, 102
Gay Men Fighting AIDS, 149
gays, 150, 269n36
Geertz, Clifford, 27, 101
Geisteswissenschaften, 10
general practitioners, 23, 74
genetics, new, 169, 259n19
Genovese, Eugene, 45
Gentile, Giovanni, 107–8
geopolitics, 186
Georgetown University, 176
"gift relationship," 180
Gilman, Sander, 81, 114, 115; and AIDS posters, 119, 127–31, 132, 134
Ginzburg, Carlo, 278n7
Gladstone, Jo, 251n*v*
Global Media Aids Initiative, 265n76

globalism, 141, 139, 153, 159, 267n12; political and epistemological function, 154, 157, 158; semantic hegemony, 154
Gluckman, André, 254n36
Goethe, Johann Wolfgang von, 53
Goffman, Erving, 82, 84, 85
governmentality, 186, 202, 275n6
Gramsci, Antonio, 26–27, 41, 86
graphic design, 268n21
Gray, Chris Hables, 192
Greenblatt, Stephen, 32
Grosz, Elizabeth, 81, 104
Guardian, The, 124
Guilhaumou, Jacques, 253n34
Gumbrecht, Hans, 107
Gutman, Herbert, 45

Habermas, Jürgen, 78
Hacking, Ian, 45, 65, 107, 180, 274n33
Hall, Stuart, 85, 258n88
Haraway, Donna, 80, 89
Hardt, Michael, 188, 197
Harvey, David, 220
Harwood, Jon, 65
Hastings Centre, 176
health: consumers, 75; education, 116; insurance, 177; resources, 182; services, 73. *See also* public health
Heath, Deborah, 191
Hegel, G. W. F., 5
Heidegger, Martin, 106
heritage industry, 78
Heron, Craig, 44
Hesse, Mary, 24
Hessen, Boris, 54
historians: *a*historicity, 7, 13; and autonomy, 242n153; blind assassins, 40; constructed thought, 11; in epistemology, 12; and ethical grip, 207; global framework, 157; image of, 16; neoconservative, 31, 210; and objectivity, 33, 64, 235n65; passive voice, 20; and postmodernism, 77; practice, 8–9, 77; rationalist agendas, 235n64; scientificity, 7; self-awareness, 16, 66; and "theory," 13; values and virtues, 7, 13, 19, 65. *See also* facts; Foucault, Michel
"historical epistemology," 4, 11, 16, 65, 110, 161, 217
"historical materialism," 80
"historical ontology," 65
historicization, 159, 226; of the visual, 112. *See also a*historicization; body; bioethics; biology; de-historicization; Foucault, Michel; Latour, Bruno; medicine
historiography, 18, 19, 64
history: academic discipline, 33, 78, 158, 181; assault on, 33–36; constructedness, 16, 159; as critical forum, 212; as critique, 207; of emotions, 31; as entertainment, 33, 78; of epidemiology, 51; epistemological foundations, 77; "from below," 99; ~/philosophy, 62; ~/social context, 62; as invention, 77; life in, 30; as metanarrative, 5, 64; philosophy of, 235n64; *vs.* psychology 25; "wars," 72, 162. *See also* Foucault, Michel
History and Theory, 106, 107
history of medicine: and medical humanism, 21; patient-centered, 22; problematized, 71; social, 32, 46, 72, 76, 82, 89, 99
history of science, 23–25
"history wars," 72, 162
History Workshop, 26
history-writing: *a*historical, 102; critical, 39, 67, 214, 228; as critique, 215, 225; eclecticism, 213; embodied, 19; "end of," 31, 36; and essentialisms, 160; as historiography, 11; as ideology, 13; irrelevance, 15; and modernity, 96; and new game of truth, 10; non-transcendent, 281n31; and politics, 66,

225; "relevance," 54; resting place, 158; theory adoption, 223; trivialized, 35; useless, 4; utilitarian, 213; and wider public, 35
Hobsbawm, Eric, 77
Holocaust, 184
Holt, Thomas, 260n31
Horkheimer, Max, 140, 246n12
Huisman, Frank, 69
human experimentation, 180
"human nature," 10, 106, 242n153; culturally open, 191; greedy, 34; inalienable, 82; neurobiological, 36, 233n48; real and fixed, 27; reconfigured, 180–81. *See also* posthuman; transhuman
humanism, 187, 192; faith in, 10, 25; opposition to, 218, 281n35; social, 192; socialist, 20. *See also* bio-humanism, posthumanism
humanists, 11, 15, 218
humanities, 10, 206, 225
"humanity," 2, 233n48
hunger, 214
Hunt, Lynn, 77, 81
Husserl, Edmund, 101
Huxley, Julian, 25
hysteria, 164

idealism, 14, 68
identity: construction, 80; human, 9; gay, 150, 269n36; neurological, 1; politics, 22, 142, 151, 223; studies, 278n7; and visual politics, 151. *See also* AID/HIV; self
ideology, 80; "end of," 78, 85; utilitarian, 34; *vs.* discourse, 115. *See also* capitalism
Iggers, Georg, 32, 34
Illich, Ivan, 21, 72, 99
imagination, 237n89
immunitas/communitas, 200, 201
impact indicators, 35

"impolitical," 202
incommensurability, 28
"individual," 96
individualism, 189, 196, 197, 224
inequality, 20
"influence," 41, 54
"informed consent," 180, 182
injustice, 221
instantiation, 82
Institut d'études politiques, 280n24
"integrative medicine," 37, 243n164
interdisciplinarity, 37, 38, 176, 223, 232n39
interiority, 80
"internal economies," 22

Jacobs, Margaret, 77
Jacyna, Stephen, 65
James, William, 14
Jameson, Fredric, 85
Jay, Martin, 27
Jefferys, Margot, 47
Jenner, Mark, 65, 91
Johnson, Barbara, 279n18
Jones, Gareth Stedman, 77
Jonsen, Albert, 176
Jordanova, Ludmilla, 24, 70, 71, 73, 81, 86–87, 119
Joseph and Rose Kennedy Institute for the Study of Human Reproduction and Bioethics, 176
Journal of the History of Medicine and Allied Sciences, 70
Joyce, Patrick, 64, 77, 82, 206, 223; against binaries, 227; and "critique," 215; and Foucault, 227; and Latour, 221; reinvigoration of history, 220; and social constructivism, 68; on "technostate," 222; on theory, 210; "What Is the Social in Social History?," 211, 223
Judovitz, Dalia, 110
justice, distributive, 182

Kant, Emmanuel, 226
Kantianism, 218
Kealey, Greg, 44
Kemp, Martin, 105–6
Kirby, David, 123, 124, 125, 128, 143, 151
Kjellé, Rudolph, 186
Knights, Mark, 69
Knorr-Cetina, Karin, 195
knowledge: "economy," 14, 36; "false" *vs.* "scientific," 52; historical production, 19; historical and historiographical, 16, 86; as mediated ideology, 41; natural, 15, 16, 41, 179, 217; reified conception, 55; as revelation, 107; uncertain status, 79
Kuhn, Thomas, 24, 40, 168, 217; *Structure of Scientific Revolutions*, 23, 84

labor: mental *vs.* manual, 60; human *vs.* human essences, 52
Lacan, Jacques, 265n73
Laclau, Ernesto, 79
laity, 82
language: gap with "reality," 261n44; non-determinist view of, 87. *See also* Foucault, Michel; Marxism; semantics
Laqueur, Thomas, 102
Lasch, Christopher, 45
Latour, Bruno, 37–38, 68, 206, 208; ANT, 216–18; "critique," 218; historicization of present, 219; *Leviathan and the Air Pump*, 67, 68, 281n31; and Marxism, 282n39; and neoliberalism, 220; and palaeontology, 167; *Pasteurization of France*, 162; and postmodernism, 218; retrospective diagnosis, 168; *Reassembling the Social*, 219; "the social," 281n34; social constructivism, 45, 217–18. *See also* Joyce, Patrick
Lawrence, Christopher 47
Legionnaires' disease, 163
Leibbrand, Werner, 43
Lemke, Thomas, 186

leprosy, 166
Leviathan and the Air Pump, 67, 68. *See also* Latour, Bruno
Lévy, Bernard-Henri, 254n36
liberalism, 215, 223
Lichterman, Boleslav, 178
"life," 6; biological reduction, 109; as life, 202; and/as politics, 187–88, 194, 198, 202; life sciences, 3, 10
"linguistic turn," 97
"literary turn," 32, 33, 64, 100, 101, 106, 179
literature, study of, 25. *See also* medicine
"looking practices," 135
Loudon, Irvine, 23, 46
Luckin, Bill, 65
Ludwig-Maximilians-Universität, 232n39
Lunganini, Hendrik, 268n27
Lupton, Deborah, 256n57
Lyotard, Jean-François, 78, 84
Lysenko, Trofim, 53

Maastricht, 29, 69
MacIntyre, Alasdair, 172
MacKenzie, Donald, 46
Maclean, Charles, 57, 58, 59
Maehle, Holger, 273n27
Making of the Modern Body, 102
malpractice, 74
Malthus, Thomas, 210
Man, Paul de, 253n34
management studies, 225
managerialism, 22, 66, 67
Manifestos for History, 215, 226
Mantel, Hilary, 33
Marcellesi, Jean-Baptiste, 253n34
marketplace. *See* medicine; neoliberalism
Marshall, T. H., 196, 197
Marwick, Arthur, 13, 77
Marx, Karl, 8, 32, 53, 77, 79, 84, 86, 99; "First Thesis on Feuerbach," 48, 49; view of history, 5
Marx, Roger, 268n19

Marxism, 77, 85, 218, 283n63; critical theory, 226; dogmatism, 253n34; economic determinism, 43; episteme, 44; and language, 64; materialism, 212; notion of power, 98–99; and poststructuralism, 254n34; sociological paradigm, 100; theoretical agendas, 134; validity, 78; "vulgar," 41. *See also* Foucault, Michel; Latour, Bruno
materiality, 31, 228; objects, 142; and "the social," 227; theoretics, 225
"material turn," 212
Mbeki, Thabo, 153
McCarthyism, 84
McKay, Ian, 44
McKeown, Thomas, 72, 86
"meaning culture," 107
meanings/materiality, 213
media, multinational, 154
mediation, 55, 247n22
medical humanism, 21, 39, 171, 174
medicalization, 76, 87, 99
medicine, Western, 74, 150; as analytic category, 76; and commerce, 134; historicization, 23; in history, 76; and humanitarianism, 134; image, 21; like religion, 73; marketplace, 22, 75; problematized, 71, 89; practitioners, 75, 99; reform process, 74; sociology of, 72; and the study of literature, 174. *See also* biomedicine; evidence-based medicine; integrative medicine; medical humanism; modernity; narrative-based medicine; translational medicine
Merleau-Ponty, Maurice, 101, 106, 201, 247n21
metaphor, 49, 88, 179; atmosphere, 58; biography, 163; body, 101; disease, 120–21; frame, 84, 85; globe, 158; networks, 216; social, 251; somatic, 88
miasma, 55, 59
microbes, 168
microbiology, 166

Milbank Quarterly, 68, 83
Mill, John Stuart, 164
Mirzoeff, Nicholas, 132
Mitchell, Joni, 100
Mitchell, W. T., 10, 115, 131
Mitterrand, François, 206
modernity, 217, 218; and medicine, 237n86; and humanism, 281n35
Mommsen, Wolfgang, 13
monetarism, 22, 30
Montgomery, David, 45
Moore, Jim, 24
Morrell, Jack, 65
Moscoso, Javier, 93
MTV Networks International, 265n76
multidisciplinarity. *See* interdisciplinarity
multinational corporations, 74
museology, 78
Museum für Kunst und Gewerbe, Hamburg, 142, 143, 144, 145, 156, 158, 268n19
Museum für Kunst und Gewerbe, Dresden, 268n19
museums, 116, 142. *See also specific museums*

Nadesan, Majia Holmer, 183
"narrative-based medicine," 74
"narrative closure," 77, 119
"narrative *fetishism*," 79
National Health Service, UK, 22, 74
National Library of Medicine, Bethesda, MD, 113
naturalism, 19, 110
natural knowledge. *See* knowledge
Natural Order, 27
"nature," 15, 96, 247n22; ~/culture, 24, 84, 221; ~/politics, 217
Naturwissenschaften, 10
Nazism, 201
Needham, Joseph, 25
negotiation, sub-cortical, 271n3
Negri, Antonio, 188, 197

neoconservatives, 36, 114. *See also* historians
neoliberalism, 37, 67, 138, 208, 224; governments, 74; marketplace, 34, 220; and postmodernity, 30. *See also* Latour, Bruno
neural patterns, 168
neurobiology, 1, 17, 206, 222, 232n39
neuroculture, 2
neuroeconomics, 34
neuroethics, 174
"neuro-turn," 28, 31, 109, 110, 174, 222
New Historicism, 7, 32, 81, 102, 103
"New philosophers," French, 78
Nietzsche, Friedrich, 3, 12, 25, 78, 84, 199, 200, 202, 209, 233n46, 236n75, 254n36
non-representational theory, 6, 7, 8, 222
normal/pathological, 73
normativity, 259n13
Novas, Carlos, 196
"novelty," 4, 190, 191, 192
Novick, Peter, 83

obesity, 164, 165
objectivity, 4, 15, 17, 19, 32, 77, 181, 239n112; in crisis, 83; "mechanical," 15; as subject matter, 18; ~/subjectivity, 24, 52, 53, 6. *See also* historians; natural knowledge
O'Connor, Erin, 88, 89
ontology, 106. *See also* historical ontology
"Orient," 87
Orwell, George, 20, 172
"other," 224
Otter, Chris, 279n21

Pagel, Julius, 19
paleopathology, 168
Palladino, Paolo, 91
Palmer, Bryan, 44, 77, 79
Papoulias, Constantina, 222, 223
Pappworth, Maurice, 274n33

"paradigm," 179
"past": as neutral space, 13; objectified, 13; as a representation, 5; unwanted, 4
paternalism, 175
"patient," 74, 180; de-skilled, 76; historical focus, 237n98; self-help groups, 190, 196; voice, 75
Pêchux, Michel, 253n34
Pelling, Margaret, 51–54; *Cholera, Fever, and English Medicine*, 46, 49
pension funds, 225
"people," 79, 180. *See also* public
perception: epistemological, 137; physiological, 146
performativity, 25, 260n32
periodization, 77
Pernick, Martin, 179
personhood, 180, 191
Peters, Rik, 107–8
pharmaceutical industry, 22, 75, 95, 142, 182, 206, 243n164
phrenology, 25, 26–27, 53, 59
Pickering, Andrew, 68, 166, 260n32
Pickstone, John, 29, 65, 73, 240n127
"pictorial turn," 132, 142
plague, 166, 167
plant-centricity, 233n46
Platt, Sir Harry, 240n127
Platt, Sir Robert, 274n33
"political," 90, 214
politics, 89; of appearance, 157; feminist, 106; and postmodernism, 225, 254n34. *See also* aesthetics; biopower; Foucault, Michel; geopolitics; history-writing; life; nature; neoliberalism; thanatopolitics
Poovey, Mary, 88, 89
Popper, Karl, 26
Porter, Dorothy, 81
Porter, Roy, 22, 24, 75, 86, 100, 139, 256n57
positivism, 52, 53, 54, 63
Post, Robert, 241n138

postcolonial studies, 228
"posthistorical age," 32–33
posthumanism, 9, 10, 19, 111, 218, 222. *See also* truth
"posthuman turn," 278n*i*
postmodernism: as *a*political, 254n34; anti-essentialist, 106; and critical theory, 225, 227; definition, 236n79; as discourse, 97; Francophone, 85; intellectual movement, 17, 135; obituary on, 211; *vs.* postmodernity, 237n82. *See also* disease; historians; Latour, Bruno; neoliberalism; politics
postnaturalist philosophy, 19
posters: and art, 122, 125, 130; attributed functions, 121; consumption, 121; German movement, 268n19; and historical narratives, 120; viewers, 125
posters of health, 114, 115, 141; AIDS, 146 (*see also* Gilman, Sander); material objects, 116; propaganda, 116
post-structuralism. *See* death; epistemology; Marxism
power, 80, 98, 115, 119, 122, 126, 134, 187; relations *vs.* relations of meaning, 203. *See also* biopower; Foucault, Michel; Sontag, Susan
Power of the Poster, The, 116
Powers, Richard, 3
pragmatism, 14, 15, 53, 85. *See also* utilitarianism
presentationalism, 97, 106–8; metaphysics, 261n44; as representation, 5
presentism, 12
Private Finance Initiative, 74
privatization, 23, 100, 142
Problem of Medical Knowledge, 46
Probyn, Elspeth, 81
professionalization, 74, 76, 87, 90
"progress," 4, 11, 75, 176
proteins, 169
proteomics, 9
Prozac, 96, 197

pseudo-science, 26, 52
psychology, 25, 222
"public," 121–22; and posters, 121; postmodern conception, 263n22; reconstituted, 270n56
public health, 71, 112–14, 116, 120–23, 126–27, 131, 134–35, 176
Pumphrey, Steve, 29
purity concept, 56
Putman, Hilary, 84, 86
Putman, Robert, 77

quarantine, 50

Rabinow, Paul, 44, 183, 189, 191, 192, 196, 198
Radical Science Collective, 47
Rancière, Jacques, 220
Ranke, Leopold von, 65
Rapp, Rayna, 191
Rawls, John, 75, 86
"reading practices," 119, 135
Reagan, Ronald, 77
realism, 15, 35, 88, 166; neo-, 8; philosophical, 217; social, 223; virtual, 80. *See also* language
Reconstructing History, 88
reductionism, 101; biological, 222, 230n16; neurobiological, 2, 36, 181
"reflective equilibrium," 75
"reflexive modernity," 75
"regulative ideals," 15
relativism, 84, 85
religionists, 79, 152
representationalism, 4, 11, 80, 103, 260n32; ~/authenticity, 106; ~/materiality, 213; ~/presentationalism, 106
representations, 78, 103, 214. *See also* AIDS/HIV; future; past; presentationalism
Representations, 81, 92, 102
research assessment exercise, 36, 220

"retrospective diagnosis," 20, 160–61, 164–65
Revetz, Jerry, 46
Rigney, Ann, 213, 214, 215
risk, 14, 75, 128, 164, 180, 182, 191
Roberts, Morley, 187
Robin, Régine, 253n34
Rockefeller Medicine, 23
Roedy, Bill, 265n76
Rorty, Richard, 252n17
Rose, Nikolas, 2, 6, 35, 44, 47, 76, 95, 182, 183, 191, 194, 195, 254n37, 256n57; *Politics of Life Itself*, 189–96
Rosen, George, 42, 82
Rosenberg, Charles, 61–62, 68, 85; and Ackerknecht, 42; bioethics, 173; *Cholera Years*, 167; "frame," 72, 83–87, 162; "the social," 71, 82
Rothman, David, 175
Royal Society of Arts, 232n39
Rudwick, Martin, 24
Runia, Eelco, 106
Ryle, John, 86

Said, Edward, 27, 80
Salter, Brian, 70
Samuel, Raphael, 77, 26, 283n60
Sappol, Michael, 278n2
Sartre, Jean-Paul, 106, 107, 193
Savage-Smith, Emily, 46
Sawday, Jonathan, 81
Schaffer, Simon, 24, 67
science: discourse, 11, 215, 223; "false," 53metanarrative of progress, 25; political epistemology, 38; practitioners, 26; rejected, 26; ~/context, 217; ; ~/humanities, 223, 228; ~/ideology, 84, 217; ~/pseudoscience, 52; ~/scientism, 84; ~/society, 24, 217; as social relations, 24; social studies, 26, 68; *vs.* morality, 195; "wars," 24, 67, 72, 83, 162; in working-class culture, 26

Scott, Joan W., 13–14, 209, 210, 214, 216, 225, 227; "experience," 45, 104
Searle, Geoffrey, 69
Sebald, W.G., 70
Secord, Ann and Jim, 24
Sedov, Leon, 244n*iii*
self, 95; as biology, 2; care of, 98; concept of, 180; corporeal, 103; historical, 19; modern, 100; technology of, 127; transformation, 27. *See also* identity
semantics, 154, 188–89, 198–204. *See also* biopower; Foucault, Michel; globalism; language
"semantic turn," 64, 97
semiotics, 79, 203, 204; Francophone, 78; visual, 130, 131
sex, 87, 128
sexually transmitted diseases (STDs), 114, 117
Shapin, Steve, 46, 246n73
Shortt, Sam, 245n*ix*
Sigerist, Henry, 23, 42, 85, 86
signifiers, 80, 115
Sloterdijk, Peter, 87
Smail, Daniel Lord, 110, 271n3
smallpox, 163
Smith, Roger, 10, 24; *Being Human*, 232n41
Smith, Southwood, 56, 58, 59, 60, 250n46, 250n49
Smith, Wesley J., 173
Snow, C. P., 25
"social," 72, 78, 79, 80, 88, 96, 100, 180, 214; category, 64, 90, 223, 260n27; environment of strategies, 192; problematized in history of medicine, 71, 81–82, 195; re-theorized, 90, 228; "social control," 76. *See also* cultural; frame; Joyce, Patrick; Latour, Bruno; materiality; metaphor; theory
Social History, 45
social history, 72; "end of," 78; "old," 211–12; *vs.* cultural history, 213

socialism, scientific, 25
"sociality," 222
social justice, 212
social medicine, 72, 86
social sciences, 195, 212, 228
society: ~/culture, 213, 217; ~/economy, 221; somatic, 95
Society for Plant Neurobiology, 233n46
Society for the Social History of Medicine, 23, 81
sociobiology, 2, 36
sociology: abandonment, 196; "of associations," 216, 219; "of criticism," 219; "of interaction," 216; "of networks," 216; and the neuro-turn, 36; non-transcendent, 281n31; of scientific knowledge, 35, 67, 77; "of translation," 280n25
Sokal, Alan, 283n60
solidarity, 40
somaticism, 82, 95, 82, 196
"somatic turn," 96, 100, 101
Sontag, Susan, 115, 157; *AIDS and Its Metaphors*, 120; AIDS global character, 270n54; *Illness as Metaphor*, 120; posters, 120–23, 124, 126, 127, 130, 133, 134, 147; and power, 122
sovereignty, 188, 199
"speech acts," 104
Stacey, Meg, 47
Stafford, Barbara, 109
Statue of Liberty, 78
Steedman, Carolyn, 278n7
Stein, Claudia, 29–30, 92, 168
Stevens, Tina, 177
structure, 79; ~/agency, 213; ~/culture, 213
Sturdy, Steve, 29, 65
subaltern studies, 225
subjectivity, 19, 20 33. *See also* objectivity
subject/object, 213
Sudhoff, Karl, 23, 168
"surplus valuelessness," 35

Swazey, Judith, 176
symmetry, 26, 68, 281n31

Taithe, Bertrand, 91
Tarde, Gabriel de, 281n34
Taussig, Karen-Sue, 191
TBWA, 149
teamwork, 37
Temkin, Owsei, 42, 82
tendentiousness, 185
Terrence Higgins Trust, 135, 149
text/context, 69
textuality, 88
thalidomide, 21
thanatopolitics, 188, 201, 202, 203
Thatcher, Margaret, 22, 77, 78, 86, 151, 152, 224
Theoretical Practice, 279n13
theory, 53, 214, 226; evolutionary, 11; medical, 54; politically associated, 235n62; and "the social," 212, 213; ~/practice, 279n8; "wars," 87. *See also* Actor-Network Theory; affect theory; critical theory; cultural; fact; historians; history-writing; Joyce, Patrick; Marxism; non-representational theory
Third Reich, 187, 200
Thompson, E. P., 26, 45, 72, 81, 99, 212; *Making of the English Working Class*, 102; *Poverty of Theory*, 27
Thrift, Nigel, 216, 223, 243n159, 281n34; *Non-Representational Theory*, 222
Titmuss, Richard, 20, 180
Toscani, Oliviero, 123, 124, 128, 134, 143, 151
Tosh, John, 234n60
Toulmin, Stephen, 172
"trained judgment," 15
transhuman, 111
"translational medicine," 37, 243n164
Treichler, Paula, 115, 126, 131–33, 135, 154, 224

"truth," 3, 121; biological, 4, 16, 17, 166; of biopower, 189; ethical, 176; "game of," 7, 10; historical, 65, 77, 165; ontological, 227; posthuman, 4; production of, 181; "regimes," 3, 5, 16, 100, 102, 132, 134, 180, 226; scientific, 23, 26, 84
"truth to nature," 15
tuberculosis, 167
Turner, Bryan, 256n57
"turns," 227. *See* cognitive; cultural; descriptive; linguistic; literary; material; neuro-turn; pictorial; posthuman; semantic; somatic; visual
Tuskegee, 22

UNAIDS, 135, 270n54
UNESCO, 174
United Colors of Benetton, 123, 143, 151
United Nations, Joint Program on HIV/AIDS. *See* UNAIDS
University College London, 70
University of East Anglia, 69
utilitarianism, 34; in culture, 38; dehumanized, 173. *See also* history-writing; ideology

values, normative, 13. *See also* epistemic virtues
Velikovsky, Immanuel, 28
venereal disease, 117, 119, 120, 167, 168. *See also* sexually transmitted diseases
Vernon, James, 212
Viagra, 95, 197
Victoria and Albert Museum, 142
visual culture (visual studies), 105, 112, 113, 130, 134, 153; and history of medicine, 114

"visual turn," 113
Vrettos, Athena, 81

Waddington, Ivan, 175
Wakley, Thomas, 249n36
"wars": cultural, 72, 224; history, 72, 162; science, 24, 67, 72, 83, 162; theory, 87
Watney, Simon, 126, 128, 132
"ways of knowing," 169
Wear, Andrew, 71
Weber, Max, 8, 21, 38, 79, 242n147
Webster, Charles, 45, 46, 47, 86
Weindling, Paul, 46
welfare medicine, 74
Wellcome Foundation, 22
Wellcome Trust, 39, 240n127, 278n*ii*; Centre for the History of Medicine, University College London, 139
Westminster Review, 57
Whig history, 48
White, Hayden, 13, 80, 214, 216, 226, 227, 242n147, 284n75; *Metahistory*, 97, 110
Whong-Barr, Michael, 177
Williams, Raymond, 42
Wilson, Adrian, 168
Wilson, Harold, 206
"wisdom," 10, 39
womanhood, 105
Woods, Andy, 69
Woolgar, Steve, 280n24
World Health Organization, 154

"young conservatives," 78
Young, Robert (Bob) M., 24, 25, 47, 217
Yoxen, Edward, 47

Zinsser, Hans, 163
Žižek, Slavoj, 32, 38
zoë, 188